The **BEST WRITING** on **MATHEMATICS**

2016

The BEST WRITING on MATHEMATICS

2016

Mircea Pitici, Editor

PRINCETON UNIVERSITY PRESS
PRINCETON AND OXFORD

Published by Princeton University Press, 41 William Street,
Princeton, New Jersey 08540
In the United Kingdom: Princeton University Press,
6 Oxford Street, Woodstock, Oxfordshire OX20 1TW

press.princeton.edu

ISBN (pbk.) 978-0-691-17529-4

This book has been composed in Perpetua

Printed on acid-free paper. ∞

Printed in the United States of America

1 3 5 7 9 10 8 6 4 2

for David W. Henderson

Contents

Introduction

Mircea Pitici

For the seventh consecutive year, we offer an anthology of recent writings on mathematics, easily accessible to general readers but intriguing enough to interest professional mathematicians curious about the reach of mathematics in the broader concert of ideas, disciplines, society, and history. With one exception (which escaped my radar two years ago), these pieces were first published during the calendar year 2015 in various venues, including professional journals, mass media, and online. In the introductions to the previous volumes and in several interviews, I detailed the motivations and the procedures that underlie the making of the books in this series; before I turn to the content of the current volume, I remind you of some of those tenets, then I point out a few constant and a few changing features that will help you, the reader, consider this enterprise in long-term perspective. I see each year's edition as the product of a convention imposed by our inevitable subservience to calendar strictures; each volume should be considered together with the other volumes, an integral part of the series. It seems to me that the reviewers of all the volumes published so far ignored this aspect. To rectify somehow this prevailing perception of discontinuity, in the near future we will make available a combined index of the pieces published so far.

The main message of the series is that there is a lot more to mathematics than formulas and learning by rote—a lot more than the stringency of proof and the rigor usually associated with mathematics (and held so dear by mathematicians). Mathematics has interpretive sides with endless possibilities, many made manifest by writing in natural language. By now this kind of literature has produced veteran practitioners who are easily recognized by the public but also talented newcomers. It is a new genre that has spun a vast literature but remains completely ignored in educational settings. I hope that *The Best Writing on Mathematics* series helps educators and perhaps policy makers see that it is worth broadening students' understanding of mathematics. Mathematics is an extraordinarily diverse phenomenon molded by the human mind but anchored in a realm both abstract and concrete at the same time. That is not easy to comprehend and to master if we restrict the learning of mathematics to the use of conventional mathematical symbols and notations. I am happy to see that in the books of this series we can bring together constituencies that are sometimes uncomfortable with each other, such as mathematicians and mathematics educators, "purists" and mavericks, entrepreneurs and artists. This series can contribute to a change in the perception of mathematics, toward a broad view that includes the anchoring of mathematics in social, cultural, historical, and intellectual phenomena that hold huge stakes in the working of contemporary humanity. For what it is worth, with its good qualities and its faults, this series undoubtedly reflects my vision, even though other people also add opinions along the way. I propose the initial selection of pieces. I choose most of the pieces myself, but occasionally I receive suggestions from other people (including their own productions). I always welcome suggestions, and I try to remain blind to my correspondents' eventual self-interest and self-promotion. I start from the premise that people are well-intended. I consider everything that comes to my attention, regardless of the manner in which it reaches me. The nontechnical literature on mathematics is immense and it is impossible for me to survey all of it; if people come forward to point out something new to me, I am grateful. A few of the pieces brought to my attention by their authors rightly made it into volumes; most of them did not.

Seizing an opportunity mediated by Steven Strogatz and accepted by Vickie Kearn, I did the first volume in this series (2010) quickly, in just two months, pulling all-nighters in the library at Cornell University; it had a slightly different structure than subsequent volumes. Starting with the second volume (2011) we settled on a template and on certain routines, schedules, and processes that work smoothly—with occasional bumps in the road, to keep us on our toes! This is a fast-paced series; there is not much room for delays and second-guessing.

For the final selection of texts, we take guidance from several professional mathematicians who grade each of about sixty to seventy pieces I initially propose for consideration; we also have to consider constraints related to length, copyright, and diversity of topics and authors. Thus, the content of the books is dependent on the changing literature available but the format is fairly stable. Yet changes in format do occur, gradually. For instance, starting with the second volume, I added the long list of also-rans, supplemented later by a list of special journal issues and now by two new lists, one of remarkable book reviews and another of interviews. The profile of the volumes in *The Best Writing on Mathematics* has shifted from anthology-only to anthology *and* reference work for additional sources about mathematics. By now, in each book we offer not only a collection of writings but also a fairly detailed list of supplementary resources, worth the attention of researchers who might study on their own the complex phenomenon of mathematics and its reverberations in contemporary life. The section containing notable writings, which I present toward the end of each volume, takes me a lot more time to prepare than the rest of the book. Many of the sources I list there are difficult to find for most readers of our series; I hope the list will stimulate you to seek at least a few other readings, in addition to those included here. In this volume another new feature is the printing of figures in full color within the pieces where they belong (not in a separate insert of the book).

I receive a lot of feedback on this series, most of it informal, most of it good, pleasing, cheery, even exalted. The more formal feedback comes in book reviews. It is fascinating and reassuring to compare the reviews. Almost every aspect criticized by some reviewer is praised by other reviewers. Where some people find faults, others see virtue. The point I am making is that regardless of what *would* be desirable to include in the volumes of this series, we are circumscribed by the material extant out there. I cannot include a certain type of article if I encounter nothing of that sort or if we cannot overcome republishing hurdles.

Inevitably, the final selection of pieces for each volume is slightly tilted each year toward certain themes. This happens despite my contrary tendency, to reach a selection as varied as possible, unbiased, illustrating as many different perspectives on mathematics as the recent literature allows. The attentive reader will notice that in this edition a group of contributions refer to the dynamic tension between the object and the practice of "pure" versus "applied" mathematics. This topic is intertwined with the millennia-old history of mathematics and has been addressed by many mathematicians, nonmathematicians, and philosophers.

Overview of the Volume

I now offer a brief outline of each contribution to the volume, with the caveat that the sophisticated arguments made by our authors and the assumptions that support those arguments are worth reading in detail, comparing, corroborating, and contrasting.

Hyman Bass reveals the learning experiences that shaped his views of mathematics and places them in a broader discussion of the complex rapport between mathematics and its pedagogy.

Daniel Silver presents H. G. Hardy's blazing ideas from that most famous exposition of a working "pure" mathematician, *A Mathematician's Apology*—and gives us a potpourri of reactions that followed its publication.

In a piece that can be seen as a genuine continuation on the same theme, Hannah Elizabeth Christenson and Stephan Ramon Garcia detail the fortuitous confluence of mathematics, cricket, and genetics—all of which conspired to make Hardy, at least for once, an *applied* mathematician.

Derek Abbott contends that the applicability of mathematics is considerably overrated, takes the decidedly anti-Platonist position that mathematics is entirely a human construct, and proposes that accepting these unpopular tenets will "accelerate progress."

Aided by suggestive and clear illustrations, Burkard Polster manages to use no formulas in a plain-language proof of a geometric result concerning circles arranged in a rectangle—and leaves us with a few challenges on similar topics.

Joshua Bowman considers the problem of periodic paths of billiard balls in pools of triangular shape; he gives some known results and mentions a few open problems.

In his second piece included in this volume, Burkard Polster teaches us how to invent our own variants of the card game *Spot It!*

Jennifer Quinn reports on a contest in mathematical virtuosity, with participants so ingenious that the victor cannot be settled on the spot;

all this comes from one of the newly popular, enormous, always sold-out arenas of mathematical-gladiatorial disputes!

In a compelling piece that combines elements of group, string, and number theories, Erica Klarreich weaves together considerations made by several mathematicians and physicists struck by the connections they find among fields of research apparently far apart.

Davide Castelvecchi describes the avatars of the proposed proof for the *abc* conjecture in numbers theory and relates them to the peculiarities of the proof's author, Shinichi Mochizuki.

Kevin Hartnett tells the decisive story of a recent breakthrough, Ciprian Manolescu's proof that some spaces of dimension higher than three cannot be subdivided.

Steven Strogatz presents a brilliantly insightful proof of the Pythagorean theorem from the young Albert Einstein.

In a semiautobiographical recollection, Brian Greene outlines the evolving speculative assumptions of the physicists who developed the mathematical underpinnings of string theory.

Tanya Khovanova, Eric Nie, and Alok Puranik describe step by step a generalization about hexagonal grids of a fractal iteration first proposed by Stanislaw Ulam.

Marc Frantz shows us beautiful fractal images and explains how he constructed them.

Joseph Dauben and Marjorie Senechal took strolls through the Metropolitan Museum in New York and discovered a treasure trove of mathematical content—some in plain view, some deftly hidden in details.

Alan Schoenfeld places the contentious Common Core Standards for School Mathematics in perspective, relating them to previous programmatic documents in U.S. education and discussing their rapport with a few important issues in mathematics instruction.

Katharine Beals and Barry Garelick take a decidedly critical stand toward the Common Core, arguing that the standards run against developmental constraints and may impede the learning of mathematics, not promote it.

David Acheson, Peter Turner, Gilbert Strang, and Rachel Levy present four visions of teaching applied mathematics, each of them emphasizing different elements they see as important in such work.

David Richeson details the very early history of one of the most widely recognized mathematical facts, the constancy of the proportion between the circumference and the diameter.

Isabel Serrano and Bogdan Suceavă reconsider the contribution of the medieval thinker Nicholas Oresme to the history of the idea of curvature and conclude that Oresme deserves more credit than he is usually given in accounts of the history of geometrical concepts.

In a similar reconsideration but with an opposite twist than Serrano and Suceavă, Viktor Blåsjö contends that Gottfried Wilhelm Leibniz receives *too much* credit for "proving" the fundamental theorem of calculus.

Amy Shell-Gellasch examines some connections between geometric curves and mechanical drawing devices, with a focus on the drawing toy Spirograph.

John Stillwell ponders what qualities might account for the "depth" of mathematical results and compares from this viewpoint a number of well-known theorems.

Drawing on their work experience, Marco Puts, Piet Daas, and Ton de Waal propose solutions for the instances in which errors find their way into large sets of data.

Brian Hayes summarizes the problematic current state of achievement in the quest to construct probabilistic (as opposed to deterministic) programming languages.

In a piece that combines psychological insights and mathematical rigor, Jorge Almeida analyzes several common misperceptions of the people who attempt to rationalize the most likely outcome of the lottery process.

Andrew Gelman explains why some research claims based on null hypothesis significance tests are spurious, especially in psychological studies.

Howard Wainer and Richard Feinberg point out that the practice of administering long tests under the uncritical pretense that they result in accurate assessments leads to colossal wastes of time, when aggregated to a social scale.

In the last piece of the volume, Ian Stewart attempts to circumscribe popular mathematics as a genre, affirms its value for the public, and advises those who want to contribute to it.

More Writings on Mathematics

In every edition of this series, I list other writings on mathematics published mainly during the previous year, to stimulate readers in their search for broad views of mathematics. The writings mentioned are either books and online resources (listed in the introduction) or articles (listed in the section of notable writings). Most of these items are not highly technical, but occasionally mathematical virtuosity or even theory does enter the picture; a clear distinction between technical and nontechnical writings on mathematics is not possible. The pretense that these bibliographies are comprehensive would be quixotic in the present state of publishing on mathematics. I list only items I have seen—courtesy of authors and publishers who sent me books (thank you!) or through the valuable services of the Cornell University Library and, in a few instances, of the Z. Smith Reynolds Library at Wake Forest University, where my wife worked for four years. For additional books I refer the reader to the monthly new books section published by *The Notices of the American Mathematical Society* and to the excellent website for book reviews hosted by the Mathematical Association of America.

Among the books that came to my attention over the past year, three collective volumes stand out: *The Princeton Companion to Applied Mathematics* edited by Nicholas Higham, *The First Sourcebook on Asian Research in Mathematics Education* edited by Bharath Sriraman, Jinfa Cai, and Kyeong-Hwa Lee, and the anthology *An Historical Introduction to the Philosophy of Mathematics* edited by Russell Marcus and Mark McEvoy.

Some books in which the authors focus on mathematics as it commingles with our lives, in its habitual, entertaining, puzzling, and gaming aspects, are *The Proof and the Pudding* by Jim Henle, *A Numerate Life* by John Allen Paulos, *Truth or Truthiness* by this year's contributor to our volume Howard Wainer, *The Magic Garden of George B and Other Logical Puzzles* by the almost centenarian Raymond Smullyan, *Problems for Metagrobologists* by David Singmaster, as well as the collective volumes *The Mathematics of Various Entertaining Subjects* edited by Jennifer Beineke and Jason Rosenhouse, *Numbers and Nerves* edited by Scott and Paul Slovic, and even *Digital Games and Mathematics Learning* edited by Tom Lowrie and Robyn Jorgensen (the last one with an emphasis on educational issues). More expository books are *Math Geek* by Raphael Rosen, *A Mathematical Space Odyssey* by Claudi Alsina and Roger Nelsen, *Beautiful, Simple, Exact, Crazy* by Apoorva Khare and Anna Lachowska, *Single Digits* by Marc Chamberland, *Math Girls Talk about Integers* and *Math Girls Talk about Trigonometry* by Hiroshi Yuki, and *Sangaku Proofs* by Marshall Unger.

An area enjoying phenomenal growth is the history of mathematics, in which I include the history of mathematical ideas and the histories of mathematical people. Here are some recent titles: *I, Mathematician,* a remarkable collection of pieces by working mathematicians, edited by Peter Casazza, Steven G. Krantz, and Randi Ruden; *The War of Guns and Mathematics* edited by David Aubin and Catherine Goldstein; *Music and the Stars* by Mary Kelly and Charles Doherty; *The Real and the Complex* by Jeremy Gray; and *Taming the Unknown* by Victor Katz and Karen Hunger Parshall. Biographical works on remarkable mathematicians are *Leonhard Euler* by Ronald Calinger, *The Scholar and the State: In Search of Van der Waerden* by Alexander Soifer, *Genius at Play: The Curious Mind of John Horton Conway* by Siobhan Roberts, *Fall of Man in Wilmslow* (a novel that starts with the death of Alan Turing) by David Lagercrantz, and *The Astronomer and the Witch: Johannes Kepler's Fight for His Mother* by Ulinka Rublack. Two collective volumes in the same biographical category are *Oxford Figures* edited by John Fauvel, Raymond Flood, and Robin Wilson; and *Lipman Bers* edited by Linda Keen, Irwin Kra, and Rubí Rodríguez. *Birth of a Theorem* by Cédric Villani is autobiographical. Collected works and new editions of old books include *The G. H. Hardy Reader* edited by Donald J. Albers, Gerald Alexanderson, and William Dunham; *Birds and Frogs* by Freeman Dyson; *A Guide to Cauchy's Calculus* by Dennis Cates; and *Tartaglia's Science of Weights and Mechanics in the Sixteenth Century* by Raffaele Pisano and Danilo Capecchi. Historical in perspective with a strong sociological component (and on an actual topic) is *Inventing the Mathematician* by Sara Hottinger.

Recent books on philosophical aspects of mathematics are *Mathematics, Substance and Surmise* edited by the son-and-father pair Ernest and Philip Davis; *G.W. Leibniz, Interrelations between Mathematics and Philosophy* edited by Norma Goethe, Philip Beeley, and David Rabouin; *Mathematical Knowledge and the Interplay of Practices* by José Ferreirós; and *The Not-Two* by Lorenzo Chiesa. Wide-ranging and difficult to categorize is *Algorithms to Live By*, by Brian Christian and Tom Griffiths.

A great number of books on mathematics education are published every year; it is not feasible for me to mention all that literature. Here are a few recent titles that came to my attention: *Confessions of a 21st Century Math Teacher* and *Math Education in the U.S.* by our contributor Barry Garelick, *What's Math Got to Do with It?* by Jo Boaler, *Mathematical Mindsets* by Jo Boaler and Carol Dweck, *More Lessons Learned from Research* edited by Edward Silver and Patricia Ann Kenney, *Assessment to Enhance Teaching and Learning* edited by Christine Suurtamm, *How to Make Data Work* by Jenny Grant Rankin, and the refreshingly iconoclastic *Burn Math Class* by Jason Wilkes.

In connections between mathematics and other domains, including directly applied mathematics and highly interdisciplinary works, recent titles are *Great Calculations* by Colin Pask, *Creating Symmetry* by Frank Farris, *Why Information Grows* by César Hidalgo, *Data Mining for the Social Sciences* by Paul Attewell and David Monaghan, *Knowledge Is Beautiful* by David McCandless, *Exploring Big Historical Data* by Shawn Graham, Ian Milligan, and Scott Weingart, *Topologies of Power* by John Allen, *Understanding and Using Statistics for Criminology and Criminal Justice* by Jonathon Cooper, Peter Collins, and Anthony Walsh, *Multilevel Modeling in Plain Language* by Karen Robson and David Pevalin, a new edition of *The Essentials of Statistics* by Joseph Healey, and *Mathematics for Natural Scientists* by Lev Kantorovich.

Other friendly books on topics in statistics are *The Seven Pillars of Statistical Wisdom* by Stephen Stigler, *The 5 Things You Need to Know about Statistics* by William Dressler, *Bayes' Rule* by James Stone, *Statistics without Mathematics* by David Bartholomew, and *Beginning Statistics* by Liam Foster, Ian Diamond and Julie Jeffries.

Highly visual is the new volume in *The Best American Infographics* series edited by Gareth Cook.

A work of fiction is *Zombies and Calculus* by Colin Adams.

I hope that you, the reader, will enjoy reading this anthology at least as much as I did while working on it. I encourage you to send comments, suggestions, and materials I might consider for (or mention in) future volumes to Mircea Pitici, P.O. Box 4671, Ithaca, NY 14852 or electronic correspondence to mip7@cornell.edu.

Books Mentioned

Adams, Colin. *Zombies and Calculus*. Princeton, NJ: Princeton University Press, 2014.

Albers, Donald J., Gerald Alexanderson, and William Dunham, eds. *The G. H. Hardy Reader*. Cambridge: Cambridge University Press, 2016.

Allen, John. *Topologies of Power: Beyond Territory and Networks*. Abingdon, UK: Routledge, 2016.

Alsina, Claudi, and Roger B. Nelsen. *A Mathematical Space Odyssey: Solid Geometry in the 21st Century*. Washington, DC: Mathematical Association of America, 2015.

Attewell, Paul, and David Monaghan. *Data Mining for the Social Sciences: An Introduction*. Oakland, CA: University of California Press, 2015.

Aubin, David, and Catherine Goldstein, eds. *The War of Guns and Mathematics: Mathematical Practices and Communities in France and Its Western Allies around World War I*. Providence, RI: American Mathematical Society, 2014.

Bartholomew, David J. *Statistics without Mathematics*. Los Angeles: SAGE, 2015.

Beineke, Jennifer, and Jason Rosenhouse, eds. *The Mathematics of Various Entertaining Subjects: Research in Recreational Math*. Princeton, NJ: Princeton University Press, 2015.

Boaler, Jo. *What's Math Got to Do with It? How Teachers and Parents Can Transform Mathematics Learning and Inspire Success*. Revised edition. New York: Penguin Books, 2015.

Boaler, Jo, and Carol Dweck. *Mathematical Mindsets: Unleashing Students' Potential through Creative Math, Inspiring Messages and Innovative Teaching*. San Francisco: Jossey-Bass, 2015.

Calinger, Ronald S. *Leonhard Euler: Mathematical Genius in the Enlightenment*. Princeton, NJ: Princeton University Press, 2016.

Casazza, Peter, Steven G. Krantz, and Randi D. Ruden, eds. *I, Mathematician*. Providence, RI: Mathematical Association of America, 2015.

Cates, Dennis M. *A Guide to Cauchy's Calculus: A Translation and Analysis of* Calcul Infinitésimal. Walnut Creek, CA: Fairview Academic Press, 2012.

Chamberland, Marc. *Single Digits: In Praise of Small Numbers*. Princeton, NJ: Princeton University Press, 2015.

Chiesa, Lorenzo. *The Not-Two: Logic and God in Lacan*. Cambridge, MA: MIT Press, 2016.

Christian, Brian, and Tom Griffiths. *Algorithms to Live By: The Computer Science of Human Decisions*. New York: Henry Holt and Co., 2016.

Cook, Gareth, ed. *The Best American Infographics 2015*. Boston: Houghton Mifflin Harcourt, 2015.

Cooper, Jonathon A., Peter A. Collins, and Anthony Walsh. *Understanding and Using Statistics for Criminology and Criminal Justice*. New York: Oxford University Press, 2016.

Davis, Ernest, and Philip J. Davis, eds. *Mathematics, Substance and Surmise: Views on the Meaning and Ontology of Mathematics*. Heidelberg, Germany: Springer Verlag, 2015.

Dressler, William W. *The 5 Things You Need to Know about Statistics: Quantification in Ethnographic Research*. Walnut Creek, CA: Left Coast Press, 2015.

Dyson, Freeman J. *Birds and Frogs: Selected Papers, 1990–2014*. Singapore: World Scientific, 2015.

Farris, Frank A. *Creating Symmetry: The Artful Mathematics of Wallpaper Patterns*. Princeton, NJ: Princeton University Press, 2015.

Fauvel, John, Raymond Flood, and Robin Wilson, eds. *Oxford Figures: Eight Centuries of the Mathematical Sciences*. Oxford, UK: Oxford University Press, 2013.

Ferreirós, José. *Mathematical Knowledge and the Interplay of Practices*. Princeton, NJ: Princeton University Press, 2016.

Foster, Liam, Ian Diamond, and Julie Jefferies. *Beginning Statistics: An Introduction for Social Scientists*. 2nd ed. London: SAGE Publications, 2015.

Garelick, Barry. *Confessions of a 21st Century Math Teacher*. CreateSpace Independent Publishing Platform, 2015.

Garelick, Barry. *Math Education in the U.S.: Still Crazy after All These Years*. CreateSpace Independent Publishing Platform, 2016.

Goethe, Norma B., Philip Beeley, and David Rabouin, eds. *G.W. Leibniz, Interrelations between Mathematics and Philosophy*. Heidelberg, Germany: Springer Verlag, 2015.

Graham, Shawn, Ian Milligan, and Scott Weingart. *Exploring Big Historical Data: The Historian's Macroscope*. London: Imperial College Press, 2016.

Gray, Jeremy. *The Real and the Complex: A History of Analysis in the 19th Century*. Heidelberg, Germany: Springer Verlag, 2015.

Healey, Joseph F. *The Essentials of Statistics: A Tool for Social Research*. Fourth ed. Boston: Cengage Learning, 2016.

Henle, Jim. *The Proof and the Pudding: What Mathematicians, Cooks, and You Have in Common*. Princeton, NJ: Princeton University Press, 2015.

Hidalgo, César. *Why Information Grows: The Evolution of Order, from Atoms to Economies*. New York: Basic Books, 2015.

Higham, Nicholas J, ed. *The Princeton Companion to Applied Mathematics*. Princeton, NJ: Princeton University Press, 2015.

Hottinger, Sara N. *Inventing the Mathematician: Gender, Race, and Our Cultural Understanding of Mathematics*. Albany, NY: State University of New York Press, 2016.

Kantorovich, Lev. *Mathematics for Natural Scientists*. Heidelberg, Germany: Springer Verlag, 2015.

Katz, Victor J., and Karen Hunger Parshall. *Taming the Unknown: A History of Algebra from Antiquity to the Early Twentieth Century*. Princeton, NJ: Princeton University Press, 2014.

Keen, Linda, Irwin Kra, and Rubí E. Rodríguez, eds. *Lipman Bers: A Life in Mathematics*. Providence, RI: American Mathematical Society, 2015.

Kelly, Mary, and Charles Doherty. *Music and the Stars: Mathematics in Medieval Ireland*. Dublin, Ireland: Four Courts Press, 2013.

Khare, Apoorva, and Anna Lachowska. *Beautiful, Simple, Exact, Crazy: Mathematics in the Real World*. New Haven, CT: Yale University Press, 2015.

Lagercrantz, David. *Fall of Man in Wilmslow*. London: Quercus Books, 2015.

Lowrie, Tom, Robyn Jorgensen (Zevenbergen), eds. *Digital Games and Mathematics Learning: Potential, Promises, and Pitfalls*. Heidelberg, Germany: Springer Verlag, 2015.

Marcus, Russell, and Mark McEvoy, eds. *An Historical Introduction to the Philosophy of Mathematics: A Reader*. London: Bloomsbury Academic, 2016.

McCandless, David. *Knowledge Is Beautiful*. New York: HarperCollins, 2014.

Pask, Colin. *Great Calculations: A Surprising Look behind 50 Scientific Inquiries*. Amherst, NY: Prometheus Books, 2015.

Paulos, John Allen. *A Numerate Life: A Mathematician Explores the Vagaries of Life, His Own and Probably Yours*. New York: Prometheus Books, 2015.

Pisano, Raffaele, and Danilo Capecchi. *Tartaglia's Science of Weights and Mechanics in the Sixteenth Century: Selections from* Quesiti et inventioni diverse. Heidelberg, Germany: Springer Verlag, 2016.

Rankin, Jenny Grant. *How to Make Data Work: A Guide for Educational Leaders*. London: Routledge, 2016.

Roberts, Siobhan. *Genius at Play: The Curious Mind of John Horton Conway*. New York: Bloomsbury, 2015.

Robson, Karen, and David Pevalin. *Multilevel Modeling in Plain Language*. Los Angeles: SAGE, 2016.

Rosen, Raphael. *Math Geek: From Klein Bottles to Chaos Theory, A Guide to the Nerdiest Math Facts, Theorems, and Equations*. Avon, MA: Adams Media, 2015.

Rublack, Ulinka. *The Astronomer and the Witch: Johannes Kepler's Fight for His Mother*. Oxford, UK: Oxford University Press, 2015.

Silver, Edward A., and Patricia Ann Kenney, eds. *More Lessons Learned from Research: Useful and Usable Research Related to Core Mathematical Practices*. Reston, VA: The National Council of Teachers of Mathematics, 2015.

Singmaster, David. *Problems for Metagrobologists: A Collection of Puzzles with Real Mathematical, Logical or Scientific Content*. Singapore: World Scientific, 2016.

Slovic, Scott, and Paul Slovic, eds. *Numbers and Nerves: Information, Emotion, and Meaning in a World of Data*. Corvallis, OR: Oregon State University Press, 2015.

Smullyan, Raymond. *The Magic Garden of George B and Other Logic Puzzles*. Singapore: World Scientific, 2015.

Soifer, Alexander. *The Scholar and the State: In Search of Van Der Waerden*. Heidelberg, Germany: Birkhäuser/Springer, 2015.

Sriraman, Bharath, Jinfa Cai, and Kyeong-Hwa Lee, eds. *The First Sourcebook on Asian Research in Mathematics Education: China, Korea, Singapore, Japan, Malaysia, and India*. Charlotte, NC: Information Age Publishing, 2015.

Stigler, Stephen M. *The Seven Pillars of Statistical Wisdom*. Cambridge, MA: Harvard University Press, 2016.

Stone, James V. *Bayes' Rule: A Tutorial Introduction to Bayesian Analysis*. [No Place]: Sebtel Press, 2013.

Suurtamm, Christine, ed. *Assessment to Enhance Teaching and Learning*. Reston, VA: The National Council of Teachers of Mathematics, 2015 [in NCTM's new series *Annual Perspectives on Mathematics Education*, the continuation of the *NCTM Yearbook*].

Unger, J. Marshall. *Sangaku Proofs: A Japanese Mathematician at Work*. Ithaca, NY: Cornell East Asia Program, 2015.

Villani, Cédric. *Birth of a Theorem: A Mathematical Adventure*. New York: Farrar, Straus and Giroux, 2015.

Wainer, Howard. *Truth or Truthiness: Distinguishing Fact from Fiction by Learning to Think Like a Data Scientist*. Cambridge: Cambridge University Press, 2016.

Wilkes, Jason. *Burn Math Class: And Reinvent Mathematics for Yourself*. New York: Basic Books, 2016.

Yuki, Hiroshi. *Math Girls Talk about Integers*. Austin, TX: Bento Books, 2014.

Yuki, Hiroshi. *Math Girls Talk about Trigonometry*. Austin, TX: Bento Books, 2014

The BEST WRITING on MATHEMATICS

2016

Mathematics and Teaching

Hyman Bass

Induction

I was not born to be a mathematician. Like many, I was drawn to mathematics by great teaching. Not that I was encouraged or mentored by supportive and caring teachers—such was not the case. It was instead that I had as teachers some remarkable mathematicians who made the highest expression of mathematical thinking visible and available to be appreciated. This was like listening to fine music with all of its beauty, charm, and sometimes magical surprise. Though not a musician, I felt that this practice of mathematical thinking was something I could pursue with great pleasure and capably so, even if not as a virtuoso. And I had the good fortune to be in a time and place where such pursuits were comfortably encouraged.

The watershed event for me was my freshman (honors) calculus course at Princeton. The course was directed by Emil Artin, with his graduate students John Tate and Serge Lang among its teaching assistants. It was essentially a Landau-style course in real analysis (i.e., one taught rigorously from first principles). Several notable mathematics research careers were launched by that course. Amid this cohort of brilliant students, I hardly entertained ideas of an illustrious mathematical future, but I reveled in this ambience of *beautiful thinking*, and I could think of nothing more satisfying than to remain a part of that world. It was only some fifteen years later that I came to realize that this had not been a more-or-less standard freshman calculus course.

Certain mathematical dispositions that were sown in that course remain with me to this day and influence both my research and my teaching. First is the paramount importance of proofs as the defining source of mathematical truth. A theorem is a distilled product of a proof, but the proof is a mine from which much more may often be profitably extracted. Proof analysis may show that the argument in fact proves much more than the theorem statement captures. Certain hypotheses may not have been used, or only weakly used, and so a stronger conclusion might be drawn from the same argument. Two proofs may be observed to be structurally similar, and so the two theorems can be seen to be special cases of a more unifying claim. The most agreeable proofs explain rather than just establish truth. And the logical narrative clearly distinguishes the illuminating turn from technical routine.

Artin himself once reflected on teaching in a review published in 1953:

> We all believe that mathematics is an art. The author of a book, the lecturer in a classroom tries to convey the structural beauty of mathematics to his readers, to his listeners. In this attempt, he must always fail. Mathematics is logical to be sure, each conclusion is drawn from previously derived statements. Yet the whole of it, the real piece of art, is not linear; worse than that, its perception should be instantaneous. We have all experienced on some rare occasion the feeling of elation in realising that we have enabled our listeners to see at a moment's glance the whole architecture and all its ramifications.

Two things of a more social character about mathematics also impressed me. First, the standards for competent and valid mathematical work appeared to be clear, objective, and (so I thought at the time) culturally neutral. The norms of mathematical rigor were for the most part universal and shared across the international mathematics community. Of course, mathematical correctness is not the whole story; there is also the question of the interest and significance of a piece of mathematical work. On this score, matters of taste and aesthetics come into play, but there is still, compared with other fields, a remarkable degree of consensus among mathematicians about such judgments. One circumstance that readily confers validation is a rigorous solution to a problem with high pedigree, meaning that it was posed long ago and has so far defied the efforts of several recognized mathematicians.

A social expression of this culture of mathematical norms stood out to me. Success in mathematics was independent of outward trappings, physical or social, of the individual. This was in striking contrast with almost every other domain of human endeavor. People, for reasons of physical appearance, affect, or personality, were often less favored in nonmathematical contexts. But, provided that they met mathematical norms, such individuals would be embraced by the mathematics community. At least so it seemed to me, and this was a feature of the mathematical world that greatly appealed to me. I have since learned that, unfortunately, many mathematicians individually compromised this cultural neutrality, allowing prejudice to discourage the entry of women and other culturally defined groups into the field.

One effect of this intellectual indifference to social norms in mathematics is that a number of accomplished mathematicians do not present the socially favored images of appearance and/or personality, and so the field is sometimes caricatured as being one of brilliant but socially maladroit and quirky individuals. On the contrary, it is a field with the full range of personality types, and it is distinctive instead for its lack of the kind of exclusion based on personality or physical appearance that infects most other domains of human performance. Perhaps the intellectual elitism and sense of aristocracy common to many mathematicians act as a counterpoint to this social egalitarianism.

The second social aspect of mathematics that stood out to me concerned communication about mathematics to nonexperts. Throughout my student years, undergraduate and graduate, I was amazed and excited by the new horizons being opened up. I enthusiastically tried to communicate some of this excitement to my nonmathematical friends, from whom I had eagerly learned so much about their own studies. These efforts were increasingly frustrated despite my efforts to put matters in analytically elementary terms. I think perhaps that I had already become too much of a mathematical formalist and considered the formal rendering of the ideas an important part of the message. This rendering was often inaccessible to my friends despite my enthusiasm. As my research career entered more abstract theoretical domains, I gradually, and with disappointment, retreated somewhat from efforts to talk to others about what I did as a mathematician. Samuel Eilenberg, my mentor when I first joined the Columbia University faculty, once said something to the effect that

> Mathematics is a performance art, but one whose only audience is fellow performers.

I remain to this day deeply interested in this communication problem, and I have admired and profited from the growing number of authors who have found the language, representations, styles, and narratives with which to communicate the nature of mathematics, its ideas, and its practices. Also, my current studies of mathematics teaching have reopened this question, but now in a somewhat different context. For twelve years of schooling, mathematics has a captive audience of young minds with a natural mathematical curiosity too often squandered. And these children's future teachers are students in the mathematics courses we teach.

Mathematical Truth and Proof

Each discipline has its notions of truth, norms for the nature and forms of allowable evidence, and warrants for claims. Mathematics has one of the oldest, most highly evolved, and well-articulated systems for certifying knowledge—deductive proof—dating from ancient Greece and eventually fully formalized in the twentieth century. There may be philosophical arguments about allowable rules of inference and about how generous an axiom base to admit, and there may be practical as well as philosophical issues about the production and verification of highly complex and lengthy, perhaps partly machine-executed, "proofs." But the underlying conceptual construct of (formal) proof is not seriously thrown into question by such productions, only whether some social or artifactual construction can be considered to legitimately support or constitute a proof.

Mathematical work generally progresses through a trajectory that I would describe as

Exploration → discovery → conjecture → proof → certification

In my work with Deborah Ball (2000, 2003), we have described the first three phases as involving *reasoning of inquiry* and the last two as involving *reasoning of justification*. The former is common to all fields of science. The latter has only a faint presence in mathematics education, even though it is the distinguishing characteristic of mathematics as a discipline.

Deductive proof accounts for a fundamental contrast between mathematics and the scientific disciplines: they honor very different epistemological gods. Mathematical knowledge tends much more to be cumulative. New mathematics builds on, but does not discard, what came before. The mathematical literature is extraordinarily stable and reliable. In science, by contrast, new observations or discoveries can invalidate previous models, which then lose their scientific currency. The contrast is sharpest in theoretical physics, which historically has been the science most closely allied with the development of mathematics. I. M. Singer is said to have once compared the theoretical physics literature to a blackboard that must be periodically erased. Some theoretical physicists—Richard Feynman, for instance—enjoyed chiding the mathematicians' fastidiousness about rigorous proofs. For the physicist, if a mathematical argument is not rigorously sound but nonetheless leads to predictions that are in excellent conformity with experimental observation, then the physicist considers the claim validated by nature, if not by mathematical logic; nature is the appropriate authority. The physicist P. W. Anderson once remarked, "We are talking here about theoretical physics, and therefore of course mathematical rigor is irrelevant."

On the other hand, some mathematicians have shown a corresponding disdain for this free-wheeling approach of the theoretical physicists. The mathematician E. J. McShane once likened the reasoning in a "physical argument" to that of "the man who could trace his ancestry to William the Conqueror, with only two gaps."

Even some mathematicians eschew heavy emphasis on rigorous proof in favor of more intuitive and heuristic thinking and of the role of mathematics to help explain the world in useful or illuminating ways. In general, they do not necessarily scorn rigorous proof, only consign it to a faintly heeded intellectual superego. But I would venture nonetheless, that, for most research mathematicians, the notion of proof and its quest are at least tacitly central to their thinking and their practice as mathematicians. And this would apply even to mathematicians who, like Bill Thurston, view what mathematicians do as not so much the production of proofs, but as "advancing mathematical understanding" (Thurston 1994). It would be hard to find anyone with the kind of mathematical understanding and function of which Thurston speaks who has not already assimilated the nature and significance of mathematical proof.

At the same time, the writing of rigorous mathematical proofs is not the work that mathematicians actually do, for the most part, or what they most cherish and celebrate. Such tributes are conferred instead on acts of creativity, of deep intuitive discovery, of insightful analysis and synthesis. André Weil (1971) said,

> If logic is the hygiene of the mathematician, it is not his source of food; the great problems furnish the daily bread on which he thrives.

Vladimir Arnold offered an even more derisory characterization:

> Proofs are to mathematics what spelling (or even calligraphy) is to poetry. Mathematical works do consist of proofs, just as poems do consist of characters.

But it is proof that finally gives mathematical achievements their pedigree.

The Proof vs. Proving Paradox

Saying that a mathematical claim is true means, for a mathematician, that there exists a proof of it. Strictly speaking, this is a theoretical concept, independent of any physical artifact and therefore also of any human agency. But *proving* is an undeniably human activity and so susceptible to human fallibility. It is an act of producing conviction about the truth of something. A mathematician, in proving a mathematical claim, typically produces a manuscript purporting to represent a mathematical proof and exhibits it for critical examination by expert peers (certification). But as Lakatos (1976) has vividly described, this process of proof certification can be errant and uneven, though ultimately robust.

The rules for proof construction are sufficiently exact that the checking of a proof should, in principle, be a straightforward and unambiguous procedure. However, formal mathematical proofs are ponderous and unwieldy constructs. For mathematical claims of substantial complexity, mathematicians virtually never produce complete formal proofs. Indeed, requiring that they do so would cause the whole enterprise to grind to a halt. The resulting license in the practices of certifying mathematical knowledge has caused some (nonmathematical) observers to conclude that mathematical truth is just another kind of social negotiation and, so, is unworthy of its prideful claims of objective certainty.

I think that this notion is based on a misunderstanding of what mathematicians are doing when they claim to be proving something. Specifically, I suggest that

> Proving a claim is, for a mathematician, an act of producing, for an audience of peer experts, an argument to convince them that a proof of the claim exists.

Two things are important to note here. First, that implicit in this description is the understanding that the peer experts possess the conceptual knowledge of the nature and significance of mathematical proof. Second, the notion of "conviction" here is operationalized to mean that

> The convinced listener feels empowered by the argument, given sufficient time, resources, and incentive, to actually construct a formal proof.

Of course, in this time of computer-aided proofs, some of this certification must be transferred to establishing the reliability of proof-checking software.

In my work with Deborah Ball, we have found this perspective helpful in studying the development of what might reasonably be called mathematical proving in the early grades. I say more about this development later.

Compression and Abstraction

Mathematical knowledge is, as I mentioned before, cumulative (nothing is discarded) and also hierarchical. What saves it then from sinking under the weight of its own relentless growth into some dense, impenetrable, massive network of ideas? It is rescued from this by a distinctive feature of mathematics that some have called *compression*. This is a process by which certain fundamental mathematical concepts or structures are characterized and named and cognitively rescaled so that they become, for the expertly initiated, as mentally manipulable as counting numbers is for a child. Think for example what a complex of ideas (and mathematical history) is packed into an expression like "complex Lie group," uttered fluently among mathematicians. Think of the years of schooling needed to invest that expression with precise meaning.

A typical form of compression arises from the unification of diverse phenomena as special cases of a single construct (for example, groups, topological spaces, Hilbert spaces, measure spaces, and categories and functors) about which enough of substance can be said in general to constitute a useful unifying theory. This leads to another salient feature of mathematics: *abstraction*. Most mathematics has its roots in science and so ultimately in the "real world." But mathematics, even that contrived to solve real-world problems, *naturally* generates its own problems, and so the process continues with these, leading to successive stages of unification through abstraction until the mathematics may be several degrees removed from its empirical origins. It is a happy miracle that this process of pursuing *natural mathematical questions* repeatedly reconnects with empirical reality in unexpected and unplanned ways. This is the "unreasonable effectiveness" of which Wigner wrote (1960) and of which Varadarajan (2015) writes eloquently in this collection of essays (Casazza et al. 2015).

Abstraction is often thought to separate mathematics from science. But even among mathematicians, there are different professed dispositions toward abstraction. Some mathematicians protest that they like to keep things "concrete," but a bit of reflection on what they consider to be concrete shows it to be far from such for an earlier generation. Indeed, I noticed while I was a graduate student that the extent to which a mathematical idea was considered abstract seemed more a measure of the mathematician's age than of the cognitive nature of the idea. As new ideas become assimilated into courses of instruction, they become the daily bread and butter of initiates, all the while remaining novel and exotic to many of their elders.

Whereas compression is an essential instrument for the ecological survival of mathematics, its very virtue presents a serious obstacle for the teaching of mathematics. The knowledgeable and skillful mathematician has assimilated and internalized years of successive compression, streamlining of ideas, and habits of mind. But a young learner of mathematics still lives and thinks in a "mathematically decompressed" world, one that has become hard for the mathematically trained person to imagine, much less remember. This phenomenon presents a special challenge to teachers of mathematics to children. And, interestingly, it requires a special kind of knowledge of mathematics itself that is neither easy for nor common among otherwise mathematically knowledgeable adults (Ball et al. 2005, 2008).

Teaching Mathematics

Among the questions to which our editors invited us to direct attention was, "How are we research mathematicians viewed by others?" For one community, (school) mathematics teachers and education researchers, I have some firsthand knowledge of this after more than a decade of interdisciplinary work in mathematics education, and what I have learned I find both interesting and important for mathematicians to understand. I began this essay with an account of how great university teaching drew me into mathematics. Now, in my close study of elementary mathematics teaching, I have a changed vision of what teaching entails.

Of course many mathematicians, like mathematics educators, are seriously interested in the mathematics education of students. But there are significant differences in how the two communities, broadly speaking, see this enterprise. The mathematicians' interests, naturally enough, are directed primarily at the graduate and undergraduate levels and perhaps somewhat at the secondary level. In contrast, the interests of the educators are predominantly aimed at the primary through secondary levels. But there is a broadening overlap in the ranges of interest of the two communities, and some fundamental aspects of what constitutes quality teaching are arguably independent of educational levels. Nonetheless, in areas of common focus there are often profound differences of perspective and understanding across the two communities about what constitutes quality mathematics instruction. I expand on this later.

There is also a difference in educational priorities between the mathematics and mathematics education communities, a difference whose importance cannot be overstated. Mathematicians are naturally interested in "pipeline" issues, the rejuvenation of the professional community, and so the induction and nurturing of talented and highly motivated students into high-level mathematical study. Mathematics educators, on the other hand, are professionally committed to the improvement of public education at scale, with the aim of *high levels of learning for all students* and with a heavy emphasis on the word "all." Although these two agendas are not intrinsically in conflict or competition, they are often seen to be so, and this can have resource and policy implications. Public education in the United States, despite long and costly interventions, continues to perform poorly in international comparisons in terms of meeting workplace demands and even in providing basic literacies. Moreover, there is a persistent achievement gap reflected in underperformance of certain subpopulations that the educational system historically has not served well. These are the "big frontline problems" of mathematics education, and they are just as compelling and urgent to mathematics educators as the big research problems are to mathematicians.

Mathematicians have an excellent tradition of nurturing students of talent. What they are less good at is identifying potential talent. The usual indices, high test scores and precocious accomplishment, are easy enough to apply, but these indices typically overlook students of mathematical promise whom the system has not encouraged or given either the expectation of or opportunity for high performance. As a result, mathematical enrichment programs, if not sensitively designed, can sometimes perpetuate the very inequities that mathematics educators are trying to mitigate.

My main focus here, however, is on teaching. Much of what I have learned I owe to work with my colleague and collaborator, Deborah Ball. To facilitate what I want to say, it is helpful to use a schematic developed by Deborah and her colleagues David K. Cohen and Stephen W. Raudenbush (2003) to describe the nature of instruction.

The "Instructional Triangle"

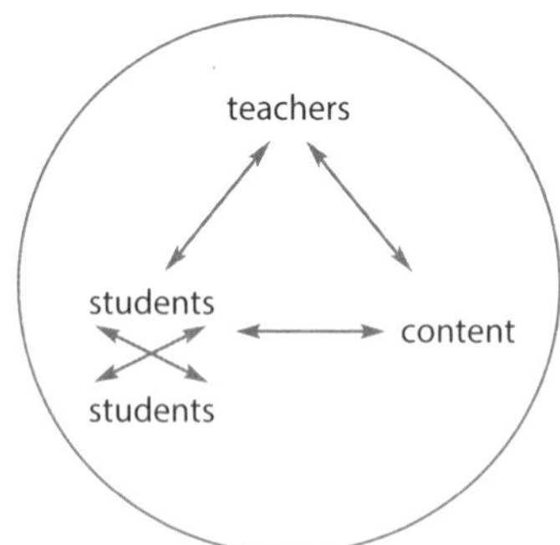

The concept proposed by this image is that instruction is about the interactions of the teacher with students (and of students with each other) around content. The double arrows emphasize that these interactions are dynamic; in particular, changing any one element of the picture significantly alters the whole picture.

Now what I suggest is that most mathematicians' conception of instruction lives primarily at the content corner of the triangle, with its school incarnation expressed in terms of curriculum (including both standards documents and curriculum materials). In this view, with high-quality curriculum in place, the mathematically competent teacher has only to implement that curriculum with intellectual fidelity, and then attentive and motivated students will learn. Of course, mathematicians have little direct influence on schoolchildren, but they often have explicit, discipline-inspired ideas about what mathematics children should learn and how. In the case of teachers, mathematicians do bear some direct responsibility for their mathematical proficiency, since many teachers learn much of the mathematics they know in university mathematics courses. Many studies have pointed to weak teacher content knowledge as a major source of underperformance in public education. Of the many remedies proposed, mathematicians have generally favored more and higher level mathematics courses as a requirement for certification. Unfortunately, none of the interventions based on these ideas so far undertaken have yet produced the desired gains in student achievement for a broad range of students. Note that all of this discussion and debate resides on the content–teacher side of the triangle, with little explicit attention yet given to the role of the students in instruction. Effective teachers teach both math and children.

An educational counterpart to the above "one-sided" (teacher–content) stance is an intellectually adventurous, but somewhat romantic, view of school teaching, this time on the teacher–student side of the triangle. This approach proposes offering tangible and somehow "real-world" related mathematical activities with which to engage learners but leaving the development of mathematical ideas to largely unrestrained student imagination and invention. In such cases, the discipline of the mathematical ideas may be softened to the point of dissipation.

The most refined understanding of mathematics teaching, of which Deborah Ball's work is exemplary, insists that teaching must coordinate attention to the integrity of the mathematics and appropriate learning goals with attention to student thinking, which needs to be honored and made an integral part of the instruction. In this view, the effective teacher has not only a deep understanding of and fascination with the mathematics being taught but also a dual knowledge of and fascination with student thinking. The underlying premise is that children have significant mathematical ideas, albeit imperfectly expressed, and that a part of the teacher's work is to recognize the presence of those ideas (for which a sophisticated knowledge of mathematics is needed!), to give them appropriate validation, and to help students shape them into a more developed articulation and understanding. It is the coordination of these dual spheres of attention, both to the mathematics and to students' thinking, that makes effective teaching the intricate and skilled work that it is.

This is a kind of professional practice that discursive rhetoric cannot adequately capture. It is best conveyed through an examination of teaching practice itself. So let me offer a vignette. Consider the teaching of mathematical proving. This teaching is typically first done in high school geometry, but there are good reasons to argue that it should be done developmentally, starting in the early grades. What might this mean or look like? After all, young children have no concept of anything approaching mathematical proof, and no one seriously argues that this should be formally taught to them. First of all, what is the intellectual imperative for proving that can be made meaningful to young children? Our view is that it is the persistent question, "Why are things true?" Deductive proof is the refined method devised by mathematicians to answer that very question. Learning the methods of deductive proving takes time, but the question "Why are things true?" is itself immediately compelling. The underlying pedagogy is that students progressively learn the methods of constructing compelling mathematical arguments in the course of repeatedly trying to convince others of things they have good reasons to believe to be true. In other words, students can be helped to construct the infrastructure of proving in the very course of proving things (to the best of their growing abilities).

To illustrate this idea, I turn to an episode from a third-grade class (eight-year-olds). The children have been exploring even and odd numbers. Although they do not yet have formal definitions, they rely on intuitive ideas of even and odd based on notions of fair-sharing, and they can identify whether particular (small) numbers are even or odd. They begin to notice addition patterns, like: even + even = even, even + odd = odd, and odd + odd = even. The teacher asks them whether they can "prove" that "Betsy's conjecture" (odd + odd = even) is *always* true. Some of the children have tested the conjecture with lots of numbers, and they are thereby convinced that it must be true. The teacher challenges them, "How do you know that someone might not come up with a new example that doesn't work?" The children are left to ponder that, some of them working collaboratively in small groups.

The next day, the teacher asks the class what they found out. Several hands go up, and she calls on Jeannie, who, speaking also for her partner, Sheena, says, "We were trying to prove that . . . you can't prove that Betsy's conjecture always works . . . because numbers go on and on forever, and that means odd numbers and even numbers go on forever and, um, so you couldn't prove that all of them work." This stunning assertion is at once insightful and revolutionary. Jeannie's comment reflects the fact that Betsy's conjecture is not just one, but rather infinitely many claims, and so not susceptible to empirical verification. Though articulated somewhat informally, she has realized that the cases of the conjecture go on and on.

A second student, Mei, objects to this argument. She objects not to the reasoning but rather to an important inconsistency. She points out that prior claims for which the class had achieved consensus were also not checked in all cases, and Jeannie and her partner had not raised any similar objection to those earlier claims. She challenges the need to make sure that the claim works for all cases, since they had not tried to do this for other conjectures.

In both of these contributions, one can see the process by which the class is beginning to construct the very norms of mathematical reasoning: rejecting empirical methods for an infinitely quantified claim in the first instance and the need for consistency of logical methods in the second. And all of this is still short of proving Betsy's conjecture.

But the next day, Betsy produces a "proof" of her conjecture. At the board, she draws a row of seven small circles and then to the right another row of seven small circles. Then she partitions the two rows into three groups of two, with one left over, and the ones left over next to each other in the middle:

o o o o o o o   o o o o o o o

Then she turns to the class and explains haltingly, "If you have an odd number and group it by twos, you have one left over, so if you have two odd numbers and group them by twos, then the ones left over can be grouped together so you have an even number." What is most significant about this oral argument is that she makes no reference to the specific numbers seven and seven that she has drawn, but rather speaks generically of "an odd number."

How does this demonstration surmount the dilemma raised by Jeannie and Sheena? Betsy has implicitly invoked a general definition of odd number—a number that when grouped by twos has one left over—and this definition, being itself infinitely quantified (it characterizes all odd numbers), can support a conclusion of similarly infinite purview. Betsy's oral argument might be algebraically expressed in the form

$$(x2 + 1) + (1 + y2) = x2 + (1 + 1) + y2 = (x + 1 + y)2.$$

(Note that, for children at this stage, $2x = x + x$ is not the same as $x_2 = 2 + 2 + \ldots + 2$.) But, although she is thinking in general terms, Betsy does not yet know how to represent her ideas generally using algebraic notation.

Notice for each of these students—Jeannie, Mei, and Betsy—the significance of the mathematical ideas, albeit sometimes expressed with difficulty. Episodes like this demonstrate what it might look like to develop the norms of proving in the course of constructing more and more well-developed arguments to convince others of the truth of claims. But this is not the whole story. This narrative resides, as does much of the education research literature, on the student–content side of the instructional triangle, focusing on accomplished student performance. The teacher seems invisible.

This kind of collective interaction and learning does not occur spontaneously, unmediated by well-informed and purposeful instruction. The question of what makes for effective teaching has to do with understanding the knowledge, skills, and sensibilities (both mathematical and pedagogical) and, above all, the practices that enable instruction that can elicit the kind of motivation and reflective engagement of students described in this episode. The kind of mathematical knowledge required includes not only a robust understanding of the mathematical terrain of the work and corresponding learning goals but also an ability to hear, in incipient and undeveloped form, significant, though often not entirely correct, mathematical ideas in student thinking. I emphasize that the latter entails a special kind of knowledge of mathematics, not just psychology. Teaching requires the skills to not only hear and validate and give space to these ideas, but also to help students reshape them in mathematically productive ways. The teacher needs to know how to instruct with questions more than with answers and to give students appropriate time, space, and resources to engage these questions. The actions to accomplish this, though purposeful, structured, and carefully calibrated to the students, are also subtle, deliberately focusing on student performance. And so, to the naive observer, it often can appear that "the teacher is not doing much." The central problem of teacher education is to provide teachers with the knowledge and skills of such effective practice.

The idea of developmental learning, such as we saw with the third graders, was also a feature of my calculus course back at Princeton in 1951. I first learned in that course what (a strange thing) a real number was (mathematically),[1] even though I had been working comfortably with real numbers throughout the upper high school grades. With regard to proving, while we learned, through witnessing stunning examples and through practice problems how to construct reasonably rigorous proofs, it was still always the case that the claims in question seemed at least reasonable and sufficiently meaningful that general mathematical intuition could guide us. But I was at first stymied by a problem on one of our take-home exams. A function $f(x)$ was defined by: $f(x) = 0$ for x irrational, and $f(x) = 1/q$ if x is rational, equal to p/q in reduced form. The question was, "Where is f continuous?" At first sight, this question seemed outrageous. How could one possibly answer it? It was impossible to sketch the graph. Intuition was useless. After some reflection, I resigned myself to the fact that all we had to work with were definitions of f and of continuity. It was a great revelation to me that these definitions and a modest amount of numerical intuition sufficed to answer the question. This was a great lesson about the nature and the power of deductive reasoning, to which my enculturation, though at a different level, was not so different from that of the third graders in our example.

The construction and timing of such a problem was itself a piece of instruction. I wish now that we had video records of Emil Artin's calculus teaching that I could show to an accomplished teacher, like Deborah, to analyze the pedagogical moves, including interaction with students, of which his practice was composed. Much as I profited from that instruction, I was not prepared at the time to see the craft of its construction.

Note

1. An equivalence class of Cauchy sequences of rational numbers.

References

Vladimir Arnold, “Proofs in Mathematics,” http://www.cut-the-knot.org/proofs/.

D. L. Ball and H. Bass, “Making believe: The collective construction of public mathematical knowledge in the elementary classroom.” *Constructivism in education: Yearbook of the National Society for the Study of Education*, D. Philips, ed., University of Chicago Press, Chicago, 2000, pp. 193–224.

D. L. Ball and H. Bass, “Making mathematics reasonable in school.” *A Research Companion to Principles and Standards for School Mathematics*, J. Kilpatrick, W. G. Martin, and D. Schifter, eds., National Council of Teachers of Mathematics, Reston, VA, 2003, pp. 27–44.

D. L. Ball, H. C. Hill, and H. Bass, “Knowing mathematics for teaching: Who knows mathematics well enough to teach third grade, and how can we decide?” *American Educator* **29(1)** (2005), 14–17, 20–22, 43–46.

D. L. Ball, M. H. Thames, and G. Phelps, “Content knowledge for teaching: What makes it special?” *Journal of Teacher Education* **59(5)** (2008), 389–407.

P. Cassazza, S. G. Krantz, and R. D. Ruden, eds., *I, Mathematician*. Mathematical Association of America, Washington, DC, 2015.

D. K. Cohen, S. W. Raudenbush, and D. L. Ball, “Resources, instruction, and research.” *Educational Evaluation and Policy Analysis* **25(2)** (2003), 119–142.

I. Lakatos, *Proofs and Refutations: The Logic of Mathematical Discovery*, Cambridge University Press, 1976.

William P. Thurston, “On proof and progress in mathematics.” *Bull. Amer. Math. Soc. (N.S.)* **30(2)** (1994), 161–177.

V. S. Varadarajan, “I am a mathematician” (essay), *I, Mathematician*. P. Cassazza, S. G. Krantz, and R. D. Ruden, eds., Mathematical Association of America, Washington, DC, 2015.

André Weil, “The future of mathematics.” *Great Currents of Mathematical Thought*, Vol. 1, F. Le Lionnais, ed., Dover Publications, New York, 1971, pp. 321–336.

Eugene Wigner, “The unreasonable effectiveness of mathematics in the natural sciences.” *Comm. Pure Appl. Math.* **13(1)** (1960), 1–14.

In Defense of Pure Mathematics

Daniel S. Silver

Godfrey Harold Hardy was one of the greatest number theorists of the twentieth century. Mathematics dominated his life, and only the game of cricket could compete for his attention. When advancing age diminished his creative power, and a heart attack at sixty-two robbed his physical strength, Hardy composed *A Mathematician's Apology*. It was an *apologia* as Aristotle or Plato would have understood it, a self-defense of his life's work.

"A mathematician," Hardy contended, "like a painter or poet, is a maker of patterns. . . . The mathematician's patterns, like the painter's or the poet's, must be beautiful; the ideas, like the colours or the words, must fit together in a harmonious way." It was a personal and profound view of mathematics for the layman, unlike anything that had appeared before. The book, which this year reaches the seventy-fifth anniversary of its original publication, is a fine and most accessible description of the world of pure mathematics.

Ever since its first appearance, *A Mathematician's Apology* has been a lightning rod, attracting angry bolts for its dismissal of applied mathematics as being dull and trivial. The shaft that lit up the beginning of a review in the journal *Nature* by Nobel laureate and chemist Frederick Soddy was particularly piercing: "This is a slight book. From such cloistral clowning the world sickens."

Hardy's opinions about the worth of unfettered thought were strong, but stated with "careful wit and controlled passion," to borrow words of the acclaimed author Graham Greene. They continue to find sympathetic readers in many creative fields. They were prescient at the time and remain highly relevant today.

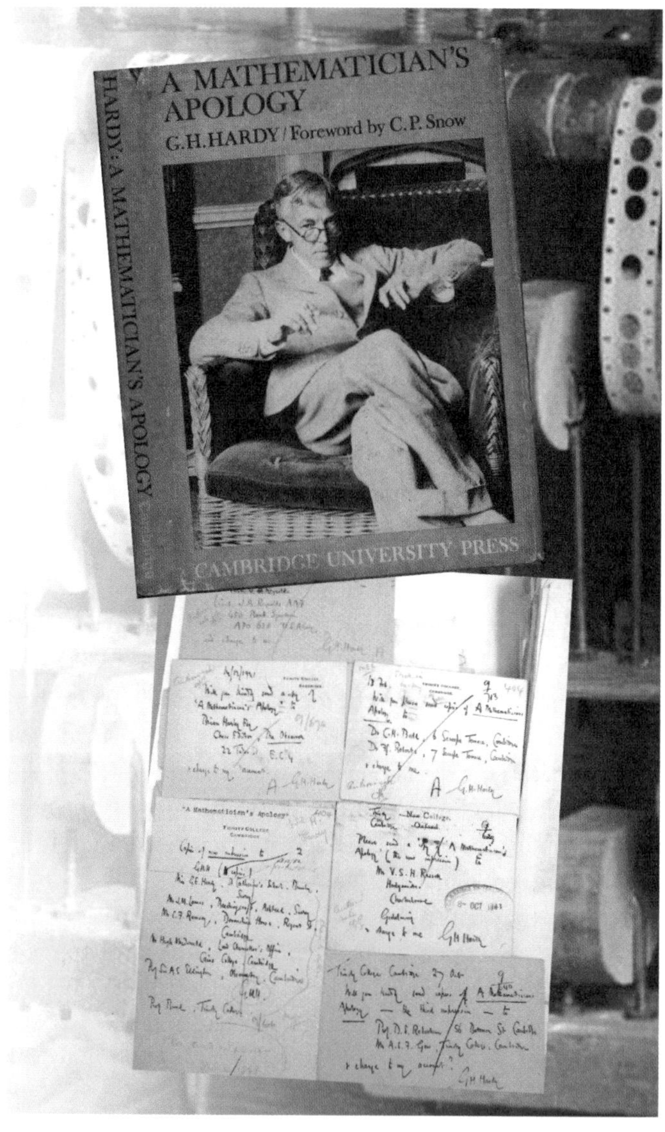

The second edition of *A Mathematician's Apology* featured Hardy's now-iconic photograph on the dust jacket (Jacket image courtesy of Cambridge University Press). For the first edition, Hardy sent postcards requesting that presentation copies be sent to colleagues, including C. D. Broad and J. E. Littlewood, the physicist Sir Arthur Eddington, chemist and novelist C. P. Snow, cricketer John Lomas (to whom he dedicated the book), and his sister Gertie. He also requested copies be sent to colleagues in the United States. Postcards photograph courtesy of Cambridge University Library.

The Art of Argument

Hardy was born on February 7, 1877, in Cranleigh, Surrey. His parents valued education, but neither had been able to afford university.

Hardy grew up to be a scholar, a sportsman, an atheist, and a pacifist, but, above all, he was an individualist. In an obituary of him, the mathematician E. C. Titchmarsh recalled: "If he dined at high table in tennis clothes it was because he liked to do so, not because he had forgotten what he was wearing."

Hardy began a famous collaboration with analyst J. E. Littlewood in 1911. Two years later, they would pore over strange, handwritten mathematical manuscripts that had been sent unsolicited by a young Indian civil servant, Srinivasa Ramanujan. Together they would decide that it was the work of a true genius. After considerable effort, Hardy succeeded in bringing Ramanujan to Cambridge, where Hardy was a professor. The "one romantic incident in my life" is how Hardy described his discovery and collaboration with his young protégé, who tragically died of illness seven years later.

In 1919, Hardy moved to Oxford University, where his eccentricities thrived. In his rooms he kept a picture of Vladimir Lenin. He shunned mechanical devices such as telephones, would not look into a mirror, rarely allowed his photograph to be taken, and was very shy about meeting people. Nevertheless, he was a superb conversationalist, able to carry on talk about many subjects (including, of course, cricket). Titchmarsh recalled: "Conversation was one of the games which he loved to play, and it was not always easy to make out what his real opinions were."

Mathematician George Pólya had a similar recollection: "Hardy liked to shock people mildly by stating unconventional views, and he liked to defend such views just for the sake of a good argument, because he liked arguing."

It is clear that Hardy enjoyed teasing his audience, something one should keep in mind when reading *A Mathematician's Apology*.

Work for Second-Rate Minds

Many reviews of *A Mathematician's Apology* appeared during the first few years of its life. Most were favorable. The author of one such review, published in *The Spectator* in 1940, was Graham Greene. Hardy must have been flattered to read: "I know no writing—except perhaps Henry James's introductory essays—which conveys so clearly and with such an absence of fuss the excitement of the creative artist."

Other reviews were less enthusiastic. Today, as then, there are several reasons to be offended by *A Mathematician's Apology*, especially if you are a scientist.

If you are the author of expository articles (such as this one), then you don't have to wait long for an insult to be hurled your way. Hardy's book began:

> It is a melancholy experience for a professional mathematician to find himself writing about mathematics. The function of a mathematician is to do something, to prove new theorems, to add to mathematics, and not to talk about what he or other mathematicians have done. . . . Exposition, criticism, appreciation, is work for second-rate minds.

Most reviewers pardoned Hardy for this assertion. Félix de Grand'Combe, professor of French at Bristol University, did not. Writing in the *Journal of Education* in August 1943, he argued that observing and reformulating can be creative and illuminating acts:

> When Linnaeus devised his wonderful classification of plants he didn't "make" anything, he merely discovered a pre-existing treasure, explaining and rendering perceptible to all eyes a series of coherent relationships actually present in Nature, but his work altered and clarified our whole conception of the vegetable world; it gave informing reason to apparent chaos, life to what the ancients had seen as a dark welter of "non-being."

If Hardy thought that exposition, criticism, and appreciation is work for second-rate minds, then he must have come to that conviction late in life. During his prime years, he wrote book reviews for *The Cambridge Review, The Times Literary Supplement, Nature*, and *The Mathematical Gazette*. He was an effective and enthusiastic lecturer, constantly in demand. His textbook *A Course in Pure Mathematics*, published in 1908 and still in print, is entirely expository.

If you are a biochemist in search of a cure for a dreadful disease, then you might be insulted by Hardy's summary of your true motives. There are three: intellectual curiosity, professional pride (including anxiety to be satisfied with one's performance), and ambition. According to Hardy:

> It may be fine to feel, when you have done your work, that you have added to the happiness or alleviated the sufferings of others, but that is not why you did it.

Writing in *The News Letter* in 1941, the English physicist and philosopher of science Norman Campbell took Hardy's assertions at face value. However, if a mathematician's principal motivation is to benefit him or herself rather than society, he asked "why should we provide . . . so many more comfortable jobs for mathematicians than for, say, poets or stamp-collectors?"

Should you be an older mathematician, you might be vexed by Hardy's reminder: "No mathematician should ever allow himself to forget that mathematics, more than any other art or science, is a young man's game."

Examples of mathematicians who have made significant discoveries in later life can be given easily. Littlewood, who remained productive well after the age of ninety, is one counterexample. Nevertheless, Hardy's belief is common today among mathematicians. It is encouraged by the fact that the Fields Medal, the highest award in mathematics, is awarded only for work done before the age of forty.

If you are a scientist whose feathers are not yet ruffled, Hardy's main contention will surely disturb your plumage. "Real" mathematics, he argued, is almost wholly "useless," whereas useful mathematics is "intolerably dull." By "real" mathematics, Hardy meant pure mathematics that tends to be abstract and general and, in Hardy's opinion, has the most aesthetic value. Opposed to it is the bulk of mathematics seen in school: arithmetic, elementary algebra, elementary geometry, differential and integral calculus, mathematics designed for computation and having the least aesthetic appeal.

Hardy was both prosecutor and defender in an imaginary trial to determine whether his life had been worthwhile:

> I have never done anything "useful." No discovery of mine has made, or is likely to make, directly or indirectly, for good or ill, the least difference to the amenity of the world. . . . I have just one chance of escaping a verdict of complete triviality, that I may be judged to have created something worth creating.

As more than one observer has noted, it is ironic that Hardy is perhaps most widely known for a discovery about genetics. A theorem that he and Wilhelm Weinberg independently proved is well known today as the Hardy–Weinberg principle.

But for Hardy, who lived through two world wars, number theory provided a retreat that was, thankfully, useless to military planners. Borrowing from his own article, "Mathematics in Wartime," published in the journal *Eureka* in 1940, Hardy wrote in *Apology* the same year that "When the world is mad, a mathematician may find in mathematics an incomparable anodyne."

According to Hardy, the mathematician's world is directly linked to reality. Theorems are non-negotiable. In contrast, he says, the scientist's reality is merely a model. "A chair may be a collection of whirling electrons, or an idea in the mind of God," he declared. "Each of these accounts of it may have merits, but neither conforms at all closely to the suggestions of common sense." The pure mathematician need not be tethered to physical facts. In Hardy's words, "'Imaginary' universes are so much more beautiful than this stupidly constructed 'real' one."

Frederick Soddy, who had helped the world understand radioactivity, was disgusted by such sentiments. In his review in *Nature*, he said that if Hardy were taken seriously, then the "real mathematician" would be a "religious maniac."

Hardy was aware of Soddy's review. He might have been amused by it. A letter to him from R. J. L. Kingsford at Cambridge University Press, dated January 1941, concluded, "I quite agree that Soddy's amazing review in *Nature* is a most valuable advertisement. I enclose a copy of the review, herewith."

Another condemnation of *A Mathematician's Apology* came from E. T. Bell, a mathematician and science fiction writer who is best remembered for his 1937 book *Men of Mathematics*. In his review, published in *The Scientific Monthly* in 1942, Bell recommended Hardy's book to "solemn young men who believe they have a call to preach the higher arithmetic to mathematical infidels." He concluded, "Congenital believers will embrace [Hardy's book] with joy, possibly as a compensation for the loss of their religious beliefs of their childhood."

It Won't Make a Nickel for Anyone

Hardy had intended to publish *A Mathematician's Apology* with Cambridge University Press at his own expense. However, Press Secretary S. C. Roberts recognized the value of the ninety-page essay and endorsed it to the Syndics, the governing body of the press. The book was reprinted twice, the second time in 1948, the year following Hardy's death. In June 1952, Hardy's sister wrote to the press:

> As *A Mathematician's Apology* is now impossible to get, both first hand and second hand, I expect that you will in time be reprinting it; I think that it would be a good idea to have a photograph of my brother in it granted that it did not make it too expensive. I have the negative of which the enclosed photograph is an enlargement; it is an amateur snap and extremely characteristic.

The photograph that she had sent would eventually appear on the dust jacket of the second edition, and has become a well-known image of her brother.

Reissuing *A Mathematician's Apology* would be difficult. Inflation in Britain had made it impossible to reprint so small a book at a reasonable sales price. Some sort of material would be needed to extend it, but none of Hardy's academic lectures seemed appropriate.

Hardy (far right) and his protégé Srinivasa Ramanujan (center) are shown with colleagues at Cambridge University. After sending Hardy many theorems in letters, Ramanujan worked closely with Hardy for five years on various aspects of number theory, including highly composite numbers, which are positive integers with more divisors than any smaller positive integer. Ramanujan received a doctorate from Cambridge for this work in 1916. Reproduced by permission of Cambridge University Press.

In 1959, eleven years later, chemist and writer C. P. Snow suggested that he might write an introduction. It seemed a superb idea. Snow had advised Cambridge University Press during the war years. He was a close friend of Hardy and had offered him advice about the book. However, Snow would not commit to a deadline. An internal memorandum from the current Secretary, R. W. David, in September 1966 complained that "we have been chasing Snow for copy at roughly yearly intervals."

After many incidents, the second edition of *A Mathematician's Apology* was finally in bookstores by the end of 1967. Snow's contribution added literary charm. It began:

> It was a perfectly ordinary night at Christ's high table, except that Hardy was dining as a guest. . . . This was 1931, and the phrase was not yet in English use, but in later days they would have said that in some indefinable way he had star quality.

As Cambridge University Press anticipated, the new edition of *A Mathematician's Apology* was received well in the United States. Byron Dobell, an author and editor in New York who helped many young writers, including Tom Wolfe and Mario Puzo, seasoned his praise with a sprinkle of caution:

> It is the kind of book you wish was being read by all your friends at the very moment when you are reading it yourself. It is one of those secret, perfect works that makes most writing seem like a mixture of lead and mush. It's the under-the-counter book we're touting this month. It has nothing to do with anything but the joy of life and mind. The price is $2.95 and, with a title like that, it won't make a nickel for anyone.

Physical Connections

The second edition of *A Mathematician's Apology* appeared as mathematics was becoming increasingly abstract. Many mathematicians rejoiced at this change of direction in their field. Others lamented. If the trend continued, some believed, mathematics would become irrelevant.

One mathematician who celebrated was University of Chicago professor Marshall Stone. His article "The Revolution in Mathematics," which first appeared in the journal *Liberal Education* in 1961 and was reprinted the same year in *American Mathematical Monthly*, saw abstraction bridging areas of mathematics that had previously been isolated islands of thought. The identification of mathematics and logic, he argued, was greatly responsible:

> Mathematics is now seen to have no necessary connections with the physical world beyond the vague and mystifying one implicit in the statement that thinking takes place in the brain. The discovery that this is so may be said without exaggeration to mark one of the most significant intellectual advances in the history of mankind.

Stone noted a paradox: Increasing abstraction was spawning new applications. He listed the mathematical theory of genetics and game theory, as well as the mathematical theory of communications with contributions to linguistics.

A very different opinion was expressed the following year in "Applied mathematics: What is needed in research and education," published in *SIAM Review*. It was the transcript of a symposium chaired by mathematician H. J. Greenberg. Its panel consisted of mathematicians George Carrier, Richard Courant, and Paul Rosenbloom, and physicist C. N. Yang. Stone's article, with its embrace of abstraction, was discussed with alarm. The panel members urged a more traditional vision of mathematics, one that draws its inspiration from science. Courant's warning sounded like a review of *A Mathematician's Apology*:

> We must not accept the old blasphemous nonsense that the ultimate justification of mathematical science is "the glory of the human mind." Mathematics must not be allowed to split and diverge towards a "pure" and an "applied" variety.

Despite Courant's warning, a line between pure and applied mathematics exists at most universities today. Too often it is a battle line, witnessing skirmishes over scant resources and bruised egos. It is a line that has perhaps been blurred a bit by pure mathematicians' widespread use of computers and technology's urgent need for sophisticated algorithms. Mathematicians who share Hardy's sentiments might feel reluctant to express them in the face of soaring costs of higher education. Students with mounting debts have become increasingly impatient with teachers who digress from material directly needed for their exams. Administrators drool over research grants in medicine and cyber-security while finding less filling the meager grants awarded in pure mathematics.

The line between pure and applied mathematics might be blurred, but it will not soon be erased. As long as it exists, G. H. Hardy's *A Mathematician's Apology* will be read and—usually—enjoyed. No finer summary can be offered than that written by J. F. Randolph in his 1942 review:

> This book is not only about mathematics, it is about ideals, art, beauty, importance, significance, seriousness, generality, depth, young men, old men, and G. H. Hardy. It is a book to be read, thought about, talked about, criticized, and read again.

G. H. Hardy: Mathematical Biologist

Hannah Elizabeth Christenson
and Stephan Ramon Garcia

Godfrey Harold Hardy (1877–1947), the magnificent analyst who "discovered" the enigmatic Ramanujan and penned *A Mathematician's Apology*, is most widely known outside of mathematics for his work in genetics. Hardy's fame stems from a condescending letter to the editor in *Science* concerning the stability of genotype distributions from one generation to the next. His result is known as the Hardy–Weinberg law, and every biology student learns it today.

How did Hardy, described by his colleague, C. P. Snow, as "the purest of the pure" [8], become one of the founders of modern genetics? What would Hardy say if he knew that he had earned scientific immortality for something so mathematically simple?

In a lecture delivered by R. C. Punnett (of Punnett square fame), the statistician Udny Yule raised a question about the behavior of the ratio of dominant to recessive traits over time. This led Punnett to question why a population does not increasingly tend toward the dominant trait. He was confused and brought the question to his colleague, G. H. Hardy, with whom he frequently played cricket (for the complete story, see references [2 and 3]).

Under certain natural assumptions, Hardy demonstrated that there is an equilibrium at which the ratio of different genotypes remains constant over time (this result was independently obtained by the German physician Wilhelm Weinberg). There is no deep mathematics involved; the derivation of the Hardy–Weinberg law involves only "mathematics of the multiplication-table type" [6]. Hardy's brief letter dismisses Yule's criticism of Mendelian genetics:

> I am reluctant to intrude in a discussion concerning matters of which I have no expert knowledge, and I should have expected

> the very simple point which I wish to make to have been familiar to biologists. . . . there is not the slightest foundation for the idea that a dominant character should show a tendency to spread over a whole population, or that a recessive should tend to die out [6].

Hardy's letter was short, tinted with contempt, and possibly unnecessary. Geneticist A. W. F. Edwards refers to the affair as "a problem that, if both parties had paid more attention to Mendel's paper itself, should never have arisen" [2]. According to the geneticist J. F. Crow, the Hardy–Weinberg law "is so self-evident that it hardly needed to be 'discovered'" [1].

This was not an argument that Hardy sought. Punnett reflected that "'Hardy's Law' owed its genesis to a mutual interest in cricket" [7]. If they had not played cricket together, Punnett probably would not have asked Hardy about the problem in the first place. Hardy certainly would never have developed an interest in it otherwise, for his aversion to applied mathematics was legendary:

> [I]s not the position of an ordinary applied mathematician in some ways a little pathetic? . . . "Imaginary" universes are so much more beautiful than this stupidly constructed "real" one [5, p. 135].

Although Titchmarsh tells us that Hardy "attached little weight to it" [9], the ubiquity of the Hardy–Weinberg law in introductory biology texts indicates the seminal nature of the result. This contradicts Hardy's bold confession:

> I have never done anything "useful." No discovery of mine has made, or is likely to make, directly or indirectly, for good or ill, the least difference to the amenity of the world. . . . Judged by all practical standards, the value of my mathematical life is nil; and outside mathematics it is trivial anyhow [5, p. 150].

However, if we scrutinize Hardy's views and personality more closely, we might gain a more nuanced perspective. Hardy did not detest applications entirely; he instead took pride in the uselessness of his work because it freed him from contributing to the terrors of war and violence:

> But here I must deal with a misconception. It is sometimes suggested that pure mathematicians glory in the uselessness of their work. If the theory of numbers could be employed for any

> practical and obviously honorable purpose, if it could be turned directly to the furtherance of human happiness or the relief of human suffering . . . then surely neither Gauss nor any other mathematician would have been so foolish as to decry or regret such applications. But science works for evil as well as for good (and particularly, of course, in time of war) . . . [5, pp. 120–121].

As an avid atheist, Hardy saw God, rather than applied scientists, as his "personal enemy" [9]. In fact, Hardy was the President of the Association of Scientific Workers from 1924–1926:

> [Hardy] said sarcastically that he was an odd choice, being "the most unpractical member of the most unpractical profession in the world." But in the important things he was not so unpractical [8].

Hardy tells us that "the noblest ambition is that of leaving behind something of permanent value." "To have produced anything of the slightest permanent interest," he says, "whether it be a copy of verses or a geometrical theorem, is to have done something utterly beyond the powers of the vast majority of men." Mathematics lends itself to this form of immortality: "Archimedes will be remembered when Aeschylus is forgotten, because languages die and mathematical ideas do not."

Hardy applauded "the permanence of mathematical achievement," regardless of its applicability to the outside world. Concerning the theorems of Euclid and Pythagoras, "each is as fresh and significant as when it was discovered—two thousand years have not written a wrinkle on either of them," despite the fact that "neither theorem has the slightest practical importance." Although he lauded the permanence of mathematical achievement, Hardy was an "anti-narcissist" who "could not endure having his photograph taken . . . He would not have any looking glass in his rooms, not even a shaving mirror." However, he clearly wanted to accomplish something everlasting, for "mathematics was his justification" [8].

G. H. Hardy achieved immortality, although his most famous accomplishment is not within his own exalted field of pure mathematics; nor is it in a field to which he attached any value. Crow conjectures,

> It must have embarrassed him that his mathematically most trivial paper is not only far and away his most widely known, but has been of such distastefully practical value. He published this paper not in the obvious place, *Nature*, but across the Atlantic in *Science*.

> Why? It has been said that he didn't want to get embroiled in the bitter argument between the Mendelists and biometricians. I would like to think that he didn't want it to be seen by his mathematician colleagues [1].

Further speculation regarding Hardy's choice of venue can be found in reference [4].

No one can know with certainty how Hardy would react now, over one hundred years later, to the impact his letter in *Science* had. There is more complexity and depth to him than can be gleaned from his writings, not even in combination with accounts from those who knew him. However interesting and revealing details may be (like those Titchmarsh provided in Hardy's obituary—he liked Scandinavia, cats, and detective stories, but not dogs, politicians, or war [9]), they will never provide a complete picture.

Reflecting on his life, Hardy considered it to be a success in terms of the happiness and comfort that he found, but the question remained as to the "triviality" of his life. He resolved it accordingly:

> The case for my life . . . is this: that I have added something to knowledge, and helped others to add more; and that these somethings have a value which differs in degree only, and not in kind, from that of the creations of the great mathematicians, or of any of the other artists, great or small, who have left some kind of memorial behind them [5, p. 151].

In all great proofs, Hardy asserted that

> there is a very high degree of unexpectedness, combined with inevitability and economy. The arguments take so odd and surprising a form; the weapons used seem so childishly simple when compared with the far-reaching results; but there is no escape from the conclusions [5, p. 113].

Extending the scope of these criteria beyond mathematics, one can argue that the Hardy–Weinberg law meets these standards.

References

[1.] James F. Crow, "Eighty Years Ago: The Beginnings of Population Genetics," *Genetics*, Volume **19,** Issue 3 (1988), pp. 473–476.

[2.] A. W. F. Edwards, "G. H. Hardy (1908) and Hardy-Weinberg Equilibrium," *Genetics*, Volume **179,** Issue 3 (2008), pp. 1143–1150.

[3.] Colin R. Fletcher, "G. H. Hardy—Applied Mathematician," *Bulletin—Institute of Mathematics and Its Applications*, Volume **16,** Issue 2–3 (1980), pp. 61–67.

[4.] Colin R. Fletcher, "Postscript To: G. H. Hardy—Applied Mathematician," *Bulletin—Institute of Mathematics and Its Applications*, Volume **16,** Issue 11–12 (1980), p. 264.

[5.] G. H. Hardy, *A Mathematician's Apology* (with a foreword by C. P. Snow), reprint of the 1967 edition, Cambridge University Press, Cambridge, U.K., 1992.

[6.] G. H. Hardy, "Mendelian Proportions in a Mixed Population," *Science*, Volume **28,** Number 706 (July 10, 1908), pp. 49–50.

[7.] R. C. Punnett, "Early Days of Genetics," *Heredity*, Volume **4,** Issue 1 (1950), pp. 1–10.

[8.] C. P. Snow, Foreword to *A Mathematician's Apology*, Cambridge University Press, London, 1967.

[9.] E. C. Titchmarsh, "Godfrey Harold Hardy. 1877–1947," *Obituary Notices of Fellows of the Royal Society*, Volume **6,** Issue 18 (1949), pp. 446–461.

The Reasonable Ineffectiveness of Mathematics

DEREK ABBOTT

The nature of the relationship between mathematics and the physical world has been a source of debate since the era of the Pythagoreans. A school of thought, reflecting the ideas of Plato, is that mathematics has its own existence. Flowing from this position is the notion that mathematical forms underpin the physical universe and are out there waiting to be discovered.

The opposing viewpoint is that mathematical forms are objects of our human imagination and we make them up as we go along, tailoring them to describe reality. In 1921, this view led Einstein to wonder, "How can it be that mathematics, being after all a product of human thought which is independent of experience, is so admirably appropriate to the objects of reality?" [1].

In 1959, Eugene Wigner coined the phrase "the unreasonable effectiveness of mathematics" to describe this "miracle," conceding that it was something he could not fathom [2]. The mathematician Richard W. Hamming, whose work has been profoundly influential in the areas of computer science and electronic engineering, revisited this very question in 1980 [3].

Hamming raised four interesting propositions that he believed fell short of providing a conclusive explanation [3]. Thus, like Wigner before him, Hamming resigned himself to the idea that mathematics is unreasonably effective. These four points are (1) we see what we look for, (2) we select the kind of mathematics we look for, (3) science in fact answers comparatively few problems, and (4) the evolution of man provided the model.

In this article, we will question the presupposition that mathematics is as effective as claimed and thus remove the quandary of Wigner's "miracle," leading to a non-Platonist viewpoint.[1] We will also revisit Hamming's four propositions and show how they may indeed largely explain that there is no miracle, given a reduced level of mathematical effectiveness.

The reader will be asked for a moment of indulgence, where we will push these ideas to the extreme, extending them to all physical law and models. Are they all truly *reified*? We will question their absolute reality and ask the question: Have we, in some sense, generated a partly anthropocentric physical and mathematical framework of the world around us?

Why should we care? Among scientists and engineers, there are those who worry about such questions and there are those who prefer to "shut up and calculate." We will attempt to explain why there might be a useful payoff in resolving our philosophical qualms and how this might assist our future calculations.

I. Mathematicians, Physicists, and Engineers

The following is anecdotal and is by no means a scientific survey. However, in my experience of interacting with mathematicians, physicists, and engineers, I would estimate that about 80% of mathematicians lean to a Platonist view.[2] Physicists, on the other hand, tend to be closeted non-Platonists. An ensemble of physicists will often appear Platonist in public, but when pressed in private I can often extract a non-Platonist confession.

Engineers by and large are openly non-Platonist. Why is that? Focusing on electrical and electronic engineering, as a key example, the engineer is well-acquainted with the art of approximation. An engineer is trained to be aware of the frailty of each model and its limits when it breaks down. For example, we know that lumped circuit models are only good for low frequencies.

An engineer is also fully aware of the artificial contrivance in many models. For example, an equivalent circuit only models the inputs and outputs of a circuit and ignores all the internal details. Moreover, the engineer knows the conditions under which these simplifications can be exploited.

An engineer often has control over his or her "universe" in that if a simple linear model does not work, the engineer, in many cases, can force a widget, by design, to operate within a restricted linear region. Thus, where an engineer cannot approximate linearity, he or she often linearizes by *fiat*.

A mathematical Platonist will often argue that number π is a real entity, claiming that a geometric circle is a reified construct that exists independently of the universe. An engineer, on the other hand, has no difficulty in seeing that there is no such thing as a perfect circle anywhere in the physical universe, and thus π is merely a useful mental construct.

In addition to the circle, many other ideal mathematical forms, such as delta functions, step functions, and sinusoids, are in an engineer's mathematical toolbox and used on a daily basis. Like the circle, the engineer sees delta functions, and for that matter all functions, as idealities that do not exist in the universe. Yet they are useful for making sufficiently accurate, yet approximate, predictions.

A physicist may have nightmares on studying a standard electronic engineering text, finding the use of negative time in the theory of noncausal filters. However, a non-Platonist engineer has no qualms about such transformations into negative spaces, as there is no ultimate reality there. These are all mental constructs and are dealt with in a utilitarian way, producing the results required for system design.

Hamming's paper marvels on how complex numbers so naturally crop up in many areas of physics and engineering, urging him to feel that "God made the universe out of complex numbers" [3]. However, for the engineer, the complex number is simply a convenience for describing rotations [7], and, of course, rotations are seen everywhere in our physical world. Thus, the ubiquity of complex numbers is not magical at all. As pointed out by Chappell et al. [8], Euler's remarkable formula $e^{-j\pi} = -1$ is somewhat demystified once one realizes it merely states that a rotation by π radians is simply a reflection or multiplication by -1.

Engineers often use interesting mathematics in entirely nonphysical spaces. For example, the support vector machine (SVM) approach to classifying signals involves transforming physical data into nonphysical higher dimensional spaces and finding the optimal hyperplanes that separate the data. In telecommunications, coding theory can also exploit higher dimensional spaces [9]. In both these examples, physically useful outcomes result from entirely mental abstractions of which there are no analogs in the physical universe.

II. Do Fractals Have Their Own Existence?

Roger Penrose, a mathematical Platonist, argues that a fractal pattern is proof of a mathematical entity having an existence of its own [6]. It is argued that the mathematician cannot foresee a beautiful fractal before applying a simple iterative equation. Therefore, a fractal pattern is not a mental construct but has its own existence on a Platonic plane waiting to be discovered.

A first objection is that there is an infinite number of ways to display the fractal data and that to "see" a fractal we have to anthropocentrically display the data in the *one* way that looks appealing to our senses. Perhaps to an alien, a random pattern based on white noise might be more beautiful?

A second objection is that out of an infinite number of possible iterative equations, perhaps only negligible numbers of them result in fractal patterns and even fewer look appealing to humans. Take the analogy of a random sequence of digits. We know any infinite random sequence encodes all the works of Shakespeare and all the world's knowledge. If we preselect appealing parts of a random sequence, we have in fact cheated.

At the end of the day, a given set of rules that turns into an elegant fractal is really no different than, say, the set of rules that form the game of chess or that generate an interesting cellular automaton. The set of moves in a game of chess is evidently interesting and richly beautiful to us, but that beauty is no evidence that chess itself has a Platonic existence of its own. Clearly, the rules of chess are purely a contrived product of the human mind and not intrinsic to nature.

A Platonist will argue that mathematical forms follow from a set of axioms, and thus exist independently of our knowledge of them. This situation is no different than our lack of foreknowledge of a fractal pattern before exercising its originating equation. What can we say of the axioms themselves? I argue that they are also mental abstractions, and an example is given in Section V to illustrate that even the simple counting of objects has its physical limits. Thus, axioms based on the assumption of simple counting are not universally real.

III. The Ineffectiveness of Mathematics

So far, we have argued that mathematics is a merely mental abstraction that serves useful purposes. A further response to answer Wigner's thought that the effectiveness of mathematics is a "miracle" is to suggest that this effectiveness might be overstated.

What we are finding in electronic engineering is that the way we mathematically model and describe our systems radically changes as we approach the nanoscale and beyond. In the 1970s, when metal oxide semiconductor field-effect transistor (MOSFET) lengths were of the order of micrometers, we were able to derive from physical first principles elegant analytical equations that described transistor behavior, enabling us to design working circuits. Today, we produce deep submicrometer transistors, and these analytical equations are no longer usable, as they are swamped with too many complicated higher order effects that can no longer be neglected at the small scale. Thus, in practice, we turn to empirical models that are embedded in today's computer simulation software for circuit design. Traditional analytical mathematics simply fails to describe the system in a compact form.

Another example is the use of Maxwell's equations for modeling integrated electromagnetic devices and structures. In modern devices, due to the complexity of design, we no longer resort to analytical calculations; instead, electromagnetic simulation programs that use numerical methods are now the standard approach.

The point here is that when we carry out engineering in different circumstances, the way we perform mathematics changes. Often the reality is that when analytical methods become too complex, we simply resort to empirical models and simulations.

The Platonist will point out that the inverse square law for gravitation is spectacularly accurate at predicting the behavior of nearby planets and distant stars across vast scales. However, is that not a self-selected case conditioned on our human fascination with a squared number? Furthermore, due to inherent stochasticity in any physical system, at the end of the day, we can only ever experimentally verify the square law to within a certain accuracy. While the Newtonian view of gravitation is a spectacularly successful model, it does not hold what we believe to be the underlying reality; it has been surpassed by the

4-D curved spaces of general relativity, and this is now the dominant viewpoint until a better theory comes along.

Note that mathematics has lesser success in describing biological systems, and even less in describing economic and social systems. But these systems have come into being and are contained within our physical universe. Could it be that they are harder to model simply because they adapt and change on human time scales, and so the search for useful invariant properties is more challenging? Could it be that the inanimate universe itself is no different, but happens to operate on a time scale so large that in our anthropocentrism we see the illusion of invariance?

An energy-harvesting device that is in thermal equilibrium cannot extract net energy or work from its environment. However, if we imagine that human lifespans are now reduced to the time scale of one thermal fluctuation, the device now has the illusion of performing work. We experience the Sun as an energy source for our planet, partly because its lifespan is much longer than human scales. If the human lifespan were as long as the universe itself, perhaps our Sun would appear to be short-lived fluctuation that rapidly brings our planet into thermal equilibrium with itself as it "blasts" into a red giant. These extreme examples show how our anthropocentric scales possibly affect how we model our physical environment.

A. Hamming's First Proposition: We See What We Look For

Hamming suggests here that we approach problems with a certain intellectual apparatus, and, thus, we anthropocentrically select out that which we can apply our tools to [3]. Our focus shifts as new tools become available. In recent years, with the emerging paradigms of complex systems and mining of so-called *big data*, traditional mathematics has a smaller role and large brute-force computing is used to search for the patterns we are looking for.

B. Hamming's Second Proposition: We Select the Kind of Mathematics We Look For

Here, Hamming points out that we tailor mathematics to the problem at hand [3]. A given set of mathematical tools for one problem does not necessarily work for another. The history of mathematics shows a continual development; for example, scalars came first, then we developed vectors, then tensors, and so on. So as fast as mathematics falls short, we invent new mathematics to fill the gap.

By contrast, a Platonist will argue for the innateness of mathematics by pointing out that we sometimes invent useful mathematics before it is needed. For example, Minkowski and Riemann developed the theory of 4-D curved spaces in the abstract, before Einstein found it of utility for general relativity. I argue that this innateness is illusory, as we have cherry-picked a successful coincidence from a backdrop of many more cases that are not as fortuitous.

C. Hamming's Third Proposition: Science Answers Comparatively Few Problems

Taking into account the entire human experience, the number of questions that are tractable with science and mathematics is only a small fraction of all the possible questions we can ask. Gödel's theorem also set limits on how much we can actually prove. Mathematics can appear to have the illusion of success if we are preselecting the subset of problems for which we have found a way to apply mathematics.

A case in point is the dominance of linear systems. Impressive progress has been made with linear systems because the ability to invoke the principle of superposition results in elegant mathematical tractability. On the other hand, developments in nonlinear systems have been arduous and much less successful. If we focus our attention on linear systems, then we have preselected the subset of problems where mathematics is highly successful.[3]

D. Hamming's Fourth Proposition: The Evolution of Man Provided the Model

A possibility is that the quest for survival has selected those who are able to follow chains of reasoning to understand local reality. This idea implies that the intellectual apparatus we use is in some way already appropriate. Hamming points out that, to some extent, we know that we are better adapted to analyzing the world at our human scale, given that we appear to have the greatest difficulties in reasoning about the very small scale and the very large scale aspects of our universe.

E. Physical Models as a Compression of Nature

There is a fifth point we might add to Hamming's four propositions, and that is that all physical laws and mathematical expressions of those laws are a compression or compact representation. They are necessarily compressed due to the limitations of the human mind. Therefore, they are compressed in a manner suited to the human intellect. The real world is inherently noisy and has a stochastic component, so physical models are idealizations with the rough edges removed.

Thus, when we "uncompress" a set of equations to solve a given problem, we will obtain an idealized result that will not entirely match reality. This can be thought of as uncompressing a video that was initially subjected to lossy compression.[4] There will always be lossy information leakage, but provided the effects we have neglected are small, our results will be useful.

F. Darwinian Struggle for the Survival of Ideas

A sixth point we can add to Hamming's list is that Wigner's sense of "magic" can be exorcised if we see that the class of successful mathematical models is preselected. Consider the millions of failed models in the minds of researchers, over the ages, which never made it on paper because they were wrong. We tend to publish the ones that have survived some level of experimental vindication. Thus, this Darwinian selection process results in the illusion of automatic success; our successful models are merely selected out from many more failed ones.

Take the analogy of a passenger on a train pulling the emergency stop lever, saving the life of a person on a railway track; this seems like a miracle. However, there is no miracle once we look at the *prior* that many more people have randomly stopped trains on other occasions saving no lives. A genius is merely one who has a great idea but has the common sense to keep quiet about his or her other thousand insane thoughts.

IV. What About the Aliens?

Mathematical Platonists often point out that a hypothetical alien civilization will most likely discover the number π and put it to good use in their alien mathematics. This is used to argue that π has its own Platonic existence, given that it is "out there" for any alien to independently discover.

Do aliens necessarily know number π? Do aliens even have the same view of physics? Given the simplicity of geometric objects such as ideal circles and squares, an alien race may indeed easily visualize them. However, this is not true of all our mathematical objects, especially for those with increased complexity. For example, an alien race may never find the Mandelbrot set and may not even pause to find it interesting if found by chance.

An alien race might happily do all its physics and engineering without the invention of a delta function. Perhaps the aliens have parameterized all their physical variables in a clever way, and if we were to compare, we would find that one of our variables was surprisingly redundant.

Perhaps not all aliens have a taste for idealizations, nor Occam's razor. Maybe all their physical equations are stochastic in nature, thereby realistically modeling all physical phenomena with inherent noise.

One might also hypothesize a superintelligent alien race with no need for long chains of analytical mathematical reasoning. Perhaps their brains are so powerful that they jump straight into performing vast numerical simulations, based on empirical models, in their heads. So the question of the effectiveness of mathematics, as we know it, has no meaning for them. This thought experiment also illustrates that human mathematics serves us to provide the necessary compression of representation required by our limited brain power.

V. One Banana, Two Banana, Three Banana, Four

I deeply share Hamming's amazement at the abstraction of integers for counting [3]. Observing that six sheep plus seven sheep make thirteen sheep is something I do not take for granted either.

A deceptively simple example to illustrate the limitations in the correspondence between the ideal mathematical world and reality is to dissect the idea of simple counting. Imagine counting a sequence of, say, bananas. When does one banana end and the next banana begin? We think we know visually, but to formally define it requires an arbitrary decision of what minimum density of banana molecules we must detect to say we have no banana.

To illustrate this to its logical extreme, imagine a hypothetical world where humans are not solid but gaseous and live in the clouds. Surely, if we evolved in such an environment, our mathematics would not so readily encompass the integers? This relates to Hamming's Fourth Proposition, where our evolution has played a role in the mathematics we have chosen.

Consider the physical limits when counting a very large number of bananas. Imagine we want to experimentally verify the one-to-one correspondence between the integer number line, for large N, with a sequence of physical bananas. We can count bananas, but for very large N, we need memory to store that number and keep incrementing it. Any physical memory will always be subject to bit errors and noise, and, therefore, there are real physical limits to counting.

An absolute physical limit is when N is so large that the gravitational pull of all the bananas draws them into a black hole.[5] Thus, the integer number line is lacking in absolute reality. Davies goes a step further and argues that real numbers are also a fiction; they cannot be reified because the universe can store at most 10^{122} bits of information [11].

VI. Strong Non-Platonism

For the purposes of this essay, we have loosely labeled mathematical Platonism as the position that ideal mathematical objects exist and they are waiting to be discovered. Similarly, physical laws are also reified.

What we loosely refer to as non-Platonism is the view that mathematics is a product of human imagination and that all our physical laws are imperfect. Nature is what it is, and by *physical law* we are, of course, referring to human compression of nature.

The reader is now asked to entertain strong non-Platonism, where all physical laws are tainted with anthropocentrism and all physical models have no real interpretative value. The interpretive value of physics is purely illusory. After all, a beam of light passing through a slit knows nothing of Fourier transforms; that is an overlaid human construct.

Imagine 3-D particles passing through a 2-D universe. A 2-D flatlander [12] can create beautiful interpretations, which may even have some predictive accuracy, regarding these mysterious particles that appear, change size, and then disappear. But these interpretations are to some extent illusory and at best incomplete.

In our world, we are trapped on human length scales, human power scales, and human time scales. We have created clever instruments that extend our reach, but we are hopelessly lacking in omnipotence.

In some cases, we knowingly build up a set of models with imaginary interpretive value purely for convenience. For example, we can measure the effective mass and drift velocity of holes in a semiconductor, knowing fully well that semiconductor holes are an imaginary artifice. We exploit them as a mental device because they provide a shortcut to giving us predictive equations with which we can engineer devices.

John von Neumann stated all this more succinctly: “The sciences do not try to explain, they hardly even try to interpret, they mainly make models. By a model is meant a mathematical construct which, with the addition of certain verbal interpretations, describes observed phenomena. The justification of such a mathematical construct is solely and precisely that it is expected to work” [13].

VII. Immutability

Another way to see the potential frailty of physical “laws” created by humans is to ask which principles in physics are sacred and immutable? I will leave this as an exercise for the reader. However, when I tried the thought experiment, I was able to stretch my imagination to permit a violation of everything we know. At some vast or small scale of any set of parameters, one can imagine breakdowns in the laws, as we know them.

Is there anything we can hold onto as inviolate under any circumstances? What about Occam’s razor? I would like to hold onto Occam’s razor as immutable, but I fear that it too may be embedded with anthropocentrism. When classifying physical data, it is known that God does not always shave with Occam’s razor [14]. Could it be that, as the human brain demands a compression of nature, Occam’s razor is our mental tool for sifting out compact representations?

VIII. A Personal Story

As this is an opinion piece, it might be pertinent to understand where my opinions come from. I have a distinct memory of being alone playing on the floor, at the age of four, with a large number of cardboard boxes strewn across the room. I counted the boxes. Then, I counted them again and obtained a different number. I repeated this a few times obtaining different numbers. This excited me because I thought it was magic and that boxes were appearing and disappearing. But the magic unfortunately disappeared, and I eventually kept obtaining a run of the same number. In a few minutes, I concluded that my initial counting was inaccurate and that there sadly never was any magic. This was my first self-taught lesson in experimental repeatability and the removal of magic from science.

At both elementary school and high school, mathematics was my favorite subject, although I spent far too many years worrying about the concept of infinity. Taking a limit to infinity was something I simply got used to, minus the desire to wildly embrace it. I struggled with accepting negative numbers, and raising numbers to the power of zero seemed absurd.[6] I remember a great sense of disappointment when I was told that vectors could not be divided. Something was not quite right, but then I could not put my finger on it. After all, complex numbers contain a direction and magnitude, yet they can be divided. The more mathematics I learned, the more it seemed like an artificial hodgepodge of disparate tools, rather than a divine order.

While I loved the beauty of mathematical proofs and the search for them, it worried me that each proof needed creative ad hoc handcrafting; there was no heavenly recipe book. The nature of proofs began to appear philosophically suspect to me; for example, how do we really know if a proof is correct if it is too long? A mathematical proof is the demonstration that a proposition is correct with a level of certainty that two mathematicians somewhere in the world understand it; that was in jest, of course, but the proof of Fermat's last theorem is arguably close to pushing that boundary.

At the age of nineteen, in my undergraduate university library, I stumbled on a textbook that changed my life. In its introduction, it stated that mathematics is a product of the human mind. Obviously, all my teachers must have been mathematical Platonists, as I had never heard such an outlandish statement before. Immediately, a great burden lifted from my shoulders, and my conversion to non-Platonism was instant. This was a road to Damascus experience for me, and my philosophical difficulties that haunted me vanished.

As Hamming aptly states, "The postulates of mathematics were not on the stone tablets that Moses brought down from Mt. Sinai" [3].

IX. Why Not Just Shut Up and Calculate?

Why should we care about the nature of mathematics? My personal story for one illustrates that there is greater freedom of thought, once we realize that mathematics is something we entirely invent as we go along. This view can move us ahead and free us from an intellectual straitjacket. With the shackles removed, we can proactively manipulate, improve, and apply mathematics at a greater rate.

If we discard the notion that mathematics is passed down to us on stone tablets, we can be more daring with it and move into realms previously thought impossible. Imagine where we could be now if the centuries of debate over negative numbers could have been resolved earlier.

Another problem with mathematics today is the lack of uniformity in the tools we use. For example, we have the Cartesian plane and the Argand plane. They are isomorphic to each other, so why must we have both? We have complex numbers and quaternions. We have scalars, vectors, and tensors. Then, we have rather clunky dot and cross products, where the cross product does not generalize to higher dimensions.

It turns out to be something of a historical accident that the vector notation, with dot and cross products, was promoted by Gibbs and Heaviside, giving us a rather mixed bag of different mathematical objects [20].

Clifford's geometric algebra, on the other hand, unifies all these mathematical forms [8], [15]–[17]. It uses Cartesian axes and replaces complex numbers, quaternions, scalars, vectors, and tensors all with one mathematical object called the *multivector*. Dot and cross products are replaced with one single operation called the *geometric product*. This new type of product is elegant and follows the elementary rules for multiplying out brackets, with the extra rule that elements do not commute. You cannot divide traditional vectors, but multivectors do not have this restriction. All the properties naturally extend to higher dimensions, and thus the limitations of the cross product are overcome. This formalism is therefore simple and powerful, and it delivers improved mathematical compression tailored for the limited human mind.

While this approach has existed since 1873, it has been largely sidelined, as Gibbs and Heaviside favored dot and cross products. However, in physics, engineering, and computer science there is an emerging interest in reviving this mathematics due to its power and simplicity. To this end, a tutorial paper on geometric algebra for electrical and electronic engineers has been written [18].

X. Conclusion

Science is a modern form of alchemy that produces wealth by producing the understanding for enabling valuable products to be made from base ingredients. Science is merely functional alchemy that has had a few incorrect assumptions fixed but has in its arrogance replaced them with more insidious ones. The real world of nature has the uncanny habit of surprising us; it has always proven to be a lot stranger than we give it credit for.

Mathematics is a product of the imagination that sometimes works on simplified models of reality. Platonism is a viral form of philosophical reductionism that breaks apart holistic concepts into imaginary dualisms. I argue that lifting the veil of mathematical Platonism will accelerate progress. In summation, Platonic ideals do not exist; however, ad hoc elegant simplifications do exist and are of utility, provided that we remain aware of their limitations.

Mathematics is a human invention for describing patterns and regularities. It follows that mathematics is then a useful tool in describing regularities we see in the universe. The reality of the regularities and invariances, which we exploit, may be a little rubbery, but as long as they are sufficiently rigid on the scales of interest to humans, then it bestows a sense of order.

Acknowledgment

This paper is based on a talk the author presented at a workshop entitled The Nature of the Laws of Physics, December 17–19, 2008, Arizona State University (ASU), Phoenix, AZ. The author would like to thank all the attendees who provided useful comments both for and against Platonism, including: Scott Aaronson of the Massachusetts Institute of Technology (MIT, Cambridge, MA); Paul C. W. Davies of ASU; George F. R. Ellis of the University of Cape Town (Cape Town, South Africa); Gregory J. Chaitin of IBM (Armonk, NY); Anthony J. Leggett of the University of Illinois at Urbana–Champaign (UIUC, Urbana, IL); N. David Mermin of Cornell University (Ithaca, NY); Leonard Susskind of Stanford University (Stanford, CA); and Steve Weinstein of the University of Waterloo (Waterloo, Ontario, Canada). The author is also grateful for a number of formative discussions on the topic, over the years, with Cosma Shalizi of Carnegie Mellon University (Pittsburgh) and Paul C. W. Davies of ASU. A special thanks goes to Anthony J. Leggett of UIUC for pointing out the problem of integer counting in the case of hypothetical gaseous beings—Tony was, in turn, inspired by Hawkins [19]. The author would also like to thank Kurt Wiesenfeld of Georgia Tech (Atlanta) for his thought experiment of an alien with computational mental powers sufficient to simulate the environment. The author asked a number of colleagues to proofread earlier drafts of this article, with strict instructions to be brutal. He is grateful to Robert E. Bogner, James M. Chappell, Paul C. W. Davies, Bruce R. Davis, George F. R. Ellis, Mark D. McDonnell, Adrian P. Flitney, Chris Mortensen, and William F. Pickard for supplying gracious brutality in the spirit of debate.

Notes

1. This explains the inverted title of the present article, "The reasonable ineffectiveness of mathematics."

2. The interested reader is referred to [4] for an entertaining view of the non-Platonist position, and [5] for a Platonist perspective.

3. One might remark that many fundamental processes rather successfully approximate linear models, and this may again seem like Wigner's magic. However, is this not self-referential? What we humans regard as "fundamental" tend to be those things that appear linear in the first place.

4. If a video suffers a loss in reproduction quality when it is uncompressed, this is due to information loss during the original compression process. This is then referred to as "lossy compression."

5. It is of interest to note here that Lloyd has exploited black holes to explore the physical limits of computation [10].

6. In retrospect, I am astonished with how my mind-set was so sixteenth century. I will argue that it is the ravages of Platonism that can lock us into that mold.

References

[1] A. Einstein, *Geometrie und Erfahrung*. Berlin: Springer-Verlag, 1921.

[2] E. P. Wigner, "The unreasonable effectiveness of mathematics in the natural sciences," *Commun. Pure Appl. Math.*, vol. XIII, pp. 1–14, 1960.

[3] R. W. Hamming, "The unreasonable effectiveness of mathematics," *Amer. Math. Monthly*, vol. 87, no. 2, pp. 81–90, 1980.

[4] D. C. Stove, *The Plato Cult and Other Philosophical Follies*. Oxford, U.K.: Blackwell, 1991.

[5] S. C. Lovatt, *New Skins for Old Wine: Plato's Wisdom for Today's World*. Boca Raton, FL: Universal, 2007.

[6] R. Penrose, *The Road to Reality: A Complete Guide to the Laws of the Universe*. New York: Knopf, 2004.

[7] E. O. Willoughby, "The operator j and a demonstration that $\cos\theta + j\sin\theta = e^{j\theta}$," *Proc. Inst. Radio Electron. Eng.*, vol. 26, no. 3, pp. 118–119, 1965.

[8] J. M. Chappell, A. Iqbal, and D. Abbott, *Geometric Algebra: A Natural Representation of Three-Space*. [Online]. Available: http:// arxiv.org/pdf/1101.3619.pdf.

[9] M. El-Hajjar, O. Alamri, J. Wang, S. Zummo, and L. Hanzo, "Layered steered space-time codes using multi-dimensional sphere packing modulation," *IEEE Trans. Wireless Commun.*, vol. 8, no. 7, pp. 3335–3340, 2009.

[10] S. Lloyd, "Ultimate physical limits to computation," *Nature*, vol. 406, no. 6799, pp. 1047–1054, 2000.

[11] P. C. W. Davies, *The Goldilocks Enigma: Why Is the Universe Just Right for Life?* London: Penguin, 2007.

[12] E. A. Abbott, *Flatland: A Romance of Many Dimensions*. London: Seeley, 1884.

[13] J. von Neumann, "Method in the physical sciences," *The Unity of Knowledge*, L. Leery, Ed. New York: Doubleday, 1955, pp. 157–164.

[14] H. Bensusan, "God doesn't always shave with Occam's razor—Learning when and how to prune," *Lecture Notes in Computer Science*. Berlin: Springer-Verlag, 1998, vol. 1398, pp. 119–124.

[15] D. Hestenes, *New Foundations for Classical Mechanics: Fundamental Theories of Physics*. New York: Kluwer, 1999.

[16] S. Gull, A. Lasenby, and C. Doran, "Imaginary numbers are not real—The geometric algebra of spacetime," *Found. Phys.*, vol. 23, no. 9, pp. 1175–1201, 1993.

[17] M. Buchanan, "Geometric intuition," *Nature Phys.*, vol. 7, no. 6, p. 442, 2011.

[18] J. M. Chappell, S. P. Drake, C. L. Seidel, L. J. Gunn, A. Iqbal, A. Allison, and D. Abbott, "Geometric algebra for electrical and electronic engineers," *Proceedings of the IEEE*, Vol. 102, no. 9, pp. 1340–1363, 2014.

[19] D. Hawkins, *The Language of Nature*. San Francisco: Freeman, 1964.

[20] J. M. Chappell, A. Iqbal, J. G. Hartnett, and D. Abbott, "The vector algebra war: A historical perspective," *IEEE Access*, Vol. 12, pp. 1997–2004, 2016.

Stacking Wine Bottles Revisited

Burkard Polster

Here are twenty-five bottles in a wine rack. We've placed the four bottles at the bottom so that the two on the far left and far right touch the vertical sides. Then we stacked the remaining bottles row by row as shown.

Because the bottles at the bottom are not equally spaced, the other rows end up not being level . . . except for the seventh row at the very top. Surprisingly, this is always the case, no matter where we place the middle two bottles in the bottom row.

More generally, if bottles are stacked like this, starting with n bottles in the bottom row that are not too widely spaced, then the $2n - 1$st row of bottles is level.

This pretty theorem was discovered by Charles Payan in 1989. The $n = 3$ case featured as a problem in Velleman and Wagon's fabulous collection of math-club problems, *Which Way Did the Bicycle Go?* (1996), and in Honsberger's *Mathematical Diamonds* (2003). The *Cut-the-Knot* website also has a number of pages with very nice interactive applets dedicated to this theorem; look for the page entitled *More Bottles in a Wine Rack* and follow the links from that page. Also part of these pages are proofs of the general theorem and some generalizations by Nathan Bowler. The only other published proof we are aware of appears in a post by David Robbins to the Math Forum under the topic *Stacking Bottles in a Crate.*

In terms of generalizations, Robbins's post covers about the same ground as Nathan Bowler does on *Cut-the-Knot.*

None of the abovementioned proofs make for light reading, and one of the aims in this note is to present easily accessible proofs of the main results mentioned on *Cut-the-Knot* and in the Math Forum post that should be very close to those "in-the-book," proofs that we hope are worthy of such pretty results. We'll also be considering a number of new natural generalizations of the bottle-stacking problem.

Here is the plan. To start with, in order to avoid getting bogged down in notation, we'll restrict ourselves to discussing the case of four bottles in the bottom row and seven rows, as in our example. The proof for this case features all the arguments that are required to understand the general case.

We first show that all our stacks have a half-turn symmetry. This symmetry is apparent in our sample stack—the picture of our stack consisting of seven rows stays unchanged when rotated one hundred eighty degrees around the center blue bottle. Because such a half-turn transforms the level bottom row into the seventh row, it follows that the seventh row has to be level, too, which is what we want to show.

Our proof is based on the following closely related result by Adam Brown: If we stack bottles in a pyramid, as shown here, the center of the blue bottle at the top is exactly halfway between the two sides of the rack (Adam Brown, A circle-stacking theorem, *Mathematics Magazine* 76 [2003], 301–302.) Here is an illustrated version of Brown's proof of this result that sets the scene for everything that follows.

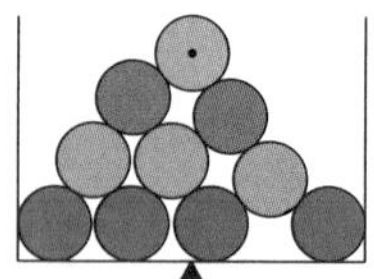

Stacks Have a Half-Turn Symmetry

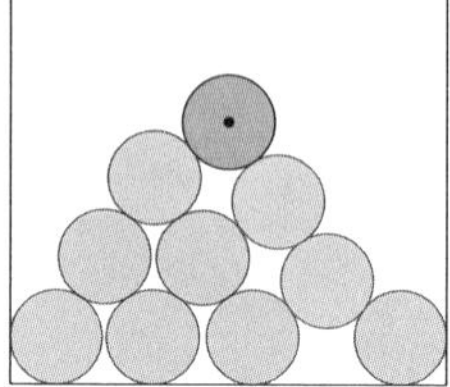

Consider the pyramid of bottles built on top of the bottom row.

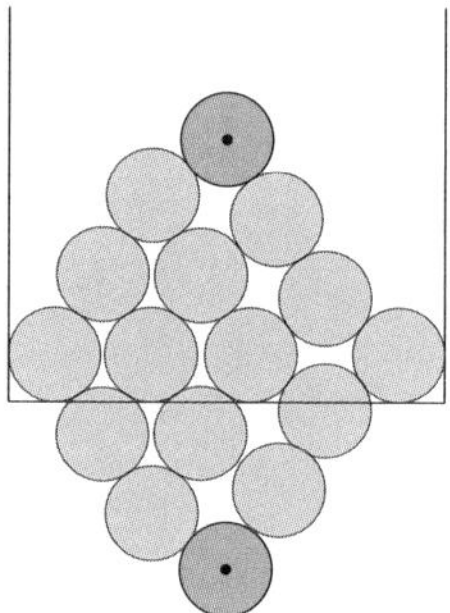

Build a second pyramid pointing down. It is the mirror image of the original pyramid.

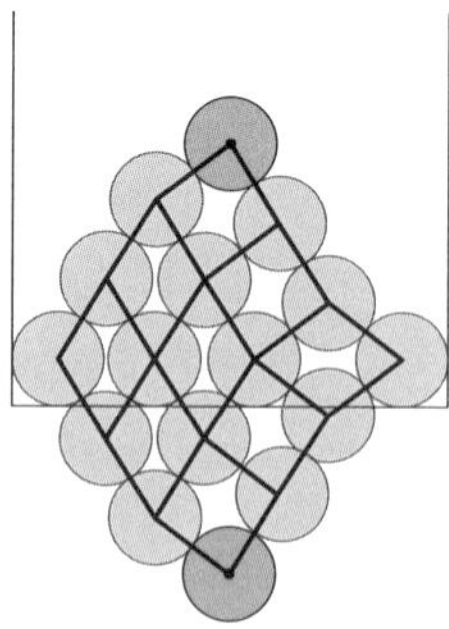

Connect the centers of touching bottles as shown. All connections have the same length and together form nine rhombi. Crucial to everything that follows is the basic fact that opposite sides of a rhombus are parallel.

Then it is clear that the three blue edges are parallel. And so are the three green and the three red ones. Hence the two yellow sides of the mesh are translates of each other.

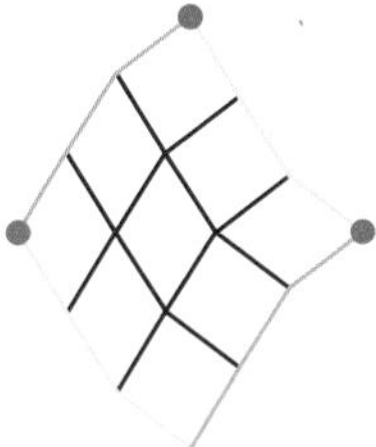

Because the orange sides are horizontal mirror images of the yellow sides, it follows that the three red points form an isosceles triangle. *Hence the blue bottle is halfway between the two sides of the rack.*

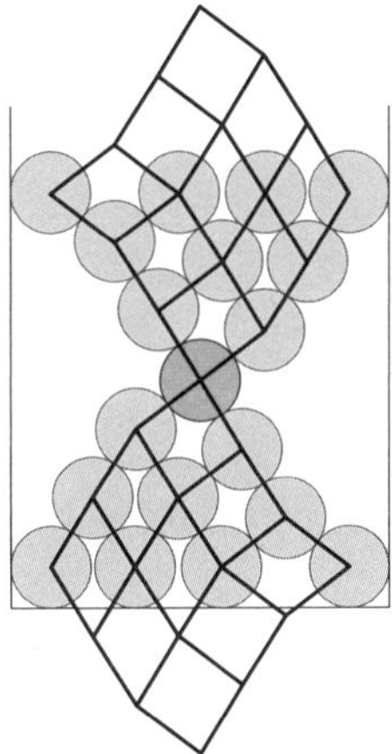

Rotate the original pyramid one hundred eighty degrees around the blue bottle to create a new pyramid balancing on its tip. Because of the halfway property of the blue bottle, the second pyramid also fits into the rack.

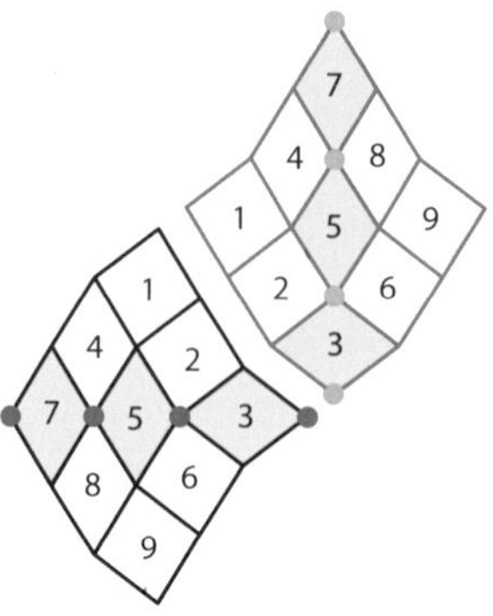

Reassemble the nine rhombi of the original mesh into a second mesh as shown. Here, corresponding rhombi in the two meshes end up being translates of each other. Therefore, because the red points are horizontally aligned, the green points are vertically aligned.

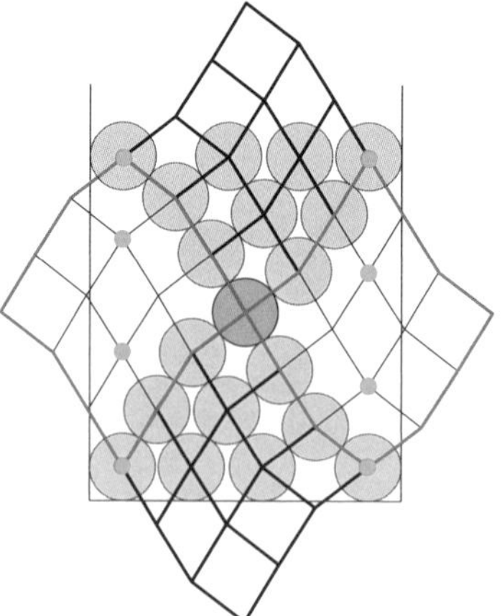

This new mesh, together with a half-turned copy, seamlessly fill the gaps between the first two meshes.

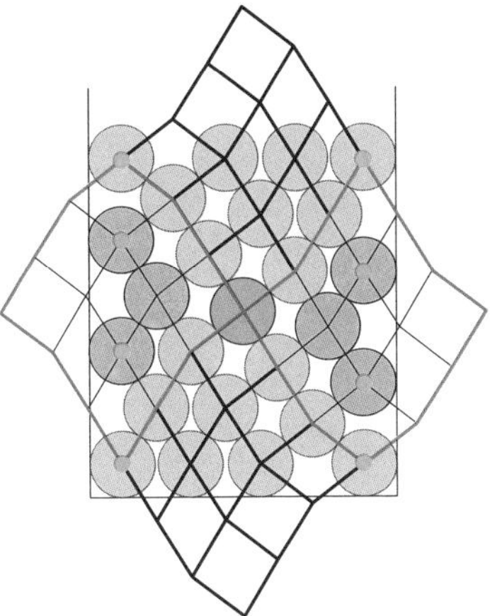

Adding bottles centered at the vertices of the new meshes clearly reconstructs the stack of wine bottles we started with and shows that it has the half-turn property.

Stacks Have a Half-Turn Symmetry Unless . . .

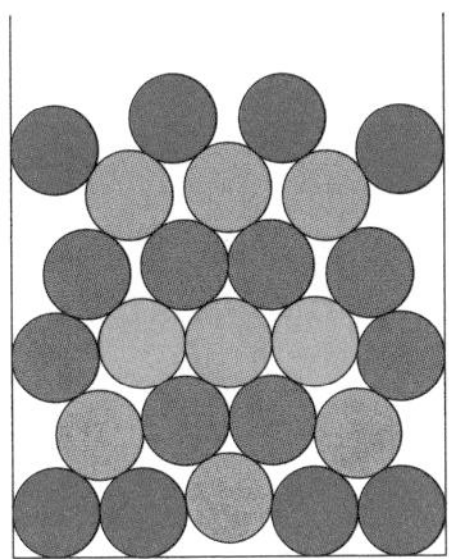

Sometimes in wine racks that almost allow five bottles to be placed in the bottom row, things go wrong and the bottles in the seventh row don't line up. This is illustrated here. Is there something wrong with our proof?

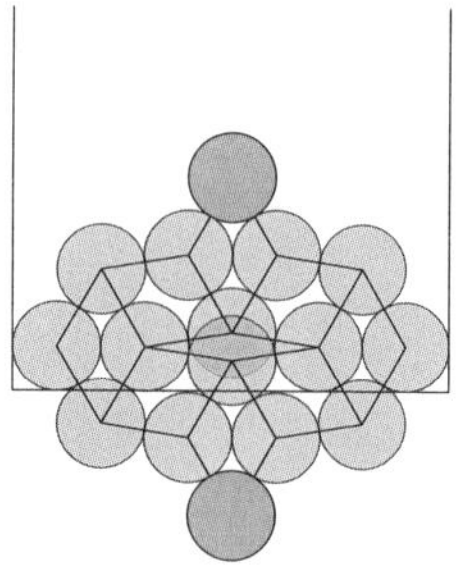

Note that in the corresponding mesh, the rhombus in the middle is too flat to accommodate the four bottles centered at its vertices without overlap. This is not a problem yet because the bottles attached at the bottom are not part of the stack.

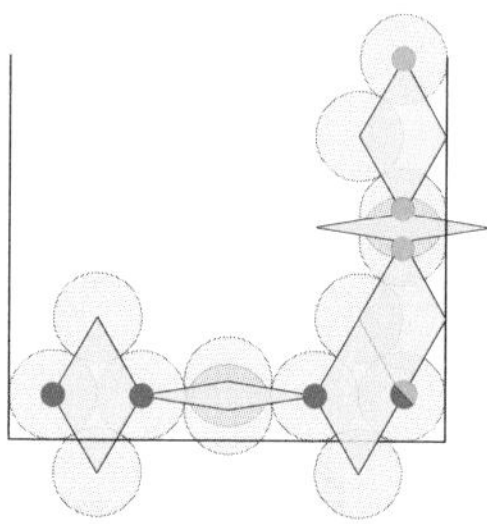

However, in our construction, the rhombi corresponding to the bottom row do occur again, as highlighted here. And in its second incarnation, the flat rhombus clearly causes our proof to break down.

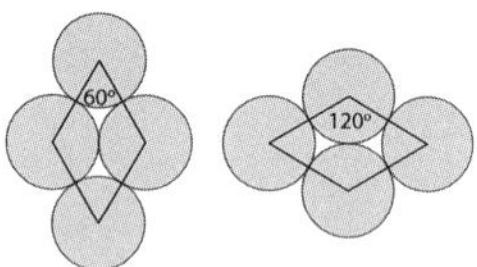

Let's call a rhombus *thin* if its top angle is less than sixty degrees and *flat* if its top angle is greater than one hundred twenty degrees. For us to be sure that our proof works, guaranteeing our level seventh row, we have to ensure that there are no thin or flat rhombi.

We now show that there won't be any thin or flat rhombi if our yellow rhombi at the bottom are not flat.

From our construction, it is clear that every edge in the big mesh we end up with is parallel to one of the edges of the yellow rhombi, as illustrated here.

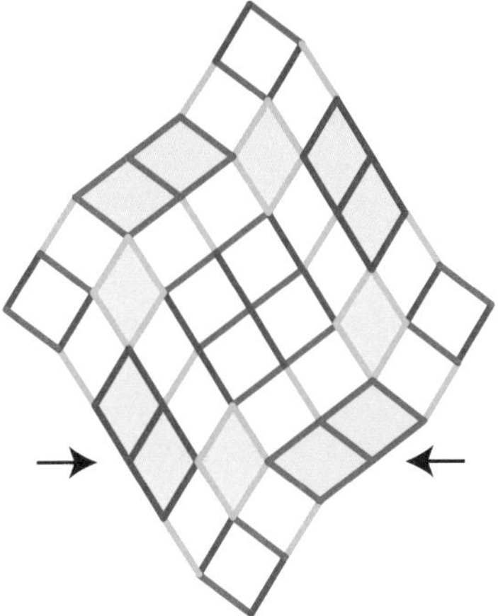

That is, segments of the same color that look parallel are parallel.

This implies that the top angle of *any* of the rhombi in the big mesh is at least as large as the smallest top angle of the yellow rhombi. As well, none of these angles is larger than the largest among the top angles of the yellow rhombi.

Because we started our construction with nonoverlapping bottles in the bottom row, we can be sure that none of the yellow rhombi is thin. Consequently, as long as none of the yellow rhombi is flat, the seventh row is level.

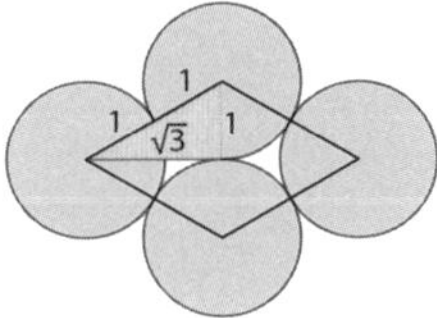

In terms of the spacing of the bottles at the bottom, this can be expressed as follows: If the bottles are of radius 1, there won't be any flat rhombi if the spacing of centers of adjacent bottles in the bottom row is never greater than $2\sqrt{3}$.

What should also be obvious at this stage is that the absence of thin or flat rhombi in a stack is equivalent to the stack "looking like" our example, that is, the stack having the following three properties:

1. The stack naturally splits into long rows and short rows, with long rows containing one more bottle than short rows, the first row being a long row, and long and short rows alternating.
2. The two outer bottles in long rows touch the sides of the rack.
3. Every bottle that touches a side touches the first or last bottles in the row above and below. All other bottles touch two consecutive bottles in the row above and two consecutive bottles in the row below.

Let's call any stack, even the fancy ones coming up, *well-behaved* if they enjoy these three properties.

Racks with Tilted Sides

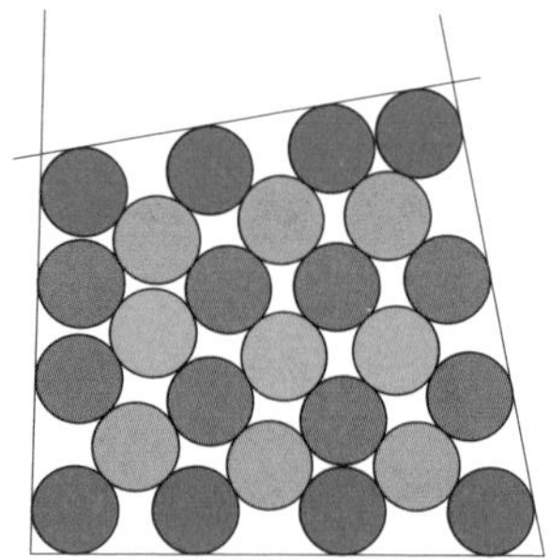

Here we'll follow the example of Nathan Bowler and David Robbins and show that even if a rack has tilted sides, any well-behaved stack inside it has a top row of bottles that is aligned.

The idea for our proof is to use the same bottom row as in the tilted rack to build a stack inside a vertical rack. Then we transform the vertical stack together with its underlying meshes into the tilted stack. The way the meshes transform shows at a glance that the stack in the tilted rack has the desired property.

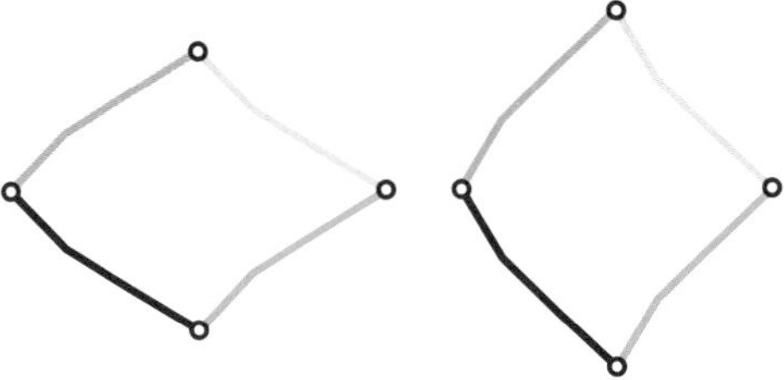

Think of the four sides of one of our meshes as a flexible frame consisting of four rigid pieces, hinged together at the corners, as indicated. It is clear that all the shapes that this frame can be flexed into have a mirror symmetry and that opposite sides of the frame are always translates of each other.

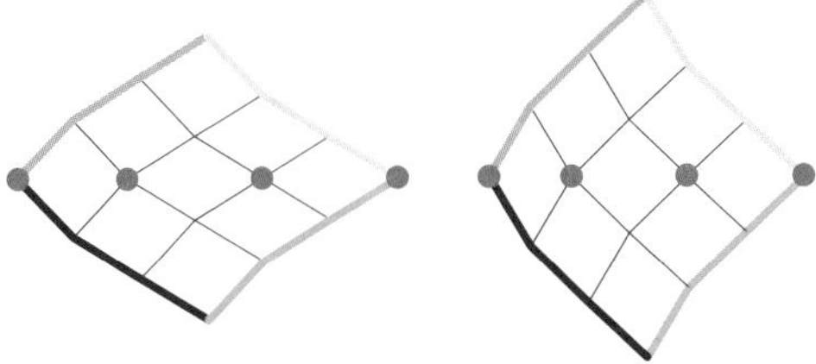

Then it is easy to see that any of these shapes spans one of our mirror-symmetric meshes, and in the following we think of the mesh flexing together with the frame. Note also that the red vertices are always aligned.

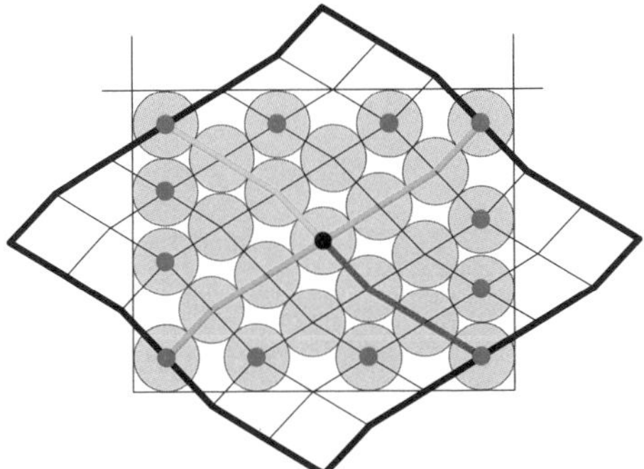

Here is our vertical stack. We are dealing with four of the special meshes glued together along the green sides. As we rotate the top two green sides around their common black endpoint, the four meshes transform, and so does the associated stack of wine bottles and the encasing rack.

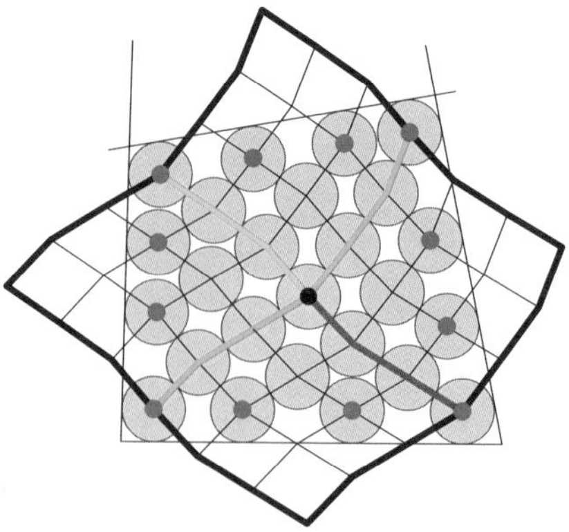

We rotate until the encasing rack is the one we are really interested in. Note that because of the way the red points line up, we can be sure that the sides as well as the top of the stack align.

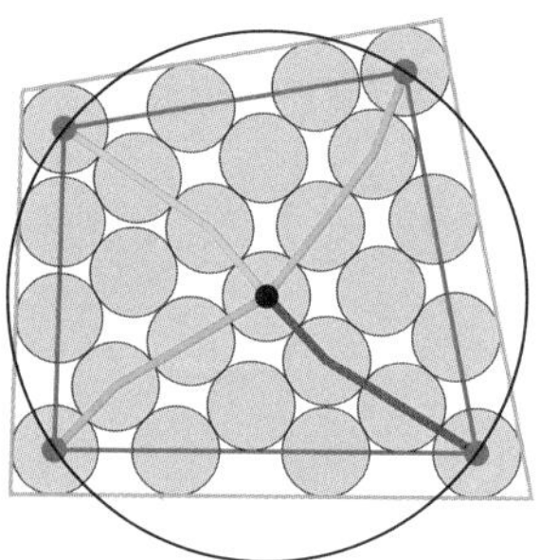

Because the four red points are at equal distance from the black point, the red quadrilateral is cyclic. This means that opposite angles in this quadrilateral add up to one hundred eighty degrees. Because the sides of the blue quadrilateral are parallel to those of the red one, it has the same angles as the red one. *This means that the lines across the tops of all well-behaved stacks inside a given rack are parallel.*

Racks with Sides That Tilt at the Same Angle

A special case of a rack with tilted sides is that of the two sides being parallel. Then the fact that opposite angles of the blue quadrilateral add up to one hundred eighty degrees implies that our stack forms an isosceles trapezium. Also, we conclude that, as in the case of racks with

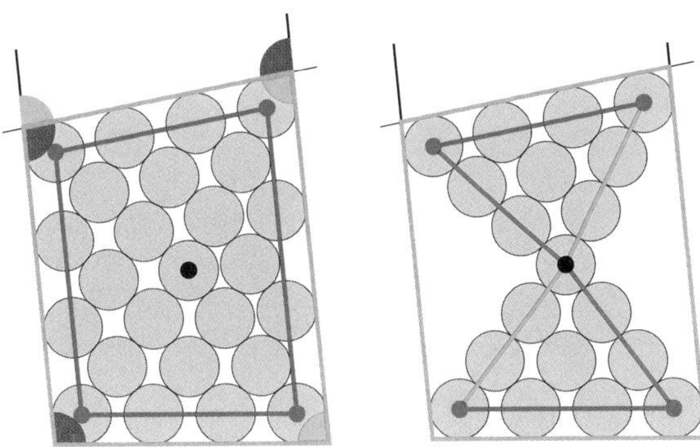

vertical sides, the top inverted pyramid in the picture on the right is just a rotated copy of the bottom pyramid. This means that if we keep stacking bottles beyond the top row, things repeat as indicated here, and we eventually end up with a horizontal row of bottles. If we start with n bottles in the bottom row, this horizontal row is row $4n - 3$. As usual, all this only works if the stack we are dealing with is well-behaved.

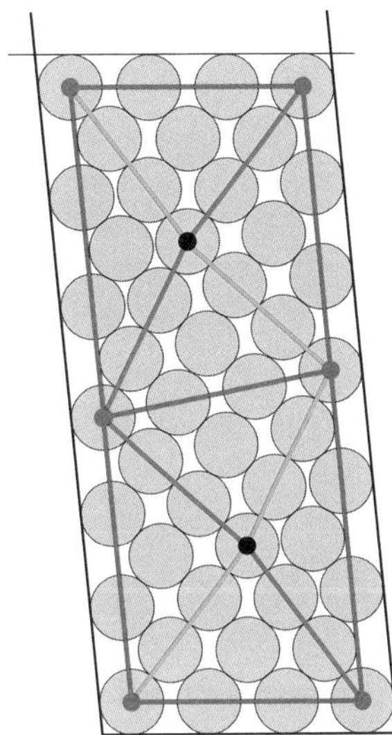

Up to this point in our account, our main contribution to the "theory" of stacking bottles has been to provide some (we hope) easily accessible explanations for why the tops of well-behaved stacks line up and to highlight the half-turn symmetry and common structures underlying well-behaved stacks.

For the rest of this article, let's discuss some things that have not been discussed elsewhere.

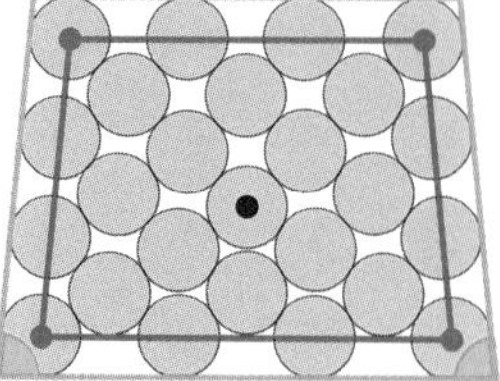

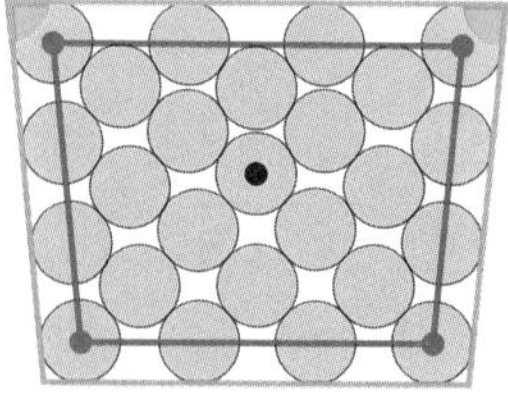

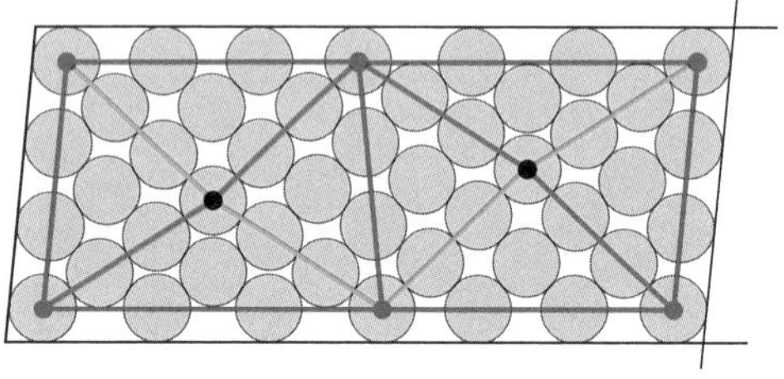

Did you notice that if you put any of our stacks on one of their sides (or upside down), you obtain another stack? If we perform this trick on a stack inside a rack with parallel sides, our previous considerations show that we get a stack inside a rack with sides that are tilted the same angle toward each other or away from each other. Also, it is clear that the top row in any such stack is actually horizontal.

And, if we perform our trick on our double stack we get something curious—a "half-stack" inside a rack with parallel sides whose top row is horizontal. Note that if there are n bottles at the bottom, then the top row of this half-stack is row n.

Meddling with the Bottom Row

In the well-behaved stacks that we've considered so far, the bottles naturally split into alternating long and short rows, with long rows containing one more bottle than short rows. Remember that we always started with two bottles in the bottom row touching the sides of the rack. Among other things, this ensures that the bottom row is a long row.

We can also build stacks in which the first row of n bottles is short with no bottle in this bottom row touching a side of the rack, as shown in the picture here.

We'd like to show that if such a stack is well-behaved (adjusting the definition of well-behaved in the obvious way), then row $2n + 1$ is a level short row.

Something similar is also true for well-behaved stacks like the one shown here.

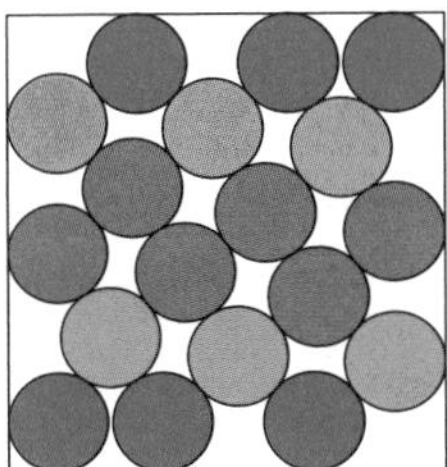

In this case, all rows contain the same number of bottles, and only one of the bottles in the bottom row touches a side of the rack. If we start with n bottles, then row $2n$ is level.

In both cases, it suffices to show that these stacks have a half-turn symmetry. We begin by streamlining our proof for the original type of stack and then indicate how this streamlined proof has to be modified to turn it into proofs for the half-turn property of the two new types of stacks. Here we go again.

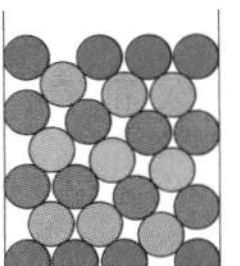
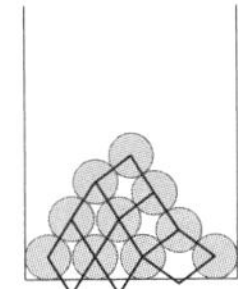
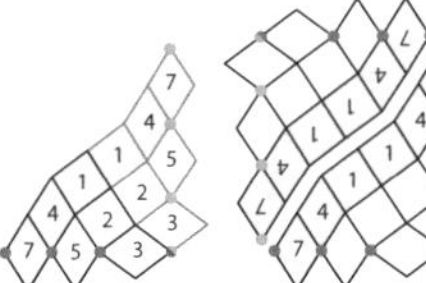
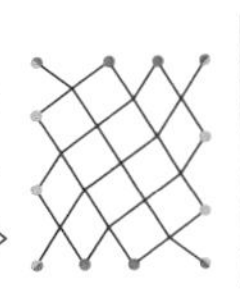
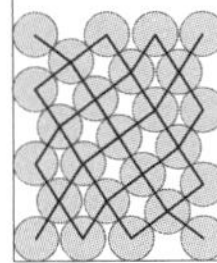

We start with the essential part of the mesh corresponding to the pyramid. We straightaway extend this mesh to one that covers half of the stack as indicated. Now it is clear that the gray band has a half-turn symmetry. We combine the mesh and a half-turned copy of this mesh. Finished!

And here is how things have to be modified for the two new types of stacks. Note that in the second case, we are overlapping the two gray bands.

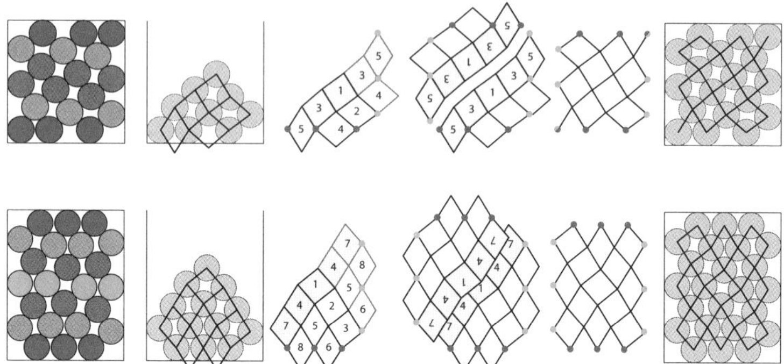

The following three diagrams show three stacks of the three different types that share the same pyramid at the bottom. Superimposed on the left diagram are the linkage and meshes that we used to show that all well-behaved stacks of the first type in tilted racks with the same bottom pyramid have a level top row. The second and third diagrams show the corresponding linkages and meshes that can be used to show that the same is true for well-behaved stacks of the second and third types.

These diagrams also illustrate how closely related the stacks of the three different types sharing the bottom pyramid are. In particular, note that both the large mesh in the middle and that on the right are made up of the same four quarter meshes as our original mesh on the left.

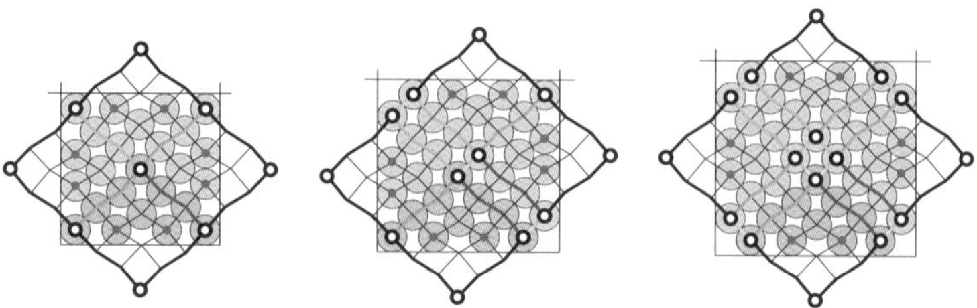

Next is an example of the middle setup in action. We start with a stack in a vertical rack and transform it into a tilted stack that shares the brown pyramid at the bottom with the vertical stack.

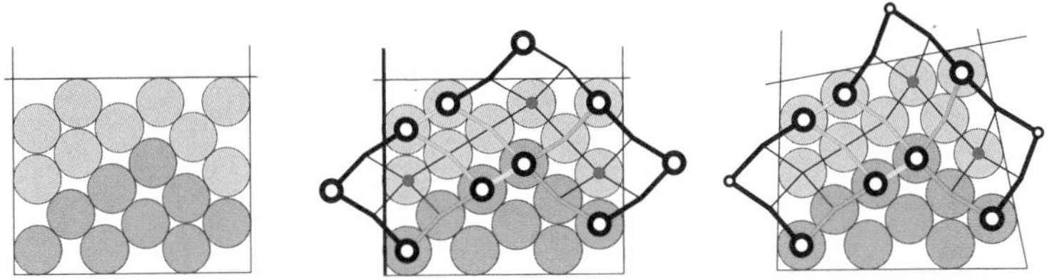

Periodic Stacking

Let's have a look at stacking bottles periodically in an infinitely long rack. Then if the stack is well-behaved and the period is n bottles, we obtain a level $n + 1$st row.

To see that this is the case, color the diagonals of bottles slanting to the right using n colors as shown. Then diagonals colored the same are horizontal translates of each other.

Now focus on two adjacent diagonals of bottles (blue and orange on the left) and, in particular, on the highlighted segments connecting the centers of adjacent bottles. Then it is clear that segments of the same color are translates of each other. From this it follows immediately that the black segments connecting the midpoints of the top and bottom bottles of diagonals, highlighted on the right side of the diagram, are horizontal translates of each other. In turn, this implies that the top row is level and that the spacing of the bottles in the top row mirrors that of the bottles in the bottom row.

Note that every one of our (vertical) well-behaved stacks can be used as a building block to create a periodic stacking. For this, interleave infinitely many copies of the stack with infinitely many horizontal mirror images of the stack as shown in the following example.

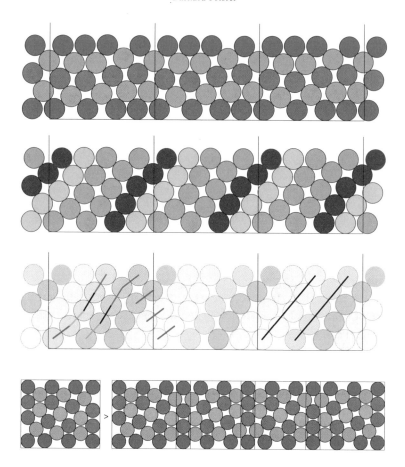

Alternatively, you can also combine infinitely many copies of a half-stack that arises from a stack in a rack with parallel sides to create a periodic stacking. On close inspection, you'll notice that this does not yield anything really new when the sides of your rack are vertical.

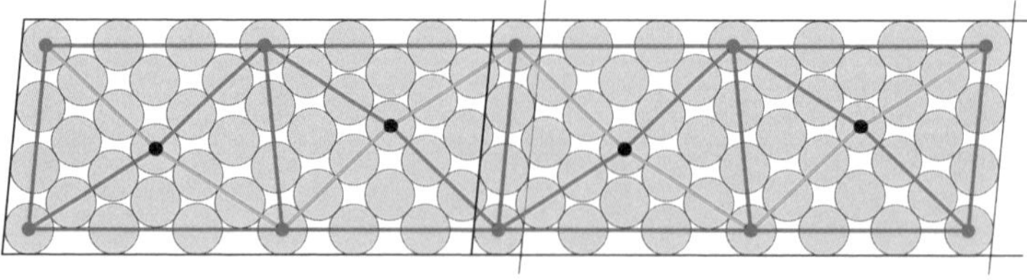

That's It

Is that really it? Is there really no more to say about stacking bottles? Not at all!

We've just considered periodic stacking. Well that's basically stacking circles on a cylinder. What about stacking circles on a Möbius strip? (Pretty easy after you know what happens on cylinders.) Or, what about stacking circles on cones? (I am just throwing this out without having given it some proper thought.) Or which well-behaved stacks in a given vertical rack have the highest or lowest top row? (Not difficult, and may come in handy when it comes to fitting your wine bottles in a wine rack that is closed at the top.) What happens if stacks are not well-behaved? (Some partial results by Nathan Bowler can be found on *Cut-the-Knot*.) What about stacking bottles of different sizes? (David Rogers notes that things still work out if we use bottles of two different sizes in long and short rows.) What about higher dimensional generalizations?

A lot of other interesting questions come to mind after you give this a little bit of thought. And we hope you do and come up with some more interesting wine-bottle stacking math.

The Way the Billiard Ball Bounces

Joshua Bowman

When it comes to real-life illustrations of mathematics in action, it's hard to beat billiards. The objects involved are simple: a table and a billiard ball (and a cue stick to get the ball moving, but because it plays no later role, we'll ignore it).

From a mathematician's perspective, there are only two basic physical principles: The ball moves in a straight line until it hits the side of the table, at which point it follows the rule that the angle of incidence equals the angle of reflection. There's no need to bother with friction, spin, or other such troublesome matters.

Carrying this simplifying process to its natural conclusion, we treat the billiard table as a polygon in the plane and the ball as a point that moves according to the previously stated rules. The course that the ball follows is called a *billiard path*.

Whereas a billiards player would be concerned with getting the ball to a particular place on the table, a mathematician might ask: If we hit the ball in such-and-such a manner, what are all the possible places it might eventually go? In this setting, billiards is a *dynamical system*. The position and direction of the ball evolve over time, and we preoccupy ourselves with the long-term behavior of its motion.

To simplify our discussion, we'll focus on the question of which billiards paths are *periodic*: That is, what starting positions and directions for the billiard ball will cause it to come back to where it started, heading in the same direction, and retrace the same path over and over?

It will probably not come as a surprise that the answer to this question depends on the shape of the table. We'll start with the most familiar shape for a billiards table, the rectangle, then progress to triangles. Finally, we will indicate some resources for further study, including the behavior of billiards in other shapes.

Billiards in Rectangles

Ordinary billiards is played on a rectangular table. The standard dimensions call for a 2:1 ratio of side lengths, but let's allow for any dimensions $a \times b$. On any such table, there are some obvious periodic paths: for example, a ball bouncing back and forth between opposite sides, striking each at right angles.

By playing around a bit, we find more periodic paths: a diamond joining the midpoints of the sides, for instance, or a crisscross pattern that hits one pair of sides three times each, and the other pair of sides twice each, as in Figure 1. Is there something common to these paths that we can generalize to find *all* periodic paths?

Any billiard path that returns to its starting position and direction must cross the rectangle in the horizontal direction some even number of times, say $2N$. (The number is even because the ball switches its left-to-right motion every time it strikes one of the vertical sides.) Likewise, it must cross the vertical direction some even number of times, $2M$. Given an arbitrary pair of nonnegative even integers $(2N,2M)$, can we find a periodic path that crosses these numbers of times?

Let's take a "through the looking-glass" approach to finding such a path. Think of the billiard path as a beam of light. Instead of taking each side of the rectangle to be a mirror, think of it as a pane of glass, and when the light reaches it, the light continues in a straight line, entering a reflected copy of the original rectangle. The path continues and eventually reaches another side inside of this copy, so we make another copy of the rectangle to continue the path in a straight line.

Each copy represents the original billiard table, but carries with it information about how many times a path has been reflected in the sides (i.e., has gone through the looking glass). In Figure 2, this information is encoded by coloring the sides of the rectangles. This process is called *unfolding* the billiard path.

FIGURE 1. A periodic billiard path in a rectangle.

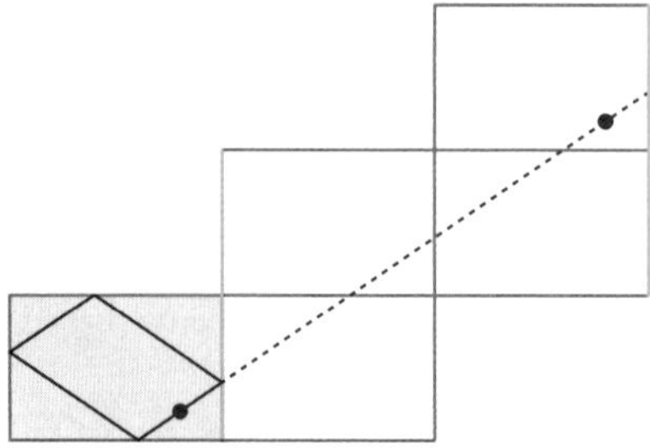

FIGURE 2. A billiard path (solid line) can be viewed as a straight line in the plane (dashed) by reflecting the table repeatedly across its sides.

A rectangle is a particularly nice polygon because by repeatedly reflecting it over its sides, we can tile, or *tessellate*, the entire plane. If we do this ahead of time, then each line in the plane corresponds to a billiard path in the rectangle (see Figure 3).

Given the pair $(2N,2M)$, we can find a periodic billiard path that reflects horizontally $2N$ times and vertically $2M$ times by drawing a straight-line path from a point in the original rectangle to the corresponding point in the rectangle located $2N$ copies to the right (or to the left) and $2M$ copies up (or down). Because we have found a relationship between periodic billiard paths and pairs of nonnegative integers $(2N,2M)$, we have the following theorem.

Theorem. *Let* R *be an* $a \times b$ *rectangle. A billiard path in* R *is periodic if, and only if, its initial slope is a rational multiple of* b/a.

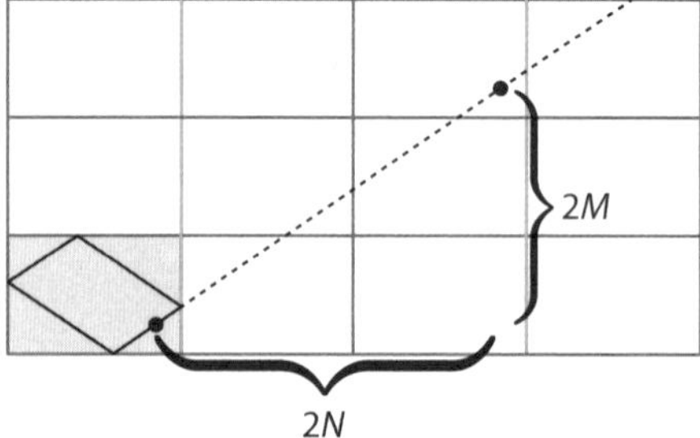

FIGURE 3. The plane, tessellated by a rectangular billiard table. Each straight line in the plane corresponds to a billiard path on the table.

Notice that the starting point plays no role in this theorem, so each periodic path belongs to a whole family of periodic paths with the same initial slope.

Billiards in Triangles

In the case of rectangles, we had two parameters, a and b, to describe the shape; after a moment's pause, we realize that only the *similarity class* of the figure matters for determining billiard paths, which reduces the family of rectangles to a single parameter: the ratio of a to b.

In the case of triangular tables, we can take any three positive numbers a, b, and c such that $a + b + c = 180$ and form a triangle with angles a, b, and c (measured in degrees), which is determined up to similarity. The equation relating a, b, and c implies that we have a two-parameter space of triangles.

In stark contrast with the case of rectangles, the question of what periodic billiard paths a given triangle has (or even whether it has any!) remains open and is an active field of research. More precisely, we know all the periodic paths for some triangles. For other triangles, we know only that there exists a periodic path. But for a great many others, *we do not know if there is a periodic path!*

Don't despair: The most familiar triangles—the equilateral, 30–60–90, and 45–45–90 triangles of high school trigonometry—are well understood.

Let's try the tactic we used with rectangles of reflecting the triangle in its sides instead of reflecting the billiard path. Conveniently, just like the rectangle, these three triangles tessellate the plane. These tessellations are shown in Figure 4; edges of the same color correspond to the same sides of the table.

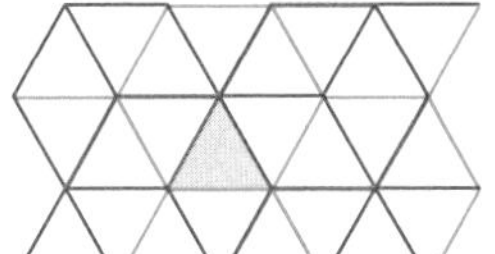
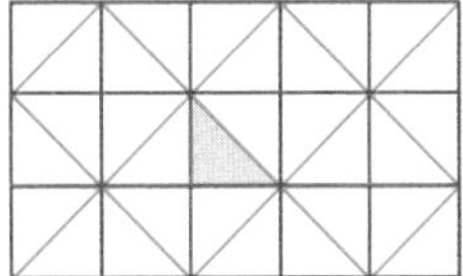
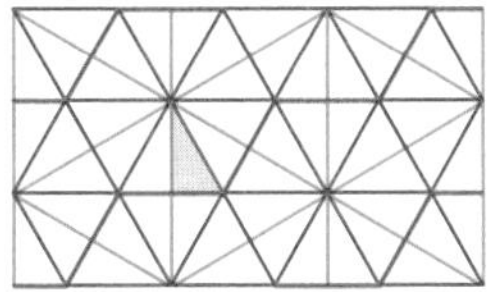

FIGURE 4. Tessellations of the plane obtained by reflecting equilateral, 45–45–90, and 30–60–90 triangles across their sides.

In exactly the way we found periodic billiard paths for the rectangle, we can connect corresponding points in two different copies of a triangle in its tessellation to construct periodic paths. Exercise: For each of these triangles, formulate the theorem analogous to that of the previous section, classifying periodic billiard paths.

Now let's cast our nets wider than these three special triangles. One of the oldest classes of periodic paths in triangles was found by Giovanni Fagnano. In 1775, he showed that every *acute* triangle T has a periodic path. (In truth, he was searching for the shortest closed path that touched all sides, not necessarily following the billiard rule of reflection, but in this case, the shortest path is also a billiard path.) The construction of the *Fagnano path* is elementary: Connect the feet of all three altitudes of the triangle (Figure 5). Exercise: Show that this triangle—called the *orthic triangle* of T—is a billiard path; that is, show that each pair of sides of the orthic triangle forms equal angles with a side of T. Why must T be acute?

From this periodic path, we get additional periodic paths for free. Unfold the path. Because it is periodic, eventually it arrives in a copy of the triangle that has the same orientation as the original, at a point corresponding to its starting location. Any path that is parallel to the original path and stays within the same sequence of unfolded triangles as the original path is also periodic. See Figure 6 for this construction.

Progress has been made on the search for periodic billiards paths in triangles, primarily in the past thirty years. Here are two theorems that illustrate what is known.

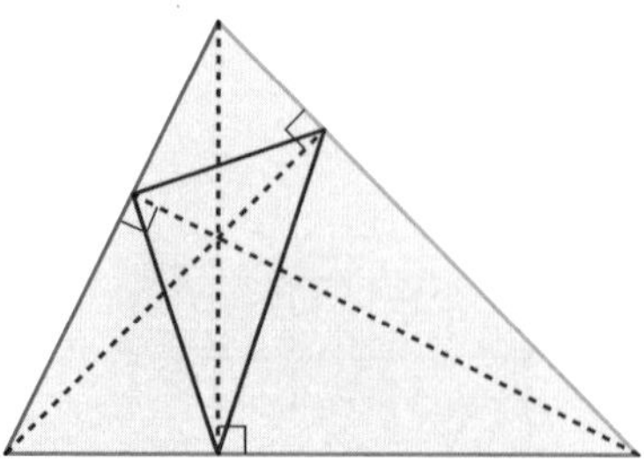

FIGURE 5. In any acute triangle, the orthic triangle is a periodic billiard path.

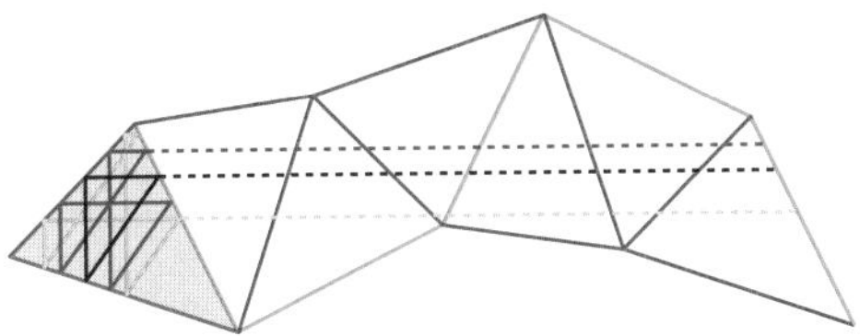

FIGURE 6. The Fagnano path is black. The purple and yellow paths run parallel to it and produce additional periodic paths.

Theorem [2]. *If all the angles of a triangle are a rational number of degrees, then the triangle has periodic billiard paths in infinitely many directions.*

Theorem [3]. *Every triangle whose angles are all less than 100° has a periodic billiard path.*

The proofs of these two theorems are radically different. The first uses advanced tools of complex analysis and topology. The second uses combinatorics and elementary real analysis and requires computer assistance.

What's Next?

To keep matters simple, we have considered only the question of which billiard paths are periodic. Classifying the behavior of nonperiodic paths requires introducing a host of notions from topology and ergodic theory.

Moreover, we have also looked only at two shapes. Billiards in general polygons, particularly those with rational angles, is undergoing the same intense study as in triangles. This remains an active area of research, with an ever-increasing arsenal of techniques, drawn from a wide variety of mathematical subjects.

Several Fields medalists have made novel and important contributions to the study of polygonal billiards, including two of the 2014 recipients, Artur Avila (the first Brazilian to receive the award) and Maryam Mirzakhani (the first woman and first Iranian to receive the award). See the sidebar.

Billiards in tables with curved boundaries (so-called *smooth billiards*) create a somewhat different flavor; circles and ellipses have nice

Billiards: On the Mathematical Frontier

Two of the four 2014 Fields Medal recipients have contributed substantially to the study of polygonal billiards. Here we give brief sketches of their work in this area. In both cases, it will help to understand the connection between polygonal billiards and a mathematical object called a *flat surface*.

From here onward, we make one additional assumption about our polygonal table: that every angle has a rational number of degrees. We build the flat surface from our polygon using the unfolding processes described in the article—by gluing each reflected copy of the polygon onto the previous copy. But when all angles are rational, the unfolding process produces only finitely many orientations of the polygon. When a reflected polygon has the same orientation as a polygon we've already created, we glue to that one instead. Do this for all edges of all reflected polygons.

In this way, we obtain an abstract surface that looks like the Euclidean plane at most points. (Exceptional points occur only at the corners.) Billiard paths in the polygon become straight-line paths on the flat surface, and vice versa. Moreover, a periodic billiard path in the polygon corresponds to a closed, straight loop in the flat surface.

As an example, consider a right triangle T with an angle of 22.5° (Figure 7). We unfold it to obtain a regular octagon. If we reflect any of these sixteen copies, we obtain one of the sixteen again. For instance, if we reflect T across its red edge, we obtain T'. So we obtain our flat surface by gluing edges A to A, B to B, C to C, and D to D.

Visually, this means that a path "leaving" the octagon on one side reenters at the corresponding point of the side with the same label. Exercise: Convince yourself that this flat surface is topologically equivalent to a two-holed torus. This construction is a generalization of the more familiar one-holed torus obtained from gluing opposite sides of a square, and it can be extended to any regular polygon with an even number of sides.

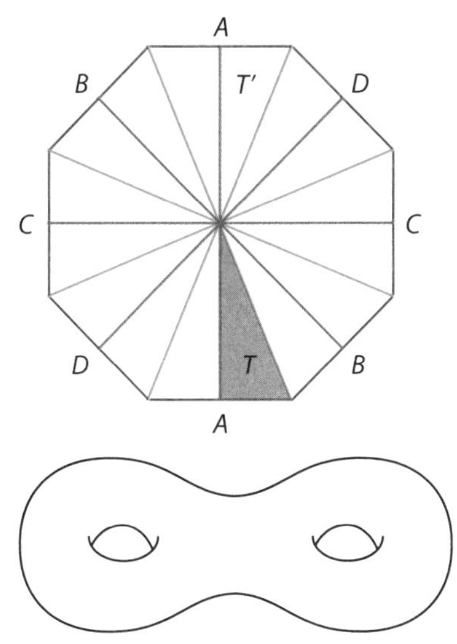

FIGURE 7. The triangle T unfolds to a regular octagon. Each pair of sides with the same label is glued together to produce a flat surface topologically equivalent to a two-holed torus.

Other flat surfaces can be made from polygons by attaching parallel edges together as we choose, not necessarily according to a reflection process. For example, we could have simply started with the octagon in Figure 7, rather than unfolding the triangle. The collection of *all* flat surfaces has a natural notion of closeness: Two surfaces are *close* if they are built out of polygons that need to be deformed by only a small amount to change from one to the other.

Artur Avila

Artur Avila's work has often sought to describe *typical* behavior in dynamical systems—that is, to answer the question of what happens "almost all the time." Periodic paths in billiards are extremely *atypical*, in the same way that rational numbers are atypical among the real numbers; most billiard paths over time approach every point of the table arbitrarily closely (in topological terms, almost every path is *dense* on the table). This is only a partial response, however; we can also ask *how quickly* the path covers the table. Answers to such questions are more easily formulated in terms of the corresponding flat surface.

Avila showed, in joint work with Giovanni Forni [2], that for almost all flat surfaces, straight-line motion is *weakly mixing*. This property may be thought of as follows: Suppose a region P covers p percent of the surface and that all points in P start moving in the same direction. Then for almost every initial direction and for any region R, on average in the long term, the points that started in P will cover p percentage of R. Thus, the balls disperse over the surface in a uniform way, even if they started out concentrated in one region. However, among flat surfaces, those that arise from billiards are also atypical, and so a priori this result might not have any direct implications for billiards.

Nonetheless, billiard tables provide an *explicit* construction of surfaces with weak mixing. As Avila showed in later work, with Vincent Delecroix [1], weak mixing is true not only for typical surfaces, but also in the specific cases of surfaces made from regular polygons (as in Figure 7), not including the triangle, square, or hexagon.

Finally, Avila showed, together with Marcelo Viana [3], that on a typical flat surface, the deviations of a typical straight-line path from its asymptotic behavior as it winds around the surface can be described in a very precise way (what is called the *Zorich phenomenon*).

More details of Avila's work on flat surfaces can be found in Forni's article "On the Brin Prize Work of Artur Avila in Teichmüller Dynamics and Interval-Exchange Transformations" [7].

Maryam Mirzakhani

Maryam Mirzakhani's work relates more directly to the ways flat surfaces can be deformed. Remember, this means deforming the polygons used to create the surface. A particular kind of such deformation changes all the polygons simultaneously by a linear transformation—that is, by applying a 2×2 invertible matrix to the polygons to produce new shapes.

Mirzakhani, in joint work with Alex Eskin and Amir Mohammadi [6], studied surfaces that can be closely approximated by applying a linear transformation to a starting surface. (This perspective underlies many of the results stated in the main article and in Avila's work.) Within the space of all flat surfaces, this subset could a priori be very complicated—a fractal, for example.

However, Mirzakhani and her collaborators showed that it is always very nice: It can (locally) be described using equations that are *linear* in the coordinates of the polygons that make up the flat surfaces. Previously the only nontrivial case in which this fact had been known was that of the two-torus, by work of Curtis McMullen (a 1998 Fields medalist) [9].

Somewhat surprisingly, conditions on what surfaces can be approximated by linear deformations of an initial surface carry information about the dynamics of straight-line paths on the initial surface [5, 13]. Thus, the work described in the previous paragraph has consequences for billiard paths—including periodic billiard paths—that are only beginning to be explored.

More details of Mirzakhani's work on surfaces can be found in McMullen's article for the International Congress of Mathematicians, "The Mathematical Work of Maryam Mirzakhani" [10].

behaviors, analogous to that of the rectangle. For details of these and other more complicated smooth billiard systems, see references [4] and [12].

Although billiards starts with simple assumptions, it has developed into a rich theory that draws from seemingly disparate areas of mathematics. This is often the hallmark of new mathematical research: It uncovers surprising connections among different areas and uses those connections to draw powerful conclusions about simple objects and processes.

Acknowledgments

Thanks to Pat Hooper and David Aulicino for several suggestions that improved the readability and accuracy of this article.

Further Reading

[1] Artur Avila and Vincent Delecroix, "Weak mixing directions in non-arithmetic Veech surfaces," *J. Am. Math. Soc.* 29 (2016): 1167–208.

[2] Artur Avila and Giovanni Forni, "Weak mixing for interval exchange transformations and translation flows," *Annals of Math.*, 165 (2007), 637–64.

[3] Artur Avila and Marcelo Viana, "Simplicity of Lyapunov spectra: Proof of the Zorich-Kontsevich conjecture," *Acta Math.* 198 (2007): 1–56.

[4] Leonid Bunimovich, "Dynamical billiards," *Scholarpedia* 2, no. 8 (2007): 1813.

[5] Laura DeMarco, "The conformal geometry of billiards," *Bull. Am. Math. Soc.* 48 (2011): 33–52.

[6] Alex Eskin, Maryam Mirzakhani, and Amir Mohammadi, "Isolation, equidistribution, and orbit closures for the SL(2,R) action on moduli space," *Annals of Math.* 182, no. 2 (2015): 673–721.

[7] Giovanni Forni, "On the Brin Prize work of Artur Avila in Teichmüller dynamics and interval-exchange transformations," *Journal of Modern Dynamics* 6, no. 2 (2012): 139–82.

[8] Howard Masur, "Closed trajectories for quadratic differentials with an application to billiards," *Duke Math. J.* 53, no. 2 (1986): 307–14.

[9] Curtis McMullen, "Dynamics of SL(2,R) over moduli space in genus two," *Annals of Math.* 165 (2007): 397–456.

[10] Curtis McMullen, "The mathematical work of Maryam Mirzakhani," *Proceedings of the International Congress of Mathematicians*, vol. 1, Seoul: Kyung Moon Sa Co., 2014.

[11] Richard Schwartz, "Obtuse triangular billiards II: 100 degrees worth of periodic trajectories," *Experimental Mathematics* 18, no. 2 (2008): 137–71.

[12] Serge Tabachnikov, *Geometry and Billiards*, Providence, RI: American Mathematical Society, 2005.

[13] Alex Wright, "From rational billiards to dynamics on moduli spaces," *Bull. Am. Math. Soc.* 53 (2016): 41–56.

The Intersection Game

Burkard Polster

Spot It! is a fun card game played with fifty-five cards, each of which carries eight pictures. What's special about this game is that any two cards have exactly one picture in common. The aim of Spot It! is to be the first player to spot this common picture. Can you spot the common picture on the two cards in Figure 1?

Now let's spot the cool mathematics at the core of this game and see if we can use it to design new variants.

Spotting Lines

How would we create a fifty-five-card Spot It! deck—a deck of cards decorated with pictures with the property that any two cards have exactly one common picture?

This task sounds tricky, but it's actually a piece of cake. What familiar mathematical objects have a similar property? Two nonparallel lines in a plane always intersect in a unique point.

So, to speed-design a deck of Spot It! cards, we simply draw fifty-five mutually nonparallel lines, label the points of intersection with

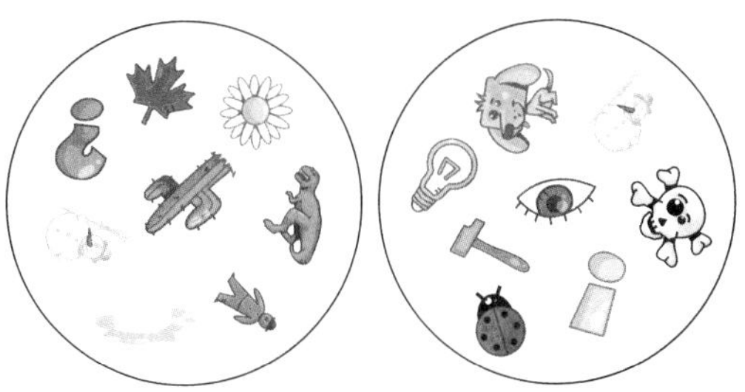

Figure 1. What picture is on both cards?

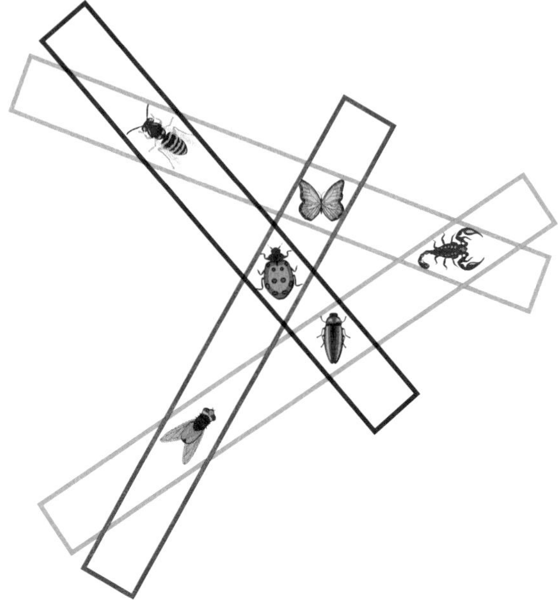

FIGURE 2. Use intersecting lines to construct a Spot It! deck with four cards.

different pictures, and gather all pictures on each line into its own card. For example, we can turn the four lines in Figure 2 into four cards, each carrying three pictures.

Pretty neat, but also pretty disappointing. Is there really no more to spot in Spot It!? Let's have a closer look at what our brand-new Spot It! deck would look like.

Using a random assortment of lines, chances are that, as in our small example, only two lines pass through each intersection point. In other words, there will be as many points of intersection as there are ways to pair up fifty-five lines, amounting to $(55 \times 54)/2 = 1{,}485$ different points, and hence 1,485 pictures. Moreover, there will be fifty-four pictures per card. That's definitely not as compact as the commercial game—there are far too many pictures in our version to be fun.

To cut down the number of pictures, we have to decrease the number of points of intersection. Let's give this a try with six lines. Our straightforward approach would result in fifteen pictures with five pictures per card. A little experimentation leads to the two sneaky, but

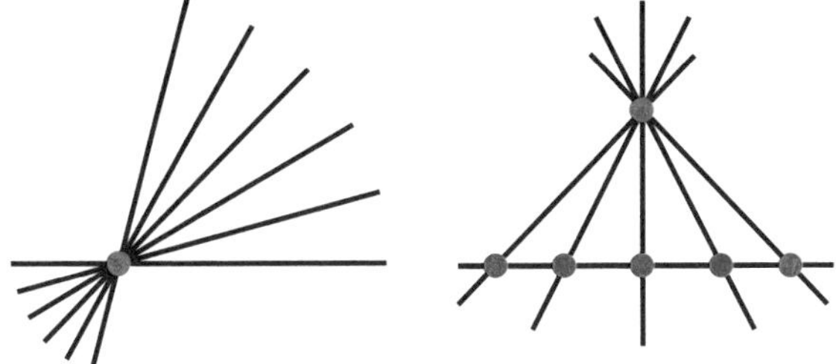

FIGURE 3. Sneaky configurations of points and lines.

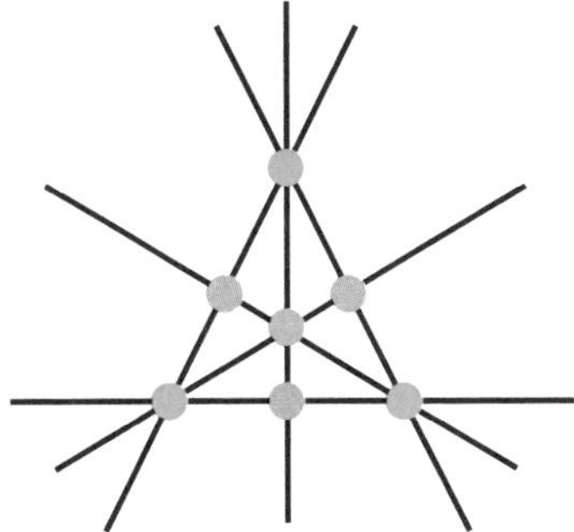

FIGURE 4. Every pair of lines intersects in one of seven points.

gamewise uninteresting, ways of keeping down the number of intersections, shown in Figure 3. More interestingly, we also spot the nicely balanced approach corresponding to seven pictures with three on every card shown in Figure 4. That's quite an improvement, and we cannot do better with six lines in a nonsneaky way.

Notice that each of the green points lies on three lines, whereas each blue point is on only two lines. So, let's collect the three blue points into a new seventh card. This can't be done with a straight line, but for the purposes of designing our game, we don't care. With this seventh card added and represented by a large circle, as in Figure 5, we arrive at a very compact seven-card version of Spot It! using only seven pictures.

To add to the fun of playing this game, we chose as pictures the seven letters in the words THEY WAR and distributed the letters as shown in the diagram. Then the seven cards consist of the words in the sentence YEA WHY TRY HER RAW WET HAT. Figure 6 shows a set of playing cards for this game.

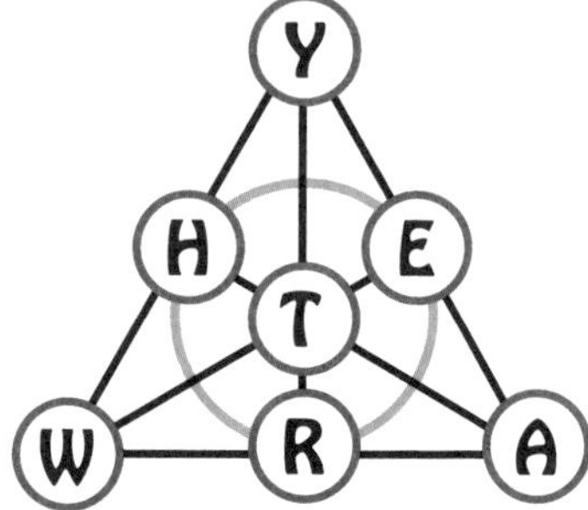

FIGURE 5. These six lines and the large circle generate a seven-card Spot it! deck.

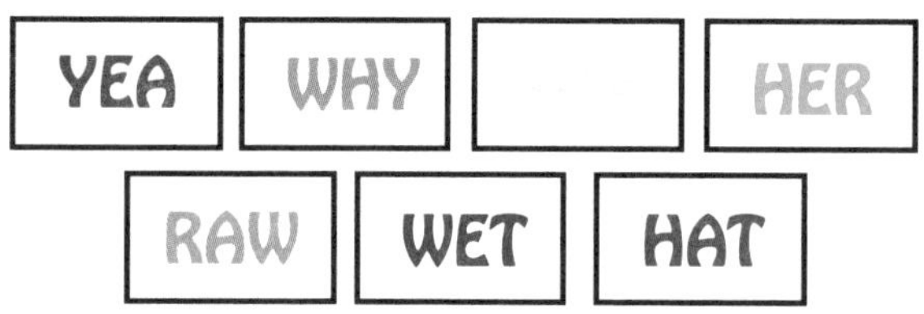

FIGURE 6. A seven-card Spot It! deck.

Spotting Projective Planes

As an abstract point-line geometry, our new seven-card version of Spot It! is actually famous. In the math biz, it is known as the *Fano plane* or the *smallest projective plane*. Let's explain what we mean by this.

The *point set* of an abstract point-line geometry can be any set whatsoever, and its *line set* is a set of subsets of the point set. In this context, we call the elements of the first set the POINTS and those of the second set the LINES of our geometry. Then it should be clear what it means for a POINT to be on a LINE, for two POINTS to be connected by a LINE, and so on.

We can interpret any deck of Spot It! cards as a point-line geometry simply by declaring the pictures to be the POINTS and the cards to be the LINES. In the case of our Fano plane, {T,H,E,Y,W,A,R} is the point set and {{Y,E,A}, {W,H,Y}, {T,R,Y}, {H,E,R}, {R,A,W}, {W,E,T}, {H,A,T}} is the line set.

Geometries obtained from a Spot It! deck are special point-line geometries called *dual linear spaces*, which are characterized by the fact that any two LINES intersect in a unique POINT.

Projective planes like the Fano plane are extra-special dual linear spaces that have the further property that any two POINTS are contained in exactly one LINE. In the case of the Fano plane, this means that any two of the letters in THEY WAR are contained simultaneously in exactly one of the words in YEA WHY TRY HER RAW WET HAT.

One of the fundamental theorems of the theory linear spaces, discovered by de Bruijn and Erdős, implies that except for some sneaky exceptions like the ones in Figure 3, Spot It! decks always have at least as many pictures as there are cards. In addition, any optimal Spot It! deck with the same number of cards and pictures is an incarnation of a finite projective plane. In this case, there are n pictures on every card, every picture appears on n cards, and there are $n^2 - n + 1$ cards and pictures, for some n.

For example, in the case of the Fano plane game, there are three pictures on every card, every picture is on three cards, and the number of cards and pictures is $3^2 - 3 + 1 = 7$.

The commercial Spot It! game has eight pictures on every card, $8^2 - 8 + 1 = 57$ pictures, but only fifty-five cards. Interestingly, the commercial Spot It! deck is a projective plane with two lines missing. (The name of the original French game is Dobble, and the double five in fifty-five had market appeal and also happened to be an ideal number of cards, taking into account manufacturing considerations.) In fact, we've reconstructed the two missing cards in our deck (we've learned that not all decks are identical); they're the ones in Figure 1.

Spot It Cubed!

What about the natural question: Is it possible to construct a deck in which any *three* cards have exactly one picture in common? The answer is "Yes."

Any three mutually nonparallel planes in space intersect in a unique point or in a line. So, we just need to take a collection of mutually nonparallel planes, no three of which pass through a common line (which is easy). We label the resulting points of intersection with different pictures and gather all pictures on the different planes into cards and voilà: instant Spot It Cubed!

It is just as easy to create cards for Spot It^4!, Spot It^5!, and so on, using intersecting hyperplanes. However, as in the case of regular Spot It!, when designing these new decks, the real trick is to keep the number of pictures small. Luckily, there are mathematical objects that can be translated into efficient decks.

Here is a nice simple one. It's a point-circle geometry. Again, this is an abstraction of ordinary geometry—CIRCLES are sets of POINTS in which every triple of POINTS lie on exactly one CIRCLE. (Here the analog is points and circles on a *sphere* rather than points and circles in the plane.)

In our geometry, the POINTS are the eight vertices of a cube and the CIRCLES are fourteen sets of four vertices each, represented by the six faces of the cube (which we'll color green), the six (orange) diagonal rectangle cuts, and the two (blue) tetrahedra inscribed in the cube.

Figure 7 shows one representative each of the three types of CIRCLES. Convince yourself that any three POINTS in our minigeometry are contained in exactly one CIRCLE.

To make this geometry work for us, we must *dualize* it: We make the CIRCLES the pictures and the POINTS the cards. The card corresponding to a POINT contains all the CIRCLES through the POINT.

Confused? Let's build the deck. Using a regular hexagon picture of the cube, we can draw the fourteen CIRCLES as in Figure 8.

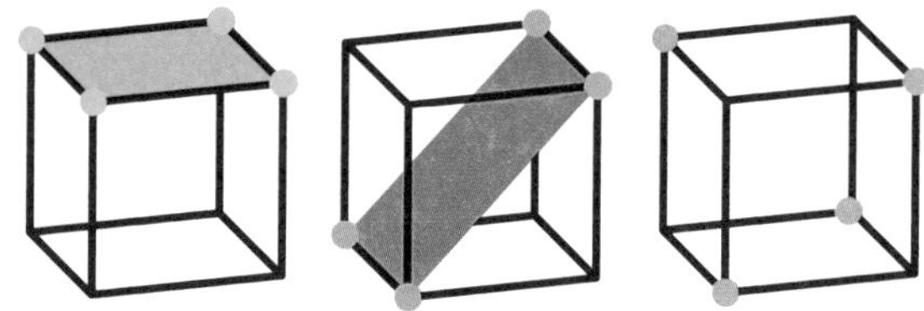

FIGURE 7. The three types of CIRCLES.

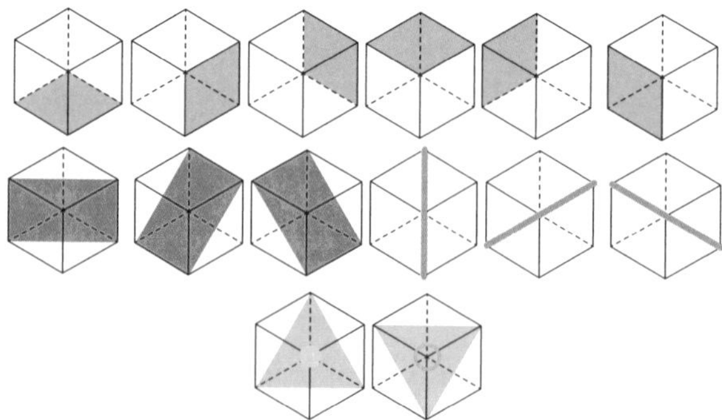

FIGURE 8. The fourteen CIRCLES.

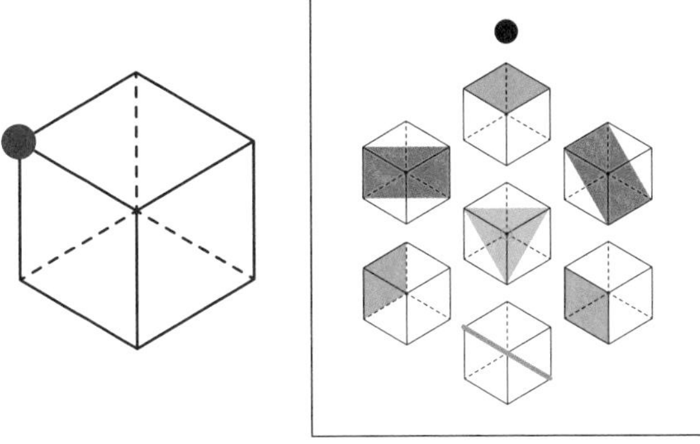

FIGURE 9. The card corresponding to the purple vertex.

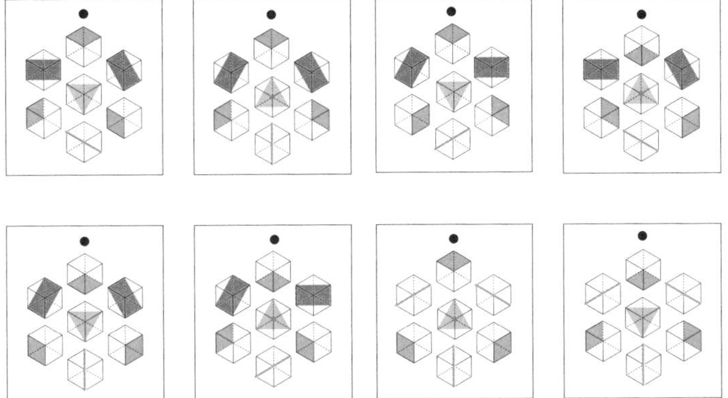

FIGURE 10. Cards corresponding to the eight POINTS.

Figure 9 shows the card corresponding to the purple POINT/vertex; it contains the pictures of all the CIRCLES that contain this vertex.

Figure 10 shows the complete set of cards. But of course we can also use pictures that don't encode any of the geometry, as in Figure 11.

Spot a Biplane!

If we want to be innovative Spot It! game designers, the mathematical objects to check out are the so-called t-(v,k,λ) *(block) designs*. A t-(v,k,λ) design has a point set consisting of v POINTS and BLOCKS consisting of k POINTS each, such that any t POINTS are contained in exactly λ BLOCKS.

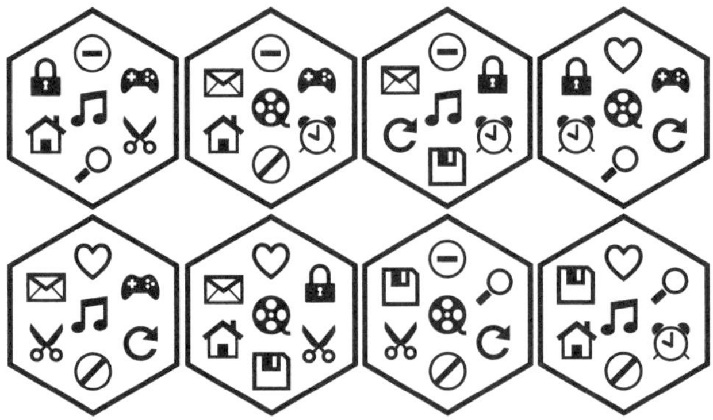

FIGURE 11. Any three cards in this deck have one picture in common.

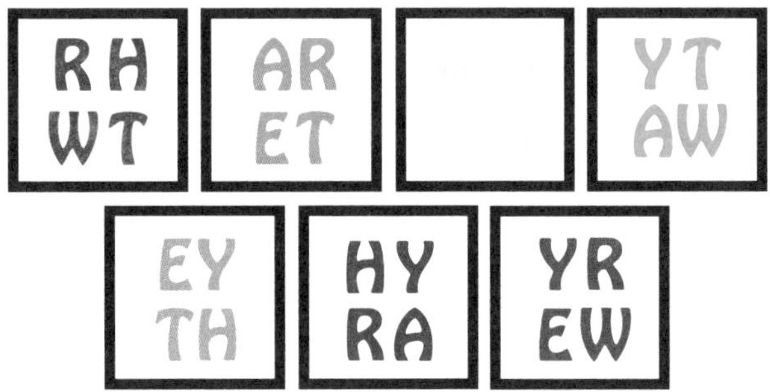

FIGURE 12. A Spot a Biplane! deck.

For example, projective planes are the 2-($n^2 - n + 1$,n,1) designs, and our geometry on the cube is a 3-(8,4,1) design. Dualizing any t-(whatever,whatever,1) design using the method we just demonstrated gives a deck of cards in which any t cards have exactly one picture in common. The problem with larger t is that no matter how hard we try, we'll always end up with a lot of pictures on our cards.

For example, starting with the amazingly compact 5-(12,6,1) Mathieu design, we'd get a deck of twelve cards with sixty-six pictures on each card out of a total of 132 pictures. Pretty mind-boggling—but the game is almost certainly not much fun to play.

Instead of increasing t, why not play with the other parameters in the definition of block designs? In particular, some of the 2-(whatever,whatever,2) designs translate into interesting decks of cards with small numbers of pictures.

These rare and exotic designs are called *biplanes*. Similar to projective planes, any two POINTS of a biplane are contained in exactly two BLOCKS and (thrown in for free!) any two BLOCKS intersect in exactly two POINTS. This intersection property means that we don't have to worry about dualizing a biplane. Just make the POINTS into the pictures and the BLOCKS into the cards, and we've got a Spot a Biplane! deck.

For example, the blocks of the unique 2-(7,4,2) biplane happen to be the complements of the LINES of the Fano plane within its point set. Figure 12 shows the corresponding Spot a Biplane! deck.

Ready to play? Then spot the two common letters on the red and green cards, and spot the two cards that contain the letters AT.

Further Reading

"The Mathematics of 'Spot It!'" by Rebekah Coggin and Anthony Meyer, *IIME Journal* 13(8), 459–467, has more details on projective planes, as do a number of articles online.

Check out Spot It! (or Dobble) for mobile devices. It incorporates (complete!) decks corresponding to projective planes with thirty-one, fifty-seven, and ninety-one cards.

If you are interested in making your own serious Spot a Biplane! deck, try Gordon Royle's site on biplanes (http://staffhome.ecm.uwa.edu.au/~00013890/remote/biplanes/index.html).

Tonight! Epic Math Battles: Counting vs. Matching

JENNIFER J. QUINN

"Ladies and gentleman, welcome to Erdős Stadium, where combinatorialists from around the world do battle, using their skills to create truly artistic mathematical identities. The winner will gain the people's ovation and fame forever."

Massive applause greeted the Chairman's statement, then he continued. "Representing counting is Erdős Stadium's resident enumerative expert. She is known to use all her fingers and all her toes. Her favorite technique is to ask a question and answer it in two different ways. You know her. You love her. I present to you . . . the Countess."

A regal woman in a flowing blue gown glided on stage. She gave a slight smile and gently inclined her head to the audience.

The Chairman continued, "If memory serves me, our challenger is known to attack in pairs. He shows no fear of negative signs. His favorite technique of 'Description—Involution—Exception' takes his challenger's strengths and makes them his own. Representing the mathematics of matching, I give you . . . Sir Match-A-Lot."

An enthusiastic man danced his way down the center aisle, playing to the audience. He waved and blew kisses until signaled to a stop.

A black-and-white-shirted official took over the microphone. "The contest will proceed in three timed rounds," she explained. "A required mystery parameter will be revealed for each round. You must prepare an identity featuring the required component and your own mathematical technique.

"Tonight's judges are: Tim Possible, known for using probabilistic arguments; Dr. X, the stadium generatingfunctionologist; and Miss Fin de Vol, a pigeon-hole-principle practitioner. I want a good clean battle. Now shake hands. May the best mathematician prevail."

The Chairman reclaimed the stage. "I raise my coffee to the Father of Combinatorics, Paul Erdős." He gestured toward a portrait hung center stage, took a sip, and bellowed, "*Allez compute*!" to signal the start of the competition. Two assistants removed drapery to reveal a chalkboard with the words *binomial coefficients*.

Round 1

While the contestants labored in isolation booths on each side of the stage, the audience listened to the judges' conversation.

"Zis means zee contestants need to present an identity zat features binomial coefficients, and ze proof must utilize zair specific technique of counting or matching, no?" commented Fin de Vol.

Tim Possible responded, "One hundred percent certain, Miss de Vol. Binomial coefficients are practically a gift. With an obvious combinatorial interpretation of $\binom{n}{k}$ as the number of subsets of size k that can be formed from an n-element set, I'd wager that both the Countess and Sir Match-A-Lot will have plenty to bring forward for judgment."

A buzzer sounded, and the contestants stepped away from their workspaces, hands held high. The official took control. "An offstage coin toss determined who presents first. Countess, what is your mathematical offering?"

The Countess stepped forward. "My motto is 'Keep it simple,' and I strove to do just that. I begin by asking the question, 'how many subsets does an n-element set contain?' The first answer to this question is 2^n by considering each element in turn and understanding its two options are to be *in* or *out* of the subset.

"The second answer directly counts subsets by cardinality. There are $\binom{n}{0}$ empty subsets, $\binom{n}{1}$ subsets with 1 element, and in general $\binom{n}{k}$ subsets with k elements. Once k is larger than n, $\binom{n}{k} = 0$. So the number of subsets is the sum of binomial coefficients $\binom{n}{k}$ as k ranges between 0 and n. Because I have answered the same question in two different ways, the answers are equal. Hence, my identity, featuring a simple sum of binomial coefficients, is

$$\binom{n}{0} + \binom{n}{1} + \binom{n}{2} + \cdots + \binom{n}{n} = 2^n."$$

Sir Match-A-Lot, eager for his chance to shine, pushed ahead. "That's nice, Countess, but I chose to add drama with some negative signs. Sure, I could have presented $\binom{n}{0} - \binom{n}{1} + \binom{n}{2} - \cdots + (-1)^n \binom{n}{n} = 0$. But that was too easy. Instead, I *generalized*. Let me break it down for you.

"Description: Choose an integer k between 0 and n. Restrict consideration to subsets of an n-element set with k or fewer members. Subsets with an even number of elements contribute +1; subsets with an odd number of elements contribute −1. The total weighted contributions from subsets of size less than or equal to k is the alternating sum of binomial coefficients:

$$\binom{n}{0} - \binom{n}{1} + \binom{n}{2} - \cdots + (-1)^k \binom{n}{k}.$$

"Involution: Start by picking your favorite of the n elements. Without loss of generality, call it Waldo. Time for the ol' switcheroo. Given any subset, match it by flipping Waldo—meaning that if Waldo is in the set, the matching subset will be the same except boot Waldo out. If Waldo is MIA, the matching subset welcomes him back. This changes the size of a subset by one, sometimes bigger, sometimes smaller, but always pairs an even subset with an odd one. Because $1 + (-1) = 0$, the matched pairs contribute a big, fat zero. Only the unmatched subsets matter.

"Exceptions: A set of size k with no Waldo cannot switcheroo. There's no elbow room. The subset size is maxed out. Anyone but Waldo could be in such a set, so there are $\binom{n-1}{k}$ exceptions. Each such set will contribute +1 or −1 to the total depending on the parity of k. This gives my kickin' identity, the alternating *partial* sum of binomial coefficients:

$$\binom{n}{0} - \binom{n}{1} + \binom{n}{2} - \cdots + (-1)^k \binom{n}{k} = (-1)^k \binom{n-1}{k}."$$

The judges conferred before Dr. X kicked off the postgame analysis: "Countess, your presentation was succinct and elegant. Simplicity of thought is admirable, but we were disappointed by the lack of complexity to your identity. You could have been true to yourself while still giving us something more substantial. Why not sums of squares or a convolution?"

"Monsieur Match-A-Lot really took us beyond," Fin de Vol pronounced, "generalizing a partner identity to zee Countess's. He is clevair, but with language like zat, no one would ever call him elegant."

Tim Possible concluded, "The binomial coefficients were prominently featured in both works, as required. So we judged on style and complexity. While Miss de Vol bristled at Match-A-Lot's nonstandard demeanor, there is no denying his raw enthusiasm. Therefore, by agreement of the judges, Round 1 is awarded to Sir Match-A-Lot."

Round 2

The second-round challenge parameter was revealed to be "Fibonacci numbers." As the competitors labored, the judges speculated on possible combinatorial interpretations.

Dr. X began, "Assuming $F_0 = 0$, $F_1 = 1$, and for $n > 1$, $F_n = F_{n-1} + F_{n-2}$, F_n is known to count binary $(n-2)$-tuples with no consecutive 0s; subsets of $\{1, 2, \ldots, n-2\}$ with no consecutive integers; or compositions of $n - 1$ using only 1s and 2s."

"Mon favorite is tilings of a $1 \times (n-1)$-board using squares and dominoes," added Fin de Vol.

Tim Possible chimed in. "There are other fascinating Fibonacci tilings too. For example, tilings of an $(n + 1)$-board with tiles of length greater than or equal to 2, or tilings of an n-board where all tiles have odd length. And that is just the tip of the iceberg."

At the conclusion of the interval, Sir Match-A-Lot presented first.

"Because you liked my previous work so much, I decided to Fibonaccize it. My offering is a formulation for an alternating sum of Fibonacci numbers, $F_1 - F_2 + F_3 - \cdots + (-1)^n F_n$. You dig? As you all surely know, F_k counts tilings of $(k-1)$-boards with squares and dominoes provided $k \geq 1$.

"Description: Count the square and domino tilings of boards with lengths between 0 and $n - 1$. Even-length boards contribute +1; odd-length boards contribute −1. The total weighted contributions from the tilings is the alternating sum of consecutive Fibonacci numbers, as promised.

"Involution: Switcheroo the last tile. If it's a domino, make it a square, and vice versa. Matched tilings add to zero.

"Exceptions: No can do a switcheroo on the empty tiling—of length zero. Zero is even, so it contributes +1. No elbow room to switcheroo on length $n - 1$ tilings that end in a square. There are F_{n-1}, and they each contribute $(-1)^n$ to the total. The take home? This sweet little ditty:

$$F_1 - F_2 + F_3 - \cdots + (-1)^n F_n = 1 + (-1)^n F_{n-1}."$$

He bowed to the audience and winked at the Countess as she made her way to the judging table. She did her best to ignore him.

"Taking your previous admonition to heart," she said to the judges, "I wanted to give you more than a sum of Fibonacci numbers. As my opponent would say, that is *too* simple. Instead, I worked to count the sum of *squares* of Fibonacci numbers by asking 'in how many ways can I tile two different boards—one of length $n - 1$ and the other of length n—with squares and dominoes?'

"You already know that there are F_n ways to tile the first and F_{n+1} ways to tile the second. So the first answer to my question is the product F_nF_{n+1}.

"For my second answer, I arrange the boards one under the other, aligned to the left. Then I count based on the rightmost position where breaks in the two tilings coincide vertically." She showed the audience Figure 1. "These locations of vertical agreement are termed *faults*. And while I prefer not to seek fault in others, the position of the rightmost (or final) fault is key.

"To its right, tilings are uniquely determined, with one board being completed by all dominoes and the other completed with a square followed by dominoes. To its left are two tilings of the same length. If the final fault occurs immediately after cell $k - 1$, there are F_k^2 possible tilings.

"Of course, a final fault always exists, although it may occur at the very beginning, which I consider *after* the 0th cell. Thus $k - 1$, the position of the final fault, ranges between 0 and $n - 1$. Summing over all possible values for k gives my second answer and leads to the identity

$$F_1^2 + F_2^2 + \cdots + F_n^2 = F_nF_{n+1}."$$

Satisfied with her presentation, the Countess took a position beside Sir Match-A-Lot and awaited the judges' response.

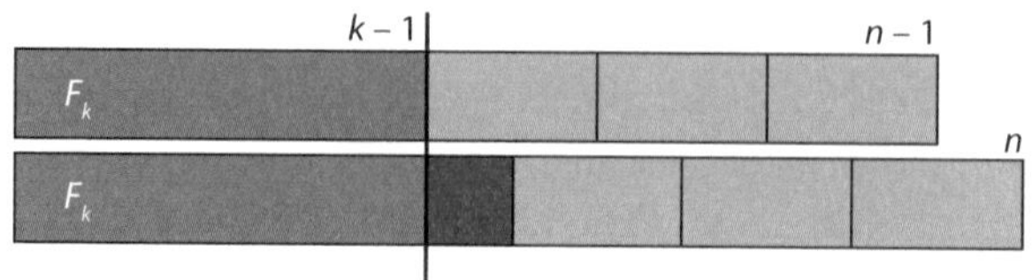

FIGURE 1. The dominoes have a fault after cell $k - 1$.

"Monsieur Match-A-Lot, zair was a symmetry between your offerings in Round 1 and Round 2. Whereas it served you well originally, it felt redundant this time," commented Fin de Vol. "Madame Countess, you 'eard our criticisms and acted on them. Well done. For complexity of identity, zis round belongs to you in my estimation."

"I don't necessarily agree with your analysis," said Dr. X. "For me, the simplicity of toggling the final tile for the matching absolutely outweighs the fault of faulty tiling pairs."

"In truth, I'm torn," said Tim Possible. "I was impressed by both presentations. So I will settle the question by flipping a coin." He removed a silver dollar from his pocket, threw it into the air, and let it land on the table. "Looks like Round 2 goes to the Countess."

Round 3

Sir Match-A-Lot was not entirely composed for the start of Round 3. He fumed about the arbitrariness of the previous decision. When the mystery parameter was revealed to be the "golden ratio," he exploded. "Hold the phone! The golden ratio is not an integer quantity. How am I supposed to use $\varphi = (1 + \sqrt{5})/2$ in a matching identity?" He paced Erdős Stadium swearing under his breath for more than half the allotted time.

The Countess, on the other hand, set to work right away, tackling the classic formula credited to Binet but known more than a century earlier to Euler, Bernoulli, and de Moivre. To the audience's delight, both had something to report at the close of the interval. The Countess presented first.

"Judges, my intention is to prove Binet's celebrated formula, $F_n = ((\varphi^n - (-\varphi)^{-n})/\sqrt{5}$ by asking a question and answering it two different ways. To do this, I use infinite tilings constructed by sequentially and independently placing tiles, selecting squares with probability $1/\varphi$ and dominoes with probability $1/\varphi^2$. Note that these are the only two options, as

$$\frac{1}{\varphi} + \frac{1}{\varphi^2} = 1 .$$

"I ask, 'what is the probability, q_n, that such an infinite tiling is breakable between cell $n - 1$ and cell n?'

"First, I compute the probability directly. Breakability after cell $n - 1$ requires that an infinite tiling begins with a square-domino segment of length $n - 1$. Each of the F_n possible initial segments occurs with probability $1/\varphi^{n-1}$. So the first answer is $q_n = F_n/\varphi^{n-1}$.

"Next, I compute the probability indirectly by considering the complementary event—when a domino covers cells $n - 1$ and n. The tilings *must* be breakable between cells $n - 2$ and $n - 1$. So the chance of being *unbreakable* after cell $n - 1$ is q_{n-1}/φ^2, leading to $q_n = 1 - q_{n-1}/\varphi^2$. Thankfully, $q_1 = 1$ and the recurrence unravels into a geometric series

$$q_n = 1 - \frac{1}{\varphi^2} + \frac{1}{\varphi^4} - \frac{1}{\varphi^6} + \cdots + \left(\frac{-1}{\varphi^2}\right)^{n-1}$$

with common ratio $-1/\varphi^2$. Answer two has the closed form

$$q_n = \frac{1 - (-1/\varphi^2)^n}{1 - (-1/\varphi^2)} = \frac{\varphi}{\sqrt{5}}(1 - (-\varphi^2)^{-n}).$$

"The final step is to set answer 1 equal to answer 2 and solve for F_n. Thus I arrive at

$$F_n = \varphi^{n-1} \cdot \frac{\varphi}{\sqrt{5}}(1 - (-\varphi^2)^{-n}) = \frac{1}{\sqrt{5}}(\varphi^n - (-\varphi)^{-n})."$$

Sir Match-A-Lot, not to be outdone, sauntered center stage and proclaimed, "That was mighty fine indeed, Countess. But your technique was probabilistic and less true to your counting roots. I, on the other hand, will prove the *exact same identity* using matching, and only matching, thereby demonstrating my superiority in this arena. Check. It. Out.

"Description: Tile a $1 \times n$ board with red squares of weight φ, orange squares of weight $\bar{\varphi}$ (aka $-\varphi^{-1}$), and dominoes of weight 1. Be warned, first tiles are special and have weights $\varphi/\sqrt{5}$, $-\bar{\varphi}/\sqrt{5}$. and 0, respectively. Compute the weight of a tiling by multiplying the individual weights. Lay your eyes on this tiling of weight $-\varphi\bar{\varphi}^5/\sqrt{5}$." He presented Figure 2.

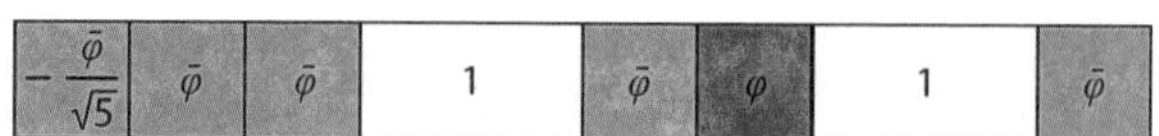

FIGURE 2. A tiling of weight $-\varphi\bar{\varphi}^5/\sqrt{5}$.

"The real deal involves all the weighted tilings of a particular length. If W_n is the sum of the weights for all n-tilings, then W_n is really the Fibonacci number F_n in disguise.

"Don't take my word for it. Do the math. Check initial values. A 1-tiling is either a red square or an orange square. Total weight $W_1 = \varphi/\sqrt{5} + (-\bar{\varphi}/\sqrt{5}) = 1$. A 2-tiling is either two red squares, a red square followed by an orange square, an orange square followed by a red square, two orange squares, or a domino. Total weight

$$W_2 = \frac{\varphi^2}{\sqrt{5}} + \frac{\varphi\bar{\varphi}}{\sqrt{5}} - \frac{\varphi\bar{\varphi}}{\sqrt{5}} - \frac{\bar{\varphi}^2}{\sqrt{5}} + 0 = \frac{\varphi^2}{\sqrt{5}} - \frac{\bar{\varphi}^2}{\sqrt{5}} = 1.$$

"For any tiling longer than 2, it ends with an ordinary tile, either a red square, orange square, or domino. Total weight

$$W_n = \varphi W_{n-1} + \bar{\varphi} W_{n-1} + W_{n-2} = W_{n-1} + W_{n-2}.$$

"You see? These weighted tilings match initial conditions and agree on the recurrence relation. So they represent. Fibonacci numbers, that is.

"Involution: The switcheroo will *usually* flip a domino (weight 1) for two squares of different colors (weight $\varphi\bar{\varphi} = -1$), or vice versa. Look for whichever comes first, a domino or a pair of consecutive squares of different colors. As long as it does not involve a leading tile of special weight, do the switch—being sure to match the first square with the color of squares that precede it." And he illustrated this by presenting Figure 3.

"The weight of the matched tilings adds to zero, just the way I like it. If a tiling begins with a red square followed by an orange square (contributing a factor of $\varphi\bar{\varphi}^5/\sqrt{5}$ to the weight), flip the colors so that it begins orange followed by red (contributing a factor of $-\varphi\bar{\varphi}^5/\sqrt{5}$ to the weight). Again, the weight of the matched tilings adds to zero. If the

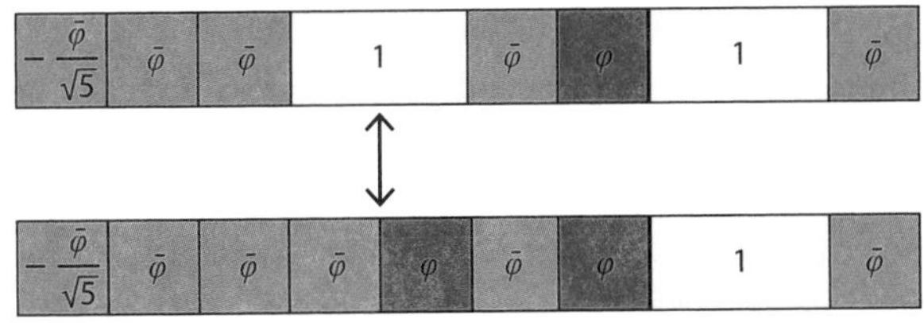

FIGURE 3. Weighted tilings are matched based on the first occurrence of a domino (weight 1) or two squares of different colors (weight $\varphi\bar{\varphi} = -1$).

tiling begins with a domino, its weight is zero by definition, so it will not matter one iota.

"Exceptions: Monochromatic squares are the only unmatched, non-zero weight tilings. There are two—all red squares and all orange squares. So,

$$W_n = (F_n =) \varphi^n/\sqrt{5} + (-\varphi^n/\sqrt{5})$$

through the beauty of matchings. I rest my case."

The Chairman led the applause. "Well done. Clearly this is a battle for the history books. Given the current score, the winner of the third and final round will be crowned Epic Math Battles champion. What do our judges have to say?"

Tim Possible was enraptured. "Countess, I could not imagine a more beautiful proof of this classic identity."

Dr. X addressed the Countess. "The balance of big picture and nitty-gritty was reasonable, but I question your choice to gloss over algebraic details. I'm taking some of your leaps on faith—particularly the simplification of the geometric series. And coming from me, that's saying a lot."

Fin de Vol addressed Sir Match-A-Lot, "I was less zan impressed by your early theatrics. Sportsmanship is essential in any competition. Ze weighted tilings seemed an overly complex combinatorial interpretation, but they swayed me when everything fell away except for ze two all-squares tilings and zair weights gave Binet's formula with no further simplification. Both presentations were truly magnifique."

Tim Possible continued, "Rather than leaving the decision to chance, we believe it best to let the audience decide. Who won? The Countess or Sir Match-A-Lot? Counting or matching?"

Cast your vote now at maa.org/mathhorizons/supplemental.htm. Voting closes on March 21. The winner will be announced on the *Math Horizons* Facebook and Twitter accounts.

Results

Decided by a margin of only five votes, *Math Horizons* announced on March 25 the Epic Math Battle: Counting vs. Matching champion to be the Countess. Overwhelmed by the award, the Countess declined to comment; her competitor vowed to return to Erdős Stadium and challenge for the title again.

Further Reading

A. T. Benjamin and J. J. Quinn, *Proofs That Really Count: The Art of Combinatorial Proof* (Washington, DC: Mathematical Association of America, 2003).

A. T. Benjamin and J. J. Quinn, "An Alternate Approach to Alternating Sums: A Method to DIE For," *College Mathematics Journal* 39, no. 3, (2008): 191–201.

A. T. Benjamin, H. Derks, and J. J. Quinn, "The Combinatorialization of Linear Recurrences," *Electronic Journal of Combinatorics* 18, no. 2, (2011): 12.

Mathematicians Chase Moonshine's Shadow

Erica Klarreich

In 1978, the mathematician John McKay noticed what seemed like an odd coincidence. He had been studying the different ways of representing the structure of a mysterious entity called the *monster group*, a gargantuan algebraic object that, mathematicians believed, captured a new kind of symmetry. Mathematicians weren't sure that the monster group actually existed, but they knew that if it did exist, it acted in special ways in particular dimensions, the first two of which were 1 and 196,883.

McKay, of Concordia University in Montreal, happened to pick up a mathematics paper in a completely different field, involving something called the *j*-function, one of the most fundamental objects in number theory. Strangely enough, this function's first important coefficient is 196,884, which McKay instantly recognized as the sum of the monster's first two special dimensions.

Most mathematicians dismissed the finding as a fluke, since there was no reason to expect the monster and the *j*-function to be even remotely related. However, the connection caught the attention of John Thompson, a Fields medalist now at the University of Florida in Gainesville, who made an additional discovery. The *j*-function's second coefficient, 21,493,760, is the sum of the first three special dimensions of the monster: 1 + 196,883 + 21,296,876. It seemed as if the *j*-function was somehow controlling the structure of the elusive monster group.

Soon, two other mathematicians had demonstrated so many of these numerical relationships that it no longer seemed possible that they were mere coincidences. In a 1979 paper called "Monstrous Moonshine," the pair—John Conway, now of Princeton University, and Simon Norton—conjectured that these relationships must result from some deep connection between the monster group and the *j*-function. "They called it moonshine because it appeared so far-fetched," said Don Zagier, a director of the Max Planck Institute for Mathematics in Bonn, Germany. "They were such wild ideas that it seemed like wishful thinking to imagine anyone could ever prove them."

It took several more years before mathematicians succeeded in even constructing the monster group, but they had a good excuse: The monster has more than 10^{53} elements, which is more than the number of atoms in a thousand Earths. In 1992, a decade after Robert Griess of the University of Michigan constructed the monster, Richard Borcherds tamed the wild ideas of monstrous moonshine, eventually earning a Fields Medal for this work. Borcherds, of the University of California, Berkeley, proved that there was a bridge between the two distant realms of mathematics in which the monster and the *j*-function live: namely, string theory, the counterintuitive idea that the universe has tiny hidden dimensions, too small to measure, in which strings vibrate to produce the physical effects we experience at the macroscopic scale.

Borcherds's discovery touched off a revolution in pure mathematics, leading to a new field known as generalized Kac-Moody algebras. But from a string theory point of view, it was something of a backwater. The twenty-four-dimensional string theory model that linked the *j*-function and the monster was far removed from the models string theorists were most excited about. "It seemed like just an esoteric corner of the theory, without much physical interest, although the math results were startling," said Shamit Kachru, a string theorist at Stanford University.

But now moonshine is undergoing a renaissance, one that may eventually have deep implications for string theory. Over the past five years, starting with a discovery analogous to McKay's, mathematicians and physicists have come to realize that monstrous moonshine is just the start of the story.

Last week, researchers posted a paper on arxiv.org presenting a numerical proof of the so-called Umbral Moonshine Conjecture, formulated in 2012, which proposes that in addition to monstrous moonshine, there are twenty-three other moonshines: mysterious correspondences between the dimensions of a symmetry group on the one hand, and the coefficients of a special function on the other. The functions in these new moonshines have their origins in a prescient letter by one of mathematics' great geniuses, written more than half a century before moonshine was even a glimmer in the minds of mathematicians.

The twenty-three new moonshines appear to be intertwined with some of the most central structures in string theory, four-dimensional objects known as K3 surfaces. The connection with umbral moonshine hints at hidden symmetries in these surfaces, said Miranda Cheng of the University of Amsterdam and France's National Center for Scientific Research, who originated the Umbral Moonshine Conjecture together with John Duncan, of Case Western Reserve University in Cleveland, Ohio, and Jeffrey Harvey, of the University of Chicago. "This is important, and we need to understand it," she said.

The new proof strongly suggests that in each of the twenty-three cases, there must be a string theory model that holds the key to understanding these otherwise baffling numerical correspondences. But the proof doesn't go so far as to actually construct the relevant string theory models, leaving physicists with a tantalizing problem. "At the end of the day when we understand what moonshine is, it will be in terms of physics," Duncan said.

Monstrous Moonshine

The symmetries of any given shape have a natural sort of arithmetic to them. For example, rotating a square ninety degrees and then flipping it horizontally is the same as flipping it across a diagonal—in other words, "ninety-degree rotation + horizontal flip = diagonal flip." During the nineteenth century, mathematicians realized that they could distill this type of arithmetic into an algebraic entity called a group. The same abstract group can represent the symmetries of many different shapes, giving mathematicians a tidy way to understand the commonalities in different shapes (Figure 1).

Over much of the twentieth century, mathematicians worked to classify all possible groups, and they gradually discovered something strange: While most simple finite groups fell into natural categories, there were twenty-six oddballs that defied categorization. Of these, the biggest, and the last to be discovered, was the monster.

Before McKay's serendipitous discovery nearly four decades ago, there was no reason to think the monster group had anything to do with the j-function, the second protagonist of the monstrous-moonshine

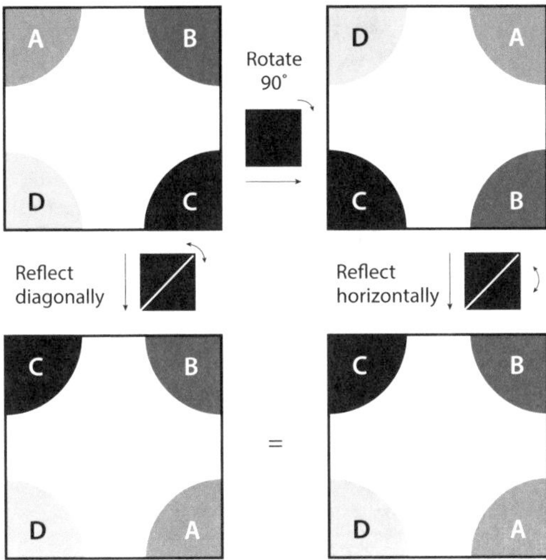

FIGURE 1. Rotating a square ninety degrees and then reflecting it horizontally has the same effect as reflecting it across a diagonal, so in the language of square symmetry arithmetic, ninety-degree rotation + horizontal reflection = diagonal reflection. Olena Shmahalo/*Quanta Magazine*

story. The *j*-function belongs to a special class of functions whose graphs have repeating patterns similar to M. C. Escher's tessellation of a disk with angels and devils, which shrink ever smaller as they approach the outer boundary (Figure 2). These "modular" functions are the heroes of number theory, playing a crucial role, for instance, in Andrew Wiles' 1994 proof of Fermat's Last Theorem. "Any time you hear about a striking result in number theory, there's a high chance that it's really a statement about modular forms," Kachru said.

As with a sound wave, the *j*-function's repeating pattern can be broken down into a collection of pure tones, so to speak, with coefficients indicating how "loud" each tone is. It is in these coefficients that McKay found the link to the monster group.

In the early 1990s, building on work by Igor Frenkel of Yale University, James Lepowsky of Rutgers University, and Arne Meurman of Lund University in Sweden, Borcherds made sense of McKay's discovery by showing that there is a particular string theory model in which the *j*-function and the monster group both play roles. The coefficients

FIGURE 2. Modular functions have repeating patterns related to the tiling in this figure. Wikimedia Commons: Tom Ruen

of the j-function count the ways strings can oscillate at each energy level. And the monster group captures the model's symmetry at those energy levels.

The finding gave mathematicians a way to study the mind-bogglingly large monster group using the j-function, whose coefficients are easy to calculate. "Math is all about building bridges where on one side you see more clearly than on the other," Duncan said. "But this bridge was so unexpectedly powerful that before you see the proof it's kind of crazy."

New Moonshine

While mathematicians explored the ramifications of monstrous moonshine, string theorists focused on a seemingly different problem: figuring out the geometry for the tiny dimensions in which strings are hypothesized to live. Different geometries allow strings to vibrate in different ways, just as tightening the tension on a drum changes its pitch. For decades, physicists have struggled to find a geometry that produces the physical effects we see in the real world.

An important ingredient in some of the most promising candidates for such a geometry is a collection of four-dimensional shapes known as K3 surfaces. In contrast with Borcherds' string theory model, Kachru said, K3 surfaces fill the string theory textbooks.

Not enough is known about the geometry of K3 surfaces to count how many ways strings can oscillate at each energy level, but physicists can write down a more limited function counting certain physical states that appear in all K3 surfaces. In 2010, three string theorists—Tohru Eguchi of Kyoto University in Japan, Hirosi Ooguri of the California Institute of Technology in Pasadena, and Yuji Tachikawa of the University of Tokyo in Japan—noticed that if they wrote this function in a particular way, out popped coefficients that were the same as some special dimensions of another oddball group, called the Mathieu 24 (M24) group, which has nearly 250 million elements. The three physicists had discovered a new moonshine.

This time, physicists and mathematicians were all over the discovery. "I was at several conferences, and all the talk was about this new Mathieu moonshine," Zagier said.

Zagier attended one such conference in Zurich in July 2011, and there, Duncan wrote in an e-mail, Zagier showed him "a piece of paper with lots of numbers on it"—the coefficients of some functions Zagier was studying called "mock modular" forms, which are related to modular functions. "Don [Zagier] pointed to a particular line of numbers and asked me—in jest, I think—if there is any finite group related to them," Duncan wrote.

Duncan wasn't sure, but he recognized the numbers on another line: They belonged to the special dimensions of a group called M12. Duncan buttonholed Miranda Cheng, and the two pored over the rest of Zagier's piece of paper. The pair, together with Jeffrey Harvey, gradually realized that there was much more to the new moonshine than just the M24 example. The clue to the full moonshine picture, they found, lay in the nearly century-old writings of one of mathematics' legendary figures.

Moonshine's Shadows

In 1913, the English mathematician G. H. Hardy received a letter from an accounting clerk in Madras, India, describing some mathematical formulas he had discovered. Many of them were old hat, and some were flat-out wrong, but on the final page were three formulas that blew Hardy's mind. "They must be true," wrote Hardy, who promptly invited the clerk, Srinivasa Ramanujan, to England, "because, if they were not true, no one would have the imagination to invent them."

Ramanujan became famous for seemingly pulling mathematical relationships out of thin air, and he credited many of his discoveries to the goddess Namagiri, who appeared to him in visions, he said. His mathematical career was tragically brief, and in 1920, as he lay dying in India at age thirty-two, he wrote Hardy another letter saying that he had discovered what he called "mock theta" functions, which entered into mathematics "beautifully." Ramanujan listed seventeen examples of these functions, but didn't explain what they had in common. The question remained open for more than eight decades, until Sander Zwegers, then a graduate student of Zagier's and now a professor at the University of Cologne in Germany, figured out in 2002 that they are all examples of what came to be known as mock modular forms (Figure 3).

After the Zurich moonshine conference, Cheng, Duncan, and Harvey gradually figured out that M24 moonshine is one of twenty-three different moonshines, each making a connection between the special dimensions of a group and the coefficients of a mock modular form—just as monstrous moonshine made a connection between the monster group and the *j*-function. For each of these moonshines, the researchers conjectured, there is a string theory like the one in monstrous moonshine, in which the mock modular form counts the string states and the group captures the model's symmetry. A mock modular form always has an associated modular function called its "shadow," so they named their hypothesis the Umbral Moonshine Conjecture—*umbra* is Latin for "shadow." Many of the mock modular forms that appear in the conjecture are among the seventeen special examples Ramanujan listed in his prophetic letter.

Curiously enough, Borcherds's earlier proof of monstrous moonshine also builds on work by Ramanujan: The algebraic objects at the core of the proof were discovered by Frenkel, Lepowsky, and Meurman as they analyzed the three formulas that had so startled Hardy in Ramanujan's first letter. "It's amazing that these two letters form the cornerstone of what we know about moonshine," said Ken Ono, of Emory University in Atlanta, Georgia. "Without either letter, we couldn't write this story."

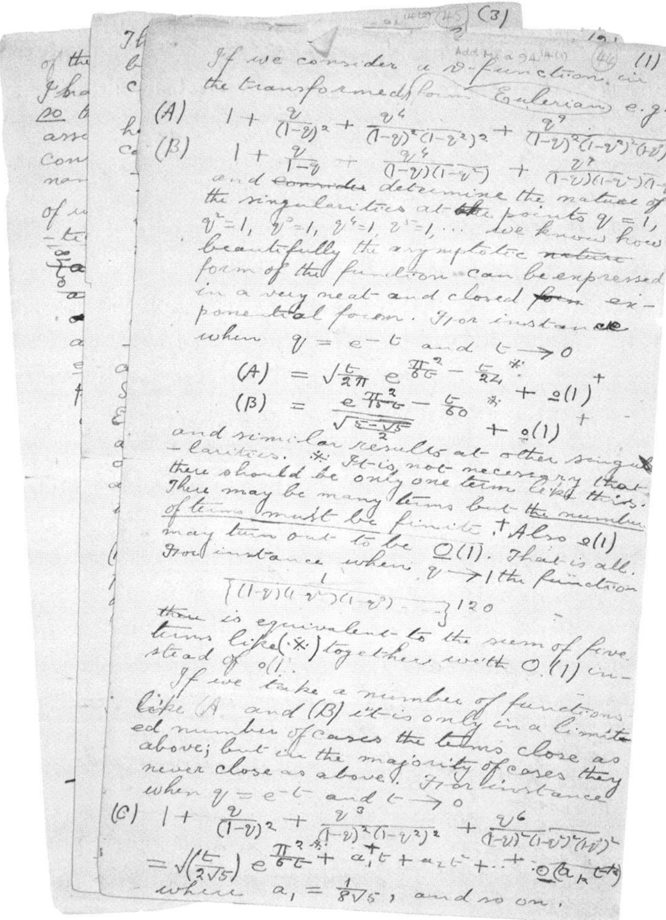

(1)

If we consider a θ-function in the transformed Eulerian form e.g.

(A) $1 + \frac{q}{(1-q)^2} + \frac{q^4}{(1-q)^2(1-q^2)^2} + \frac{q^9}{(1-q)^2(1-q^2)^2(1-q^3)^2}$

(B) $1 + \frac{q}{1-q} + \frac{q^4}{(1-q)(1-q^2)} + \frac{q^9}{(1-q)(1-q^2)(1-q^3)}$

and determine the nature of the singularities at the points $q=1$, $q^2=1$, $q^3=1$, $q^4=1$, $q^5=1$, … we know how beautifully the asymptotic form of the function can be expressed in a very neat and closed exponential form. For instance when $q = e^{-t}$ and $t \to 0$

$$(A) = \sqrt{\tfrac{t}{2\pi}}\, e^{\frac{\pi^2}{6t} - \frac{t}{24}}\ * + o(1)^{+}$$

$$(B) = \frac{e^{\frac{\pi^2}{15t} - \frac{t}{60}}}{\sqrt{\frac{5-\sqrt{5}}{2}}}\ * + o(1)^{+}$$

and similar results at other singularities. * It is not necessary that there should be only one term like this. There may be many terms but the number of terms must be finite. † Also $o(1)$ may turn out to be $O(1)$. That is all. For instance when $q \to 1$ the function

$$\frac{1}{\{(1-q)(1-q^2)(1-q^3)\cdots\}^{120}}$$

is equivalent to the sum of five terms like (*) together with $O(1)$ instead of $o(1)$.

If we take a number of functions like (A) and (B) it is only in a limited number of cases the terms close as above; but in the majority of cases they never close as above. For instance when $q = e^{-t}$ and $t \to 0$

(C) $1 + \frac{q}{(1-q)^2} + \frac{q^3}{(1-q)^2(1-q^2)^2} + \frac{q^6}{(1-q)^2(1-q^2)^2(1-q^3)^2}$

$$= \sqrt{\left(\tfrac{t}{2\sqrt{5}}\right)}\, e^{\frac{\pi^2}{6t} + a_1 t + a_2 t^2 + \cdots + O(a_k t^k)}$$

where $a_1 = \frac{1}{8\sqrt{5}}$, and so on.

FIGURE 3. Srinivasa Ramanujan's final letter to G. H. Hardy in 1920, explaining his discovery of what he called "mock theta" functions. Courtesy of Ken Ono

Finding the Beast

In the new paper posted on arxiv.org, Duncan, Ono, and Ono's graduate student Michael Griffin have come up with a numerical proof of the Umbral Moonshine Conjecture (one case of which—the M24 case—had already been proven by Terry Gannon, of the University of Alberta in Edmonton, Canada). The new analysis provides only hints of where physicists should look for the string theories that will unite the groups and the mock modular forms. Nevertheless, the proof confirms that the conjecture is on the right track, Harvey said. “We had all this structure, and it was so intricate and compelling that it was hard not to think there was some truth to it,” he said. “Having a mathematical proof makes it a solid piece of work that people can think seriously about.”

The string theory underlying umbral moonshine is likely to be “not just any physical theory, but a particularly important one,” Cheng said. “It suggests that there’s a special symmetry acting on the physical theory of K3 surfaces.” Researchers studying K3 surfaces can’t see this symmetry yet, she said, suggesting that “there is probably a better way of looking at that theory that we haven’t found yet.”

Physicists are also excited about a highly conjectural connection between moonshine and quantum gravity, the as-yet-undiscovered theory that will unite general relativity and quantum mechanics. In 2007, the physicist Edward Witten, of the Institute for Advanced Study in Princeton, New Jersey, speculated that the string theory in monstrous moonshine should offer a way to construct a model of three-dimensional quantum gravity, in which 194 natural categories of elements in the monster group correspond to 194 classes of black holes. Umbral moonshine may lead physicists to similar conjectures, giving hints of where to look for a quantum gravity theory. “That is a big hope for the field,” Duncan said.

The new numerical proof of the Umbral Moonshine Conjecture is “like looking for an animal on Mars and seeing its footprint, so we know it’s there,” Zagier said. Now researchers have to find the animal—the string theory that would illuminate all these deep connections. “We really want to get our hands on it,” Zagier said.

The Impenetrable Proof

Davide Castelvecchi

Sometime on the morning of August 30, 2012, Shinichi Mochizuki quietly posted four papers on his website.

The papers were huge—more than 500 pages in all—packed densely with symbols, and the culmination of more than a decade of solitary work. They also had the potential to be an academic bombshell. In them, Mochizuki claimed to have solved the *abc* conjecture, a twenty-seven-year-old problem in number theory that no other mathematician had even come close to solving. If his proof was correct, it would be one of the most astounding achievements of mathematics this century and would completely revolutionize the study of equations with whole numbers.

Mochizuki, however, did not make a fuss about his proof. The respected mathematician, who works at Kyoto University's Research Institute for Mathematical Sciences (RIMS) in Japan, did not even announce his work to peers around the world. He simply posted the papers and waited for the world to find out.

Probably the first person to notice the papers was Akio Tamagawa, a colleague of Mochizuki's at RIMS. He, like other researchers, knew that Mochizuki had been working on the conjecture for years and had been finalizing his work. That same day, Tamagawa e-mailed the news to one of his collaborators, number theorist Ivan Fesenko of the University of Nottingham, U.K. Fesenko immediately downloaded the papers and started to read. But he soon became "bewildered," he says. "It was impossible to understand them."

Fesenko e-mailed some top experts in Mochizuki's field of arithmetic geometry, and word of the proof quickly spread. Within days, intense chatter began on mathematical blogs and online forums (see *Nature,* http://doi.org/725; 2012). But for many researchers, early elation about the proof quickly turned to skepticism. Everyone—even those whose area of expertise was closest to Mochizuki's—was just as flummoxed by the papers as Fesenko had been. To complete the proof, Mochizuki had invented a new branch of his discipline, one that is astonishingly abstract even by the standards of pure math. "Looking at it, you feel a bit like you might be reading a paper from the future, or from outer space," number theorist Jordan Ellenberg, of the University of Wisconsin—Madison, wrote on his blog a few days after the paper appeared.

Three years on, Mochizuki's proof remains in mathematical limbo—neither debunked nor accepted by the wider community. Mochizuki has estimated that it would take a math graduate student about ten years to be able to understand his work, and Fesenko believes that it would take even an expert in arithmetic geometry some five hundred hours. So far, only four mathematicians say that they have been able to read the entire proof.

Adding to the enigma is Mochizuki himself. He has so far lectured about his work only in Japan, in Japanese, and despite being fluent in English, he has declined invitations to talk about it elsewhere. He does not speak to journalists; several requests for an interview for this story went unanswered. Mochizuki has replied to e-mails from other mathematicians and has been forthcoming to colleagues who have visited him, but his only public input has been sporadic posts on his website. In December 2014, he wrote that to understand his work, there was a "need for researchers to deactivate the thought patterns that they have installed in their brains and taken for granted for so many years." To mathematician Lieven Le Bruyn of the University of Antwerp in Belgium, Mochizuki's attitude sounds defiant. "Is it just me," he wrote on his blog earlier this year, "or is Mochizuki really sticking up his middle finger to the mathematical community?"

Now, that community is attempting to sort the situation out. In December 2015, the first workshop on the proof outside of Asia took place in Oxford, U.K. Mochizuki was not there in person, but he answered questions from the workshop through Skype. The organizers hope that the discussion will motivate more mathematicians to invest the time to familiarize themselves with his ideas—and potentially move the needle in Mochizuki's favor.

In his latest verification report, Mochizuki wrote that the status of his theory with respect to arithmetic geometry "constitutes a sort of faithful miniature model of the status of pure mathematics in human society." The trouble that he faces in communicating his abstract work to his own discipline mirrors the challenge that mathematicians as a whole often face in communicating their craft to the wider world.

Primal Importance

The *abc* conjecture refers to numerical expressions of the type $a + b = c$. The statement, which comes in several slightly different versions, concerns the prime numbers that divide each of the quantities *a, b,* and *c*. Every whole number, or integer, can be expressed in an essentially unique way as a product of prime numbers—those that cannot be further factored out into smaller whole numbers: for example, $15 = 3 \times 5$ or $84 = 2 \times 2 \times 3 \times 7$. In principle, the prime factors of *a* and *b* have no connection to those of their sum, *c*. But the *abc* conjecture links them together. It presumes, roughly, that if a lot of small primes divide *a* and *b,* then only a few, large ones divide *c*.

This possibility was first mentioned in 1985, in a rather offhand remark about a particular class of equations by French mathematician Joseph Oesterlé during a talk in Germany. Sitting in the audience was David Masser, a fellow number theorist now at the University of Basel in Switzerland, who recognized the potential importance of the conjecture and later publicized it in a more general form. It is now credited to both, and is often known as the Oesterlé-Masser conjecture.

A few years later, Noam Elkies, a mathematician at Harvard University in Cambridge, Massachusetts, realized that the *abc* conjecture, if true, would have profound implications for the study of equations concerning whole numbers—also known as Diophantine equations after Diophantus, the ancient Greek mathematician who first studied them.

Elkies found that a proof of the *abc* conjecture would solve a huge collection of famous and unsolved Diophantine equations in one stroke. That is because it would put explicit bounds on the size of the solutions. For example, *abc* might show that all the solutions to an equation must be smaller than 100. To find those solutions, all one would have to do would be to plug in every number from 0 to 99 and calculate which ones work. Without *abc*, by contrast, there would be infinitely many numbers to plug in.

Elkies's work meant that the *abc* conjecture could supersede the most important breakthrough in the history of Diophantine equations:

confirmation of a conjecture formulated in 1922 by the U.S. mathematician Louis Mordell, which said that the vast majority of Diophantine equations either have no solutions or have a finite number of them. That conjecture was proved in 1983 by German mathematician Gerd Faltings, who was then twenty-eight and within three years would win a Fields Medal, the most coveted mathematics award, for the work. But if *abc* is true, you don't just know how many solutions there are, Faltings says, "you can list them all."

Soon after Faltings solved the Mordell conjecture, he started teaching at Princeton University in New Jersey. And before long, his path crossed with that of Mochizuki.

Born in 1969 in Tokyo, Mochizuki spent his formative years in the United States, where his family moved when he was a child. He attended an exclusive high school in New Hampshire, and his precocious talent earned him an undergraduate spot in Princeton's mathematics department when he was barely sixteen. He quickly became legend for his original thinking and moved directly into a Ph.D. program.

People who know Mochizuki describe him as a creature of habit with an almost supernatural ability to concentrate. "Ever since he was a student, he just gets up and works," says Minhyong Kim, a mathematician at the University of Oxford, U.K., who has known Mochizuki since his Princeton days. After attending a seminar or colloquium, researchers and students would often go out together for a beer—but not Mochizuki, Kim recalls. "He's not introverted by nature, but he's so much focused on his mathematics."

Faltings was Mochizuki's adviser for his senior thesis and for his doctoral one, and he could see that Mochizuki stood out. "It was clear that he was one of the brighter ones," he says. But being a Faltings student couldn't have been easy. "Faltings was at the top of the intimidation ladder," recalls Kim. He would pounce on mistakes, and when talking to him, even eminent mathematicians could often be heard nervously clearing their throats.

Faltings's research had an outsized influence on many young number theorists at universities along the U.S. Eastern Seaboard. His area of expertise was algebraic geometry, which since the 1950s had been transformed into a highly abstract and theoretical field by Alexander Grothendieck—often described as the greatest mathematician of the twentieth century. "Compared to Grothendieck," says Kim, "Faltings didn't have as much patience for philosophizing." His style of math required "a lot of abstract background knowledge—but also tended to have as a goal very concrete problems. Mochizuki's work on *abc* does exactly this."

Single-Track Mind

After his Ph.D., Mochizuki spent two years at Harvard and then in 1994 moved back to his native Japan, aged twenty-five, to a position at RIMS. Although he had lived for years in the United States, "he was in some ways uncomfortable with American culture," Kim says. And, he adds, growing up in a different country may have compounded the feeling of isolation that comes from being a mathematically gifted child. "I think he did suffer a little bit."

Mochizuki flourished at RIMS, which does not require its faculty members to teach undergraduate classes. "He was able to work on his own for twenty years without too much external disturbance," Fesenko says. In 1996, he boosted his international reputation when he solved a conjecture that had been stated by Grothendieck; and in 1998, he gave an invited talk at the International Congress of Mathematicians in Berlin—the equivalent, in this community, of an induction to a hall of fame.

But even as Mochizuki earned respect, he was moving away from the mainstream. His work was reaching higher levels of abstraction, and he was writing papers that were increasingly impenetrable to his peers. In the early 2000s, he stopped venturing to international meetings, and colleagues say that he rarely leaves the Kyoto prefecture any more. "It requires a special kind of devotion to be able to focus over a period of many years without having collaborators," says number theorist Brian Conrad of Stanford University in California.

Mochizuki did keep in touch with fellow number theorists, who knew that he was ultimately aiming for *abc*. He had next to no competition: most other mathematicians had steered clear of the problem, deeming it intractable. By early 2012, rumors were flying that Mochizuki was getting close to a proof. Then came the August news: he had posted his papers online.

The next month, Fesenko became the first person from outside Japan to talk to Mochizuki about the work he had quietly unveiled. Fesenko was already due to visit Tamagawa, so he went to see Mochizuki too. The two met on a Saturday in Mochizuki's office, a spacious room offering a view of nearby Mount Daimonji, with neatly arranged books and papers. It is "the tidiest office of any mathematician I've ever seen in my life," Fesenko says. As the two mathematicians sat in leather armchairs, Fesenko peppered Mochizuki with questions about his work and what might happen next.

Fesenko says that he warned Mochizuki to be mindful of the experience of another mathematician: the Russian topologist Grigori Perelman, who shot to fame in 2003 after solving the century-old Poincaré conjecture (see *Nature* **427**, 388; 2004) and then retreated and became increasingly estranged from friends, colleagues, and the outside world. Fesenko knew Perelman, and saw that the two mathematicians' personalities were very different. Whereas Perelman was known for his awkward social skills (and for letting his fingernails grow unchecked), Mochizuki is universally described as articulate and friendly—if intensely private about his life outside of work.

Normally after a major proof is announced, mathematicians read the work—which is typically a few pages long—and can understand the general strategy. Occasionally, proofs are longer and more complex, and years may then pass for leading specialists to fully vet it and reach a consensus that it is correct. Perelman's work on the Poincaré conjecture became accepted in this way. Even in the case of Grothendieck's highly abstract work, experts were able to relate most of his new ideas to mathematical objects they were familiar with. Only once the dust has settled does a journal typically publish the proof.

But almost everyone who tackled Mochizuki's proof found themselves floored. Some were bemused by the sweeping—almost messianic—language with which Mochizuki described some of his new theoretical instructions: he even called the field that he had created "inter-universal geometry." "Generally, mathematicians are very humble, not claiming that what they are doing is a revolution of the whole Universe," says Oesterlé, at the Pierre and Marie Curie University in Paris, who made little headway in checking the proof.

The reason is that Mochizuki's work is so far removed from anything that had gone before. He is attempting to reform mathematics from the ground up, starting from its foundations in the theory of sets (familiar to many as Venn diagrams). And most mathematicians have been reluctant to invest the time necessary to understand the work because they see no clear reward: it is not obvious how the theoretical machinery that Mochizuki has invented could be used to do calculations. "I tried to read some of them and then, at some stage, I gave up. I don't understand what he's doing," says Faltings.

Fesenko studied Mochizuki's work in detail over the past year [2014], visited him at RIMS again in the autumn of 2014, and says that he has now verified the proof. (The other three mathematicians who say they have corroborated it have also spent considerable time working alongside Mochizuki in Japan.) The overarching theme of inter-universal geometry, as Fesenko describes it, is that one must look at whole numbers in a different light—leaving addition aside and seeing the multiplication structure as something malleable and deformable. Standard multiplication would then be just one particular case of a family of structures, just as a circle is a special case of an ellipse. Fesenko says that Mochizuki compares himself to the mathematical giant Grothendieck—and it is no immodest claim. "We had mathematics before Mochizuki's work—and now we have mathematics after Mochizuki's work," Fesenko says.

But so far, the few who have understood the work have struggled to explain it to anyone else. "Everybody who I'm aware of who's come close to this stuff is quite reasonable, but afterwards they become incapable of communicating it," says one mathematician who did not want his name to be mentioned. The situation, he says, reminds him of the *Monty Python* skit about a writer who jots down the world's funniest joke. Anyone who reads it dies from laughing and can never relate it to anyone else.

And that, says Faltings, is a problem. "It's not enough if you have a good idea: you also have to be able to explain it to others." Faltings says that if Mochizuki wants his work to be accepted, then he should reach out more. "People have the right to be eccentric as much as they want to," he says. "If he doesn't want to travel, he has no obligation. If he wants recognition, he has to compromise."

Edge of Reason

Some communities began to take proactive steps to validate Mochizuki's work in late 2015, when the Clay Mathematics Institute hosted the long-awaited workshop in Oxford. Leading figures in the field attended, including Faltings. Kim, who along with Fesenko is one of the organizers, says that he did not expect a few days of lectures to be enough to expose the entire theory. But some of the participants say that the workshop helped them to begin to see at least the strategy of the proof.

Most mathematicians expect that it will take many more years to find some resolution. (Mochizuki has said that he has submitted his papers to a journal, where they are presumably still under review.) Eventually, researchers hope, someone will be willing not only to understand the work, but also to make it understandable to others—the problem is, few want to be that person.

Looking ahead, researchers think that it is unlikely that future open problems will be as complex and intractable. Ellenberg points out that theorems are generally simple to state in new mathematical fields, and the proofs are quite short.

The question now is whether Mochizuki's proof will edge toward acceptance, as Perelman's did, or find a different fate. Some researchers see a cautionary tale in that of Louis de Branges, a well-established mathematician at Purdue University in West Lafayette, Indiana. In 2004, de Branges released a purported solution to the Riemann hypothesis, which many consider the most important open problem in math. But mathematicians have remained skeptical of that claim; many say that they are turned off by his unconventional theories and his idiosyncratic style of writing, and the proof has slipped out of sight.

For Mochizuki's work, "it's not all or nothing," Ellenberg says. Even if the proof of the *abc* conjecture does not work out, his methods and ideas could still slowly percolate through the mathematical community, and researchers might find them useful for other purposes. "I do think, based on my knowledge of Mochizuki, that the likelihood that there's interesting or important math in those documents is pretty high," Ellenberg says.

But there is still a risk that it could go the other way, he adds. "I think it would be pretty bad if we just forgot about it. It would be sad."

Addendum

After this article was published, the Oxford workshop took place, attended by Faltings as well as several other of the mathematicians who were quoted here. Many of the participants came away as confused as ever about Mochizuki's work, although some said that the workshop helped them to glimpse at least the beginning of an outline of the proof (doi:10.1038/nature.2015.19035).

When this book went to print, mathematicians were due to gather at RIMS in Kyoto for another workshop in July 2016, this one featuring a series of lectures by Mochizuki himself. Many researchers, however, say that they remain pessimistic that the community is making significant progress toward validating Mochizuki's proof.

A Proof That Some Spaces Can't Be Cut

Kevin Hartnett

The question is deceptively simple: Given a geometric space—a sphere, perhaps, or a doughnut-like torus—is it possible to divide it into smaller pieces? In the case of the two-dimensional surface of a sphere, the answer is clearly yes. Anyone can tile a mosaic of triangles over any two-dimensional surface. Likewise, any three-dimensional space can be cut up into an arbitrary number of pyramids (Figure 1).

But what about spaces in higher dimensions? Mathematicians have long been interested in the general properties of abstract spaces, or manifolds, which exist in every dimension. Could every four-dimensional manifold survive being sliced into smaller units? What about a five-dimensional manifold, or one with an arbitrarily large number of dimensions?

Subdividing a space in this way, a process known as triangulation, is a basic tool that topologists can use to tease out the properties of manifolds. And the triangulation conjecture, which posits that all manifolds can be triangulated, is one of the most famous problems in topology.

Ciprian Manolescu remembers hearing about the triangulation conjecture for the first time as a graduate student at Harvard University in the early 2000s. Though Manolescu was considered a phenomenon when he entered Harvard as an undergraduate—he had distinguished himself as the only person, then or since, to notch three consecutive perfect scores in the International Mathematical Olympiad—trying to prove a century-old conjecture isn't the sort of project that a wise student takes on for a doctoral thesis. Manolescu instead wrote a well-regarded dissertation on the separate topic of Floer homology and spent most of the first decade of his professional career giving the triangulation conjecture little thought. "It sounded like an unapproachable problem, so I didn't pay much attention to it," he wrote recently in an e-mail.

FIGURE 1. Every two-dimensional surface can be tiled over by a mosaic of triangles. But higher dimensional spaces can't always be triangulated in this way. Courtesy of Glen Faught

Others kept working on the problem, however, clawing toward a solution that remained stubbornly out of reach. Then in late 2012, Manolescu, now a professor at the University of California, Los Angeles, had an unexpected realization: The theory he'd constructed in his thesis eight years earlier was just what was needed to clear the final hurdle that had tripped up every previous attempt to answer the conjecture.

Building on this insight, Manolescu quickly proved that not all manifolds can be triangulated. In doing so, he not only elevated himself to the top of his field, but also created a tool with enormous potential to answer other long-standing problems in topology.

Perfect Cuts

The nineteenth-century French polymath Henri Poincaré was one of the first mathematicians to think about manifolds as combinations of simple pieces glued together. For example, a two-dimensional sphere (which means just the surface of a solid ball) can be approximated by gluing together two-dimensional triangles, and a three-dimensional sphere can be approximated by gluing together three-dimensional

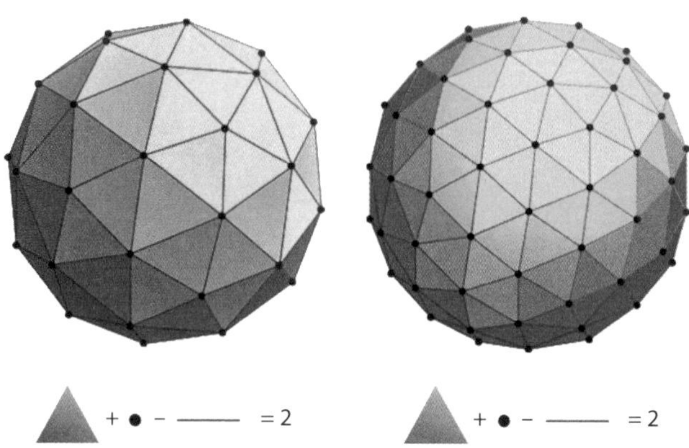

FIGURE 2. An "invariant" is a tool mathematicians use to compare spaces, or manifolds. One famous example is the Euler characteristic, shown here. To calculate it for any two-dimensional manifold, first carve the manifold into polygons (here we use triangles). Next, add the number of faces to the number of vertices and subtract the number of edges. Every sphere will have an Euler characteristic of 2, no matter how the manifold is carved up. Olena Shmahalo/*Quanta Magazine*. Source: Geodesic Spheres, Wikipedia

tetrahedra. Triangles and tetrahedra are examples of more general shapes called simplices, which can be defined in any dimension.

Triangulating a manifold is useful in a number of ways. The triangulation offers a concrete way of visualizing spaces that are difficult to see. It also provides researchers with a starting point for computing an important mathematical tool called an *invariant*.

Mathematicians use invariants to determine whether two spaces are fundamentally equivalent. If you compute the invariants of two manifolds and get unequal results, you know the manifolds are topologically distinct. (The converse doesn't always hold true—two distinct manifolds might share the same invariant.)

A simple way to see this is by computing an invariant called the Euler characteristic (Figure 2). To find the Euler characteristic of a two-dimensional surface, first divide it into any number of polygons. Now count the number of vertices, subtract the number of edges, then add the number of faces. The resulting integer will come out the same no matter how many polygons you use to triangulate the manifold. The Euler characteristic of a sphere is 2; that of a torus is 0. In two dimensions, any two manifolds with the same Euler characteristic are topologically equivalent.

Manifolds in higher dimensions also possess an Euler characteristic, but there, things don't work out quite so neatly. In three dimensions, for instance, there are infinitely many distinct manifolds with a given Euler characteristic. Even so, triangulation remains a useful tool, which naturally gives rise to the question: Can all manifolds, in any dimension, be triangulated?

The question was first posed in the early twentieth century, and the affirmative answer came to be known as the *triangulation conjecture*. Initially, mathematicians assumed that the triangulation conjecture had to be true, and by the 1950s they'd proven that it held for all manifolds in the first, second, and third dimensions. As the twentieth century wore on, however, mathematicians discovered that higher dimensional spaces lack many nice properties of lower dimensional manifolds. This discovery led mathematicians to suspect that the triangulation conjecture was likely false in higher dimensions, even though no one had been able to refute it with a proof.

The idea that spaces exist that can't be divided into smaller units seems bizarrely counterintuitive. But one way to think about it is to lay triangles on a two-dimensional sphere, going around until the entire sphere is covered. Even in this simple case, it's not immediately clear how to dovetail the final triangles with the original ones. Careful planning would be key. Add more dimensions to the scenario, and the problem of matching first simplices to last simplices becomes much more complicated.

In 1982, Michael Freedman, then at the University of California, San Diego, constructed four-dimensional manifolds that didn't allow for a natural kind of triangulation, an accomplishment that helped propel him to a Fields Medal. A few years later, Andrew Casson of Yale University proved that these particular manifolds couldn't be triangulated at all. Yet Freedman's and Casson's work didn't reveal whether triangulation is possible for all manifolds in five or more dimensions. That answer would have to wait another three decades, when Manolescu picked up the puzzle.

A Four-Dimensional Snag

Manolescu specializes in low-dimensional topology, which means that he works on problems of three- and four-dimensional manifolds. The question of whether manifolds in five or more dimensions can be triangulated would seem to lie outside his area of expertise. But in the 1970s, three mathematicians proved that solving the triangulation conjecture in higher dimensions was equivalent to answering a different question in lower dimensions. This transformation of one question into another is common in mathematics and can often provide a new perspective on a seemingly intractable situation. In 1994, when Andrew Wiles proved Fermat's Last Theorem, what he actually solved was a different problem, the semistable case of the Taniyama-Shimura-Weil conjecture, which earlier had been shown to imply the proof to Fermat's question.

Similarly, in the 1970s, a pair of mathematicians at the Institute for Advanced Study in Princeton, New Jersey—Ronald J. Stern and David Galewski—and a third mathematician working independently, Takao Matumoto, "reduced" the triangulation conjecture from a question in high dimensions to one in low dimensions.

To see how it works on a conceptual level, first imagine a two-dimensional sphere and the two-dimensional triangles that need to be glued together in order to triangulate it. One way to glue the triangles is to start with the highest dimensional part of their boundary—their one-dimensional edges—and move on to their next-highest-dimensional part—their zero-dimensional vertices.

Now consider, say, a seven-dimensional manifold. You'll need seven-dimensional simplices to try and triangulate it. To begin, you might take the same approach you took with the triangles, first gluing together the highest dimensional part of the boundaries of these simplices (six-dimensional "edges") and working your way down from there.

What Galewski, Stern, and Matumoto showed is that this gluing process goes quite well at first but snags on the border between dimensions four and three. The nature of this snag roughly boils down to a question about a topological space known as a "homology 3-sphere," which forms on the boundary between these dimensions. And deciding that question required a new kind of invariant, one that Manolescu would eventually find in his work in Floer homology.

The Big Break

Floer homology is a mathematical tool kit developed in the 1980s by Andreas Floer, a brilliant young German mathematician who died in 1991 at the age of thirty-four. It has turned out to be an incredibly successful way of thinking about manifolds, and it is now more a subfield of topology than a specific operation. Since Floer first proposed it as a way of working with three-dimensional manifolds, other mathematicians have created dozens of varieties of Floer homology, each suited to solving different kinds of problems. In the 1990s, Peter Kronheimer, Manolescu's dissertation adviser at Harvard, and Tomasz Mrowka, a topologist at the Massachusetts Institute of Technology, combined equations that originated in quantum physics with Floer homology to construct a powerful invariant of three-dimensional manifolds. In his dissertation, Manolescu created a simplified version of their theory.

"Ciprian [created] a simple, less technical way to define this Floer homology, and because it's less technical, it allows you to become more creative," said Mrowka. "You don't need to carry this big toolbox around to get the job done."

Manolescu turned Floer homology into a lighter, nimbler instrument in his dissertation, but neither he nor anyone else was immediately sure what to do with it. So there it sat, an impressive piece of work with no clear application.

In the meantime, Casson and the Norwegian mathematician Kim Frøyshov had each come up with invariants that did part of the job of solving the triangulation conjecture. But neither advance was enough on its own. "You need two properties for the invariant, and Casson had one and Frøyshov had another," Manolescu said.

Manolescu began to think about Casson's and Frøyshov's failed attempts in late 2012, and he had two important insights in quick succession. First, he realized that a major limitation of Frøyshov's invariant was that it didn't take advantage of a certain kind of symmetry known as Pin(2) symmetry. Next, he realized that his work on Floer homology eight years before was perfectly suited to incorporating that symmetry into the proof.

"There were just two [missing] ideas," Manolescu said. "In retrospect they seem straightforward, but somehow they were missed."

Once Manolescu understood the connection between his dissertation and the triangulation conjecture, he moved quickly. "I was very excited, and I wanted to write it down as fast as possible," he said. "I was kind of working around the clock." It took him a month to lay out his full refutation of the triangulation conjecture. He created a new invariant, which he named "beta," and used it to create a proof by contradiction. Here's how it works: As we have seen, the triangulation conjecture is equivalent to asking whether there exists a homology 3-sphere with certain characteristics. One characteristic is that the sphere has to have a certain property—a Rokhlin invariant of 1. Manolescu showed that when a homology 3-sphere has a Rokhlin invariant of 1, the value of beta has to be odd. At the same time, other necessary characteristics of these homology 3-spheres require beta to be even. Since beta cannot be both even and odd at the same time, these particular homology 3-spheres do not exist. Thus, the triangulation conjecture is false.

A New Set of Tools

On March 10, 2013, Manolescu posted a preprint of his paper to the online repository arXiv.org; the paper is currently under review at the *Journal of the American Mathematical Society*. Stern calls Manolescu's proof "the best result in the last couple years" in four-dimensional topology. The finding has prompted speculation that Manolescu will win the Veblen Prize in Geometry, which is given every three years for outstanding work in geometry or topology. (Casson, Freedman, Kronheimer, and Mrowka are all former winners.)

"Nobody, certainly not I, thought about using that version of Floer homology to solve the problem," said Stern. "The ingeniousness of it was just the approach [Manolescu] took." Stern added that he maintains an informal "bucket list" of problems he'd like to solve, and the triangulation conjecture was on it. "I would either like to have solved it or I'd like to know the answer," he said, "and now I know the answer."

But the more significant consequence of Manolescu's proof is the way it elevates his version of Floer homology. "For whatever reason," Mrowka said, "people haven't picked it up as much as they should have." Now, with the refutation of the triangulation conjecture, mathematicians are rushing to learn how to use Manolescu's powerful tool.

Currently, topologists at the California Institute of Technology and the University of Texas, Austin, are running seminars on Manolescu's thesis. Mrowka has two graduate students working on proving Manolescu's results again and refining his methods for other uses. Manolescu's techniques could be useful for answering questions in four-dimensional topology and other important topological subdisciplines. No one knows exactly what they will end up being useful for.

"It seems hard to believe that having a wonderful new invariant isn't going to have applications to problems that have been around in related areas," said Mrowka. "But as to what, who knows? That's what research is for."

Einstein's First Proof

Steven Strogatz

On November 26, 1949, Albert Einstein published an essay in the *Saturday Review of Literature* in which he described two pivotal moments in his childhood. The first involved a compass that his father showed him when he was four or five. Einstein recalled his sense of wonderment that the needle always pointed north, even though nothing appeared to be pulling it in that direction. He came to a conclusion, then and there, about the structure of the physical world: "Something deeply hidden had to be behind things." The second moment occurred soon after he turned twelve, when he was given "a little book dealing with Euclidean plane geometry." The book's "lucidity," he wrote—the idea that a mathematical assertion could "be proved with such certainty that any doubt appeared to be out of the question"—provoked "wonder of a totally different nature." Pure thought could be just as powerful as geomagnetism.

This month, we celebrate the hundredth anniversary of Einstein's general theory of relativity, one of his many ideas that brought lucidity to the deeply hidden. With all the surrounding hoopla, it would be nice if we could fathom something of what he actually accomplished and how he did it. That turns out to be a tall order, because general relativity is tremendously complex (http://www.newyorker.com/tech/elements/the-space-doctors-big-idea-einstein-general-relativity). When Arthur Eddington—the British astrophysicist who led the team that confirmed Einstein's predictions, during a solar eclipse in 1919—was asked if it was really true that only three people in the world understood the theory, he said nothing. "Don't be so modest, Eddington!" his questioner said. "On the contrary," Eddington replied. "I'm just wondering who the third might be."

A physicist's genius turns up in his boyhood geometry. Illustration by Tomi Um

Fortunately, we can study an earlier, simpler example of Einstein's thinking. Even before he received the little geometry book, he had been introduced to the subject by his uncle Jakob, an engineer. Einstein became particularly enamored of the Pythagorean theorem and—"after much effort," he noted in the *Saturday Review*—he wrote his own mathematical proof of it. It is my intention to lead you through that proof, step by logical step. It's Einstein's first masterpiece, and certainly his most accessible one. This little gem of reasoning foreshadows the man he became, scientifically, stylistically, and temperamentally. His instinct for symmetry, his economy of means, his iconoclasm, his tenacity, his penchant for thinking in pictures—they're all here, just as they are in his theory of relativity.

You may once have memorized the Pythagorean theorem as a series of symbols: $a^2 + b^2 = c^2$. It concerns right triangles, meaning triangles that have a right (ninety-degree) angle at one of their corners. The theorem says that if a and b are the lengths of the triangle's legs (the sides that meet at the right angle), then the length of the hypotenuse (the side opposite the right angle) is given by c, according to the formula above.

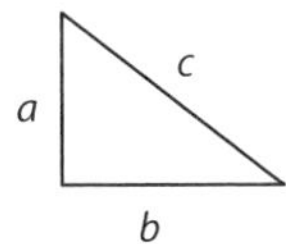

Teenagers have this rule drummed into them by the millions, year after year, in schools around the world, but most don't give it much thought. Maybe you never did, either. Once you do, though, the questions start coming. What makes it true? How did anyone ever come up with it? Why?

For a clue to that last question, consider the etymology of the word *geometry*. It derives from the Greek roots *gē* (meaning "earth" or "land") and *metria* ("measurement"). It's easy to imagine ancient peoples and their monarchs being concerned with the measurement of fields or plots of land. Officials needed to assess how much tax was to be paid, how much water they would need for irrigation, and how much wheat, barley, and papyrus the farmers could produce.

Imagine a rectangular field, thirty yards by forty.

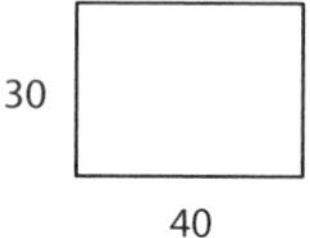

How much land is that? The meaningful measure would be the area of the field. For a thirty-by-forty lot, the area would be thirty times forty, which is twelve hundred square yards. That's the only number the tax assessor cares about. He's not interested in the precise shape of your land, just how much of it you have.

Surveyors, by contrast, do care about shapes, and angles, and distances, too. In ancient Egypt, the annual flooding of the Nile sometimes erased the boundaries between plots, necessitating the use of accurate surveying to redraw the lines. Four thousand years ago, a surveyor somewhere might have looked at a thirty-by-forty rectangular plot and wondered, How far is it from one corner to the diagonally opposite corner?

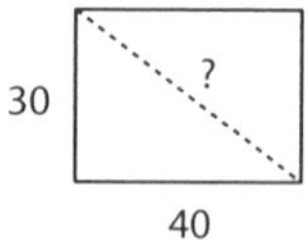

The answer to that question is far less obvious than the earlier one about area, but ancient cultures around the world—in Babylon, China, Egypt, Greece, and India—all discovered it. The rule that they came up with is now called the Pythagorean theorem, in honor of Pythagoras of Samos, a Greek mathematician, philosopher, and cult leader who lived around 550 B.C. It asks us to imagine three fictitious square plots of land—one on the short side of the rectangle, another on the long side, and a third on its diagonal.

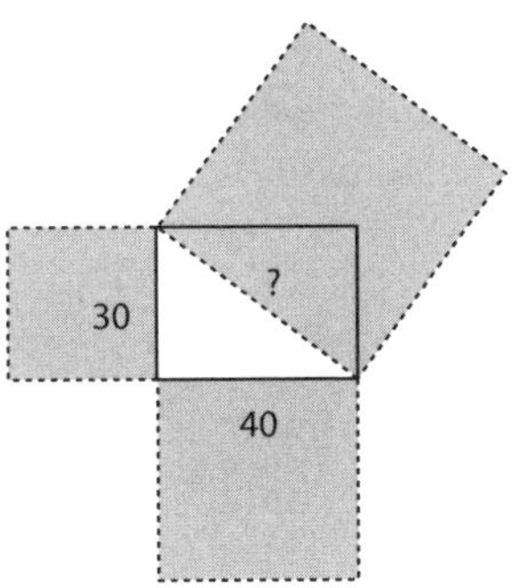

Next, we are instructed to calculate the area of the square plots on the sides and add them together. The result, 900 + 1,600 = 2,500, is, according to the Pythagorean theorem, the same as the area of the square on the diagonal. This recipe yields the unknown length that we are trying to calculate: fifty yards, since 50 × 50 = 2,500.

The Pythagorean theorem is true for rectangles of any proportion—skinny, blocky, or anything in between. The squares on the two sides always add up to the square on the diagonal. (More precisely, the *areas* of the squares, not the squares themselves, add up. But this simpler phrasing is less of a mouthful, so I'll continue to speak of squares adding up when I really mean their areas.) The same rule applies to right triangles, the shape you get when you slice a rectangle in half along its diagonal.

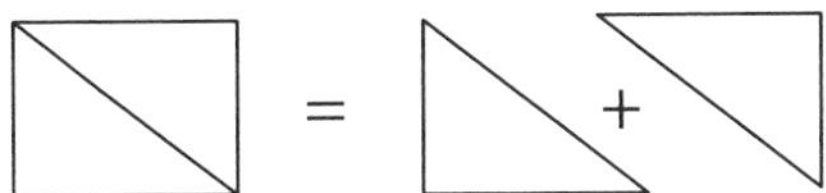

The rule now sounds more like the one you learned in school: $a^2 + b^2 = c^2$. In pictorial terms, the squares on the sides of a right triangle add up to the square on its hypotenuse.

But why is the theorem true? What's the logic behind it? Actually, hundreds of proofs are known today. There's a marvelously simple one attributed to the Pythagoreans and, independently, to the ancient

Chinese. There's an intricate one given in Euclid's *Elements*, which schoolchildren have struggled with for the past twenty-three hundred years, and which induced in the philosopher Arthur Schopenhauer "the same uncomfortable feeling that we experience after a juggling trick." There's even a proof by President James A. Garfield, which involves the cunning use of a trapezoid.

Einstein, unfortunately, left no such record of his childhood proof. In his *Saturday Review* essay, he described it in general terms, mentioning only that it relied on "the similarity of triangles." The consensus among Einstein's biographers is that he probably discovered, on his own, a standard textbook proof in which similar triangles (meaning triangles that are like photographic reductions or enlargements of one another) do indeed play a starring role. Walter Isaacson, Jeremy Bernstein, and Banesh Hoffman all come to this deflating conclusion, and each of them describes the steps that Einstein would have followed as he unwittingly reinvented a well-known proof.

Twenty-four years ago, however, an alternative contender for the lost proof emerged. In his book *Fractals, Chaos, Power Laws*, the physicist Manfred Schroeder presented a breathtakingly simple proof of the Pythagorean theorem whose provenance he traced to Einstein. Schroeder wrote that the proof had been shown to him by a friend of his, the chemical physicist Shneior Lifson, of the Weizmann Institute, in Rehovot, Israel, who heard it from the physicist Ernst Straus, one of Einstein's former assistants, who heard it from Einstein himself. Though we cannot be sure the following proof is Einstein's, anyone who knows his work will recognize the lion by his claw.

It helps to run through the proof quickly at first, to get a feel for its overall structure.

Step 1: Draw a perpendicular line from the hypotenuse to the right angle. This partitions the original right triangle into two smaller right triangles.

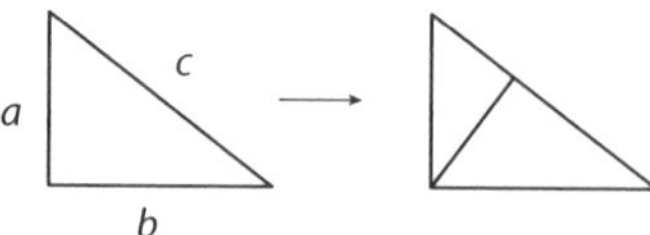

Step 2: Note that the area of the little triangle plus the area of the medium triangle equals the area of the big triangle.

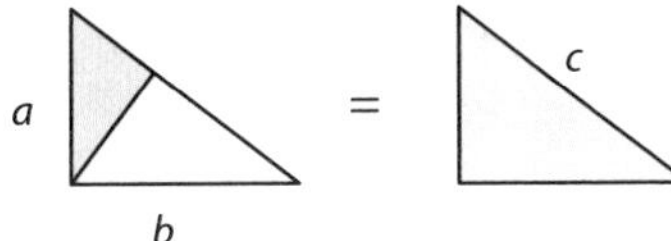

Step 3: The big, medium, and little triangles are similar in the technical sense: their corresponding angles are equal and their corresponding sides are in proportion. Their similarity becomes clear if you imagine picking them up, rotating them, and arranging them like so, with their hypotenuses on the top and their right angles on the lower left:

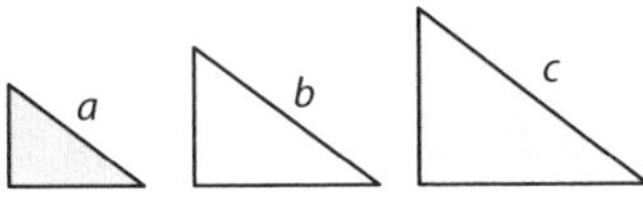

Step 4: Because the triangles are similar, each occupies the same fraction f of the area of the square on its hypotenuse. Restated symbolically, this observation says that the triangles have areas fa^2, fb^2, and fc^2, as indicated in the diagram.

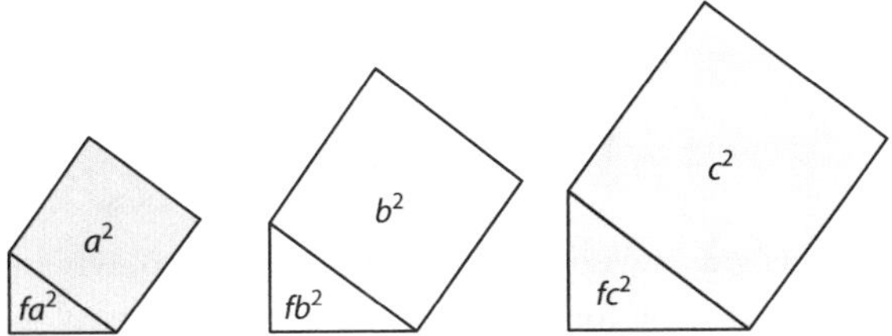

(Don't worry if this step provokes a bit of head-scratching. I'll have more to say about it below, after which I hope it'll seem obvious.)

Step 5: Remember, from Step 2, that the little and medium triangles add up to the original big one. Hence, from Step 4, $fa^2 + fb^2 = fc^2$.

Step 6: Divide both sides of the equation above by f. You will obtain $a^2 + b^2 = c^2$, which says that the areas of the squares add up. That's the Pythagorean theorem.

The proof relies on two insights. The first is that a right triangle can be decomposed into two smaller copies of itself (Steps 1 and 3). That's a peculiarity of right triangles. If you try instead, for example, to decompose an equilateral triangle into two smaller equilateral triangles, you'll find that you can't. So Einstein's proof reveals why the

Pythagorean theorem applies only to right triangles: they're the only kind made up of smaller copies of themselves. The second insight is about additivity. Why do the squares add up (Step 6)? It's because the triangles add up (Step 2), and the squares are proportional to the triangles (Step 4).

The logical link between the squares and triangles comes via the confusing Step 4. Here's a way to make peace with it. Try it out for the easiest kind of right triangle, an isosceles right triangle, also known as a 45–45–90 triangle, which is formed by cutting a square in half along its diagonal.

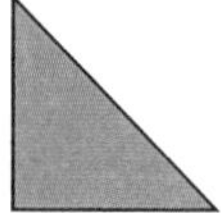

As before, erect a square on its hypotenuse.

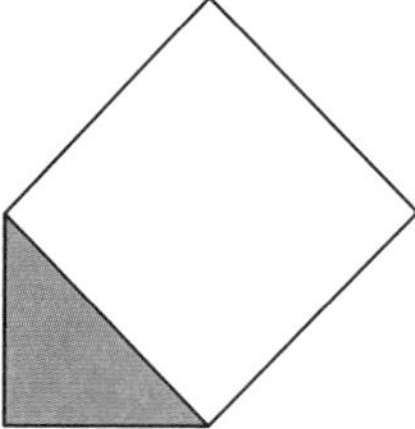

If we draw dashed lines on the diagonals of that newly built square, the picture looks like the folding instructions for an envelope.

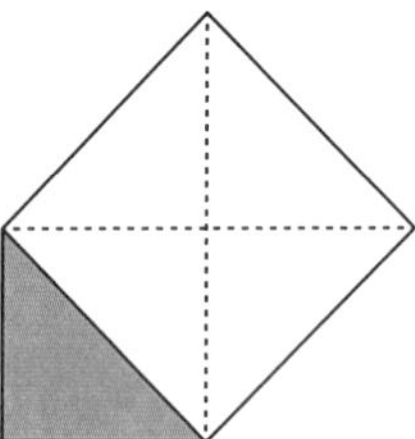

As you can see, four copies of the triangle fit neatly inside the square. Or, said the other way around, the triangle occupies exactly a quarter of the square. That means that $f = 1/4$, in the notation above.

Now for the cruncher. We never said how big the square and the isosceles right triangle were. The ratio of their areas is *always* one to four, for any such envelope. It's a property of the envelope's shape, not its size.

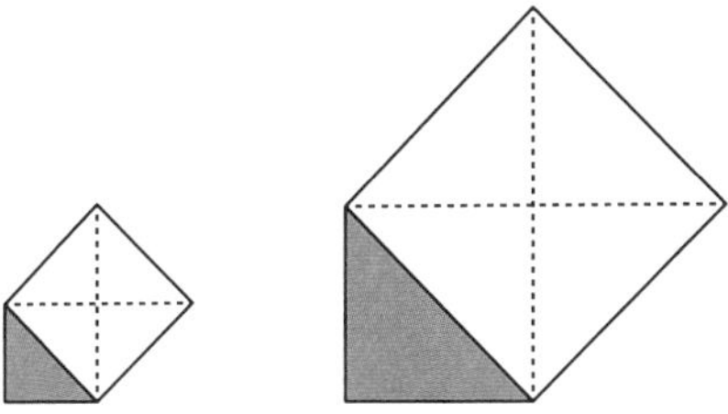

That's the thrust of Step 4. It's obvious when you think of it like that, no?

The same thing works for any right triangle of any shape. It doesn't have to be isosceles. The triangle always occupies a certain fraction, f, of the square on its hypotenuse, and that fraction stays the same no matter how big or small they both are. To be sure, the numerical value of f depends on the proportions of the triangle; if it's a long, flat sliver, the square on its hypotenuse will have a lot more than four times its area, and so f will be a lot less than 1/4. But that numerical value is irrelevant. Einstein's proof shows that f disappears in the end anyway. It enters stage right, in Step 4, and promptly exits stage left, in Step 6.

What we're seeing here is a quintessential use of a symmetry argument. In science and math, we say that something is symmetrical if some aspect of it stays the same despite a change. A sphere, for instance, has rotational symmetry; rotate it about its center and its appearance stays the same. A Rorschach inkblot has reflectional symmetry: its mirror image matches the original. In Step 4 of his proof, Einstein exploited a symmetry known as scaling. Take a right triangle with a square on its hypotenuse and rescale both of them by the same amount, as if on a photocopier. That rescaling changes some of their features (their areas and side lengths) while leaving others intact (their angles, proportions, and area ratio). It's the constancy of the area ratio that undergirds Step 4.

Throughout his career, Einstein would continue to deploy symmetry arguments like a scalpel, getting to the hidden heart of things. He opened his revolutionary 1905 paper on the special theory of relativity by noting an asymmetry in the existing theories of electricity and magnetism: "It is known that Maxwell's electrodynamics—as usually understood at the present time—when applied to moving bodies, leads to asymmetries which do not appear to be inherent in the phenomena." Those asymmetries, Einstein sensed, must be a clue to something rotten at the core of physics as it was then formulated. In his mind, everything else—space, time, matter, energy—was up for grabs, but not symmetry. Think of the courage it required to reformulate nearly all of physics from the ground up, even if it meant revising Newton and Maxwell along the way.

Both special and general relativity are also profoundly geometrical theories. They conceive of the universe as having a dimension beyond the usual three; that fourth dimension is time. Rather than considering the distance between two points (a measure of space), the special-relativistic counterpart of the Pythagorean theorem considers the interval between two events (a measure of space-time). In general relativity, where space-time itself becomes warped and curved by the matter and energy within it, the Pythagorean theorem still has a part to play; it morphs into a quantity called the metric, which measures the space-time separation between infinitesimally close events, for which curvature can temporarily be overlooked. In a sense, Einstein continued his love affair with the Pythagorean theorem all his life.

The style of his Pythagorean proof, elegant and seemingly effortless, also portends something of the later scientist. Einstein draws a single line in Step 1, after which the Pythagorean theorem falls out like a ripe avocado. The same spirit of minimalism characterizes all of Einstein's adult work. Incredibly, in the part of his special-relativity paper where he revolutionized our notions of space and time, he used no math beyond high-school algebra and geometry.

Finally, although the young Einstein made his proof of the Pythagorean theorem look easy, it surely wasn't. Remember that, in his *Saturday Review* essay, he says that it required "much effort." Later in life, this tenacity—what Einstein referred to as his stubbornness—would serve him well. It took him years to come up with general relativity, and he often felt overwhelmed by the abstract mathematics that the theory required. Although he was mathematically powerful, he was not among the world's best. ("Every boy in the streets of Göttingen understands more about four-dimensional geometry than Einstein," one of his contemporaries, the mathematician David Hilbert, remarked.)

Many years after his Pythagorean proof, Einstein shared this lesson with another twelve-year-old who was wrestling with mathematics. On January 3, 1943, a junior high school student named Barbara Lee Wilson wrote to him for advice. "Most of the girls in my room have heroes which they write fan mail to," she began. "You + my uncle who is in the Coast Guard are my heroes." Wilson told Einstein that she was anxious about her performance in math class: "I have to work longer in it than most of my friends. I worry (perhaps too much)." Four days later, Einstein sent her a reply. "Until now I never dreamed to be something like a hero," he wrote. "But since you have given me the nomination I feel that I am one." As for Wilson's academic concerns? "Do not worry about your difficulties in mathematics," he told her. "I can assure you that mine are still greater."

Why String Theory Still Offers Hope We Can Unify Physics

Brian Greene

In October 1984, I arrived at Oxford University, trailing a large steamer trunk containing a couple of changes of clothing and about five dozen textbooks. I had a freshly minted bachelor's degree in physics from Harvard, and I was raring to launch into graduate study. But within a couple of weeks, the more advanced students had sucked the wind from my sails. Change fields now while you still can, many said. There's nothing happening in fundamental physics.

Then, just a couple of months later, the prestigious (if tamely titled) journal *Physics Letters B* published an article that ignited the first superstring revolution, a sweeping movement that inspired thousands of physicists worldwide to drop their research in progress and chase Einstein's long-sought dream of a unified theory. The field was young, the terrain fertile, and the atmosphere electric. The only thing I needed to drop was a neophyte's inhibition to run with the world's leading physicists. I did. What followed proved to be the most exciting intellectual odyssey of my life.

That was thirty years ago this month, making the moment ripe for taking stock: Is string theory revealing reality's deep laws? Or, as some detractors have claimed, is it a mathematical mirage that has sidetracked a generation of physicists?

Unification has become synonymous with Einstein, but the enterprise has been at the heart of modern physics for centuries. Isaac Newton united the heavens and Earth, revealing that the same laws governing the motion of the planets and the Moon described the trajectory of a spinning wheel and a rolling rock. About 200 years later, James Clerk

Maxwell took the unification baton for the next leg, showing that electricity and magnetism are two aspects of a single force described by a single mathematical formalism.

The next two steps, big ones at that, were indeed vintage Einstein. In 1905, Einstein linked space and time, showing that motion through one affects passage through the other, the hallmark of his special theory of relativity. Ten years later, Einstein extended these insights with his general theory of relativity, providing the most refined description of gravity, the force governing the likes of stars and galaxies. With these achievements, Einstein envisioned that a grand synthesis of all of nature's forces was within reach.

But by 1930, the landscape of physics had thoroughly shifted. Niels Bohr and a generation of intrepid explorers ventured deep into the microrealm, where they encountered quantum mechanics, an enigmatic theory formulated with radically new physical concepts and mathematical rules. Though spectacularly successful at predicting the behavior of atoms and subatomic particles, the quantum laws looked askance at Einstein's formulation of gravity. This problem set the stage for more than a half-century of despair as physicists valiantly struggled, but repeatedly failed, to meld general relativity and quantum mechanics, the laws of the large and small, into a single all-encompassing description.

Such was the case until December 1984, when John Schwarz, of the California Institute of Technology, and Michael Green, then at Queen Mary College, published a once-in-a-generation paper showing that string theory could overcome the mathematical antagonism between general relativity and quantum mechanics, clearing a path that seemed destined to reach the unified theory.

The idea underlying string unification is as simple as it is seductive. Since the early twentieth century, nature's fundamental constituents have been modeled as indivisible particles—the most familiar being electrons, quarks, and neutrinos—that can be pictured as infinitesimal dots devoid of internal machinery. String theory challenges this notion by proposing that at the heart of every particle is a tiny, vibrating stringlike filament. And, according to the theory, the differences between one particle and another—their masses, electric charges, and, more esoterically, their spin and nuclear properties—all arise from differences in how their internal strings vibrate.

Much as the sonorous tones of a cello arise from the vibrations of the instrument's strings, the collection of nature's particles would arise from the vibrations of the tiny filaments described by string theory. The long list of disparate particles that had been revealed over a century of experiments would be recast as harmonious "notes" comprising nature's score.

Most gratifying, the mathematics revealed that one of these notes had properties precisely matching those of the "graviton," a hypothetical particle that, according to quantum physics, should carry the force of gravity from one location to another. With this, the worldwide community of theoretical physicists looked up from their calculations. For the first time, gravity and quantum mechanics were playing by the same rules, at least in theory.

I began learning the mathematical underpinnings of string theory during an intense period in the spring and summer of 1985. I wasn't alone. Graduate students and seasoned faculty alike became swept up in the potential of string theory to be what some were calling the "final theory" or the "theory of everything." In crowded seminar rooms and flyby corridor conversations, physicists anticipated the crowning of a new order.

But the simplest and most important question loomed large. Is string theory right? Does the math explain our universe? The description I've given suggests an experimental strategy. Examine particles, and if you see little vibrating strings, you're done. It's a fine idea in principle, but string theory's pioneers realized that it was useless in practice. The math set the size of strings to be about a million billion times smaller than even the minute realms probed by the world's most powerful accelerators. Save for building a collider the size of the galaxy, strings, if they're real, would elude brute force detection.

Making the situation seemingly more dire, researchers had come upon a remarkable but puzzling mathematical fact. String theory's equations require that the universe has extra dimensions beyond the three of everyday experience—left/right, back/forth, and up/down. Taking the math to heart, researchers realized that their backs were to the wall. Make sense of extra dimensions—a prediction that's grossly at odds with what we perceive—or discard the theory.

String theorists pounced on an idea first developed in the early years of the twentieth century. Back then, theorists realized that there might be two kinds of spatial dimensions: those that are large and extended, which we directly experience, and others that are tiny and tightly wound, too small for even our most refined equipment to reveal. Much as the spatial extent of an enormous carpet is manifest, but you have to get down on your hands and knees to see the circular loops making up its pile, the universe might have three big dimensions that we all navigate freely, but it might also have additional dimensions so minuscule that they're beyond our observational reach.

In a paper submitted for publication a day after New Year's 1985, a quartet of physicists—Philip Candelas, Gary Horowitz, Andrew Strominger, and Edward Witten—pushed this proposal one step further, turning vice to virtue. Positing that the extra dimensions were minuscule, they argued, would not only explain why we haven't seen them, but could also provide the missing bridge to experimental verification.

Strings are so small that when they vibrate they undulate not just in the three large dimensions, but also in the additional tiny ones. And much as the vibrational patterns of air streaming through a French horn are determined by the twists and turns of the instrument, the vibrational patterns of strings would be determined by the shape of the extra dimensions. Since these vibrational patterns determine particle properties like mass, electric charge, and so on—properties that can be detected experimentally—the quartet had established that if you know the precise geometry of the extra dimensions, you can make predictions about the results that certain experiments would observe.

For me, deciphering the paper's equations was one of those rare mathematical forays bordering on spiritual enlightenment. That the geometry of hidden spatial dimensions might be the universe's Rosetta Stone , embodying the secret code of nature's fundamental constituents—well, it was one of the most beautiful ideas I'd ever encountered. It also played to my strength. As a mathematically oriented physics student, I'd already expended great effort studying topology and differential geometry, the very tools needed to analyze the mathematical form of extradimensional spaces.

And so, in the mid-1980s, with a small group of researchers at Oxford, we set our sights on extracting string theory's predictions. The quartet's paper had delineated the category of extradimensional spaces allowed by the mathematics of string theory and, remarkably, only a handful of candidate shapes were known. We selected one that seemed most promising and embarked on grueling days and sleepless nights, filled with arduous calculations in higher dimensional geometry and fueled by grandiose thoughts of revealing nature's deepest workings.

The final results that we found successfully incorporated various established features of particle physics and so were worthy of attention (and, for me, a doctoral dissertation), but were far from providing evidence for string theory. Naturally, our group and many others turned back to the list of allowed shapes to consider other possibilities. But the list was no longer short. Over the months and years, researchers had discovered ever larger collections of shapes that passed mathematical muster, driving the number of candidates into the thousands, millions, billions, and then, with insights spearheaded in the mid-1990s by Joe Polchinski, into numbers so large that they've never been named.

Against this embarrassment of riches, string theory offered no directive regarding which shape to pick. And as each shape would affect string vibrations in different ways, each would yield different observable consequences. The dream of extracting unique predictions from string theory rapidly faded.

From a public relations standpoint, string theorists had not prepared for this development. Like the Olympic athlete who promises eight gold medals but wins "only" five, theorists had consistently set the bar as high as it could go. That string theory unites general relativity and quantum mechanics is a profound success. That it does so in a framework with the capacity to embrace the known particles and forces makes the success more than theoretically relevant. Seeking to go even further and uniquely explain the detailed properties of the particles and forces is surely a noble goal, but one that lies well beyond the line dividing success from failure.

Nevertheless, critics who had bristled at string theory's meteoric rise to dominance used the opportunity to trumpet the theory's demise, blurring researchers' honest disappointment of not reaching hallowed ground with an unfounded assertion that the approach had crashed. The cacophony grew louder still with a controversial turn articulated most forcefully by one of the founding fathers of string theory, the Stanford University theoretical physicist Leonard Susskind.

❦

In August 2003, I was sitting with Susskind at a conference in Sigtuna, Sweden, discussing whether he really believed the new perspective he'd been expounding or was just trying to shake things up. "I do like to stir the pot," he told me in hushed tones, feigning confidence, "but I do think this is what string theory's been telling us."

Susskind was arguing that if the mathematics does not identify one particular shape as the right one for the extra dimensions, perhaps there isn't a single right shape. That is, maybe all of the shapes are right shapes in the sense that there are many universes, each with a different shape for the extra dimensions.

Our universe would then be just one of a vast collection, each with detailed features determined by the shape of their extra dimensions. Why, then, are we in this universe instead of any other? Because the shape of the hidden dimensions yields the spectrum of physical features that allow us to exist. In another universe, for example, the different shape might make the electron a little heavier or the nuclear force a little weaker, shifts that would cause the quantum processes that power stars, including our Sun, to halt, interrupting the relentless march toward life on Earth.

Radical though this proposal may be, it was supported by parallel developments in cosmological thinking that suggested that the Big Bang may not have been a unique event, but was instead one of innumerable bangs spawning innumerable expanding universes, called the multiverse. Susskind was suggesting that string theory augments this grand cosmological unfolding by adorning each of the universes in the multiverse with a different shape for the extra dimensions.

With or without string theory, the multiverse is a highly controversial schema, and deservedly so. It not only recasts the landscape of reality, but also shifts the scientific goalposts. Questions once deemed profoundly puzzling—why do nature's numbers, from particle masses to force strengths to the energy suffusing space, have the particular values they do?—would be answered with a shrug. The detailed features we observe would no longer be universal truths; instead, they'd be local bylaws dictated by the particular shape of the extra dimensions in our corner of the multiverse.

Most physicists, string theorists among them, agree that the multiverse is an option of last resort. Yet, the history of science has also convinced us to not dismiss ideas merely because they run counter to expectation. If we had, our most successful theory, quantum mechanics, which describes a reality governed by wholly peculiar waves of probability, would be buried in the trash bin of physics. As Nobel laureate Steven Weinberg has said, the universe doesn't care about what makes theoretical physicists happy.

❧

This spring, after nearly two years of upgrades, the Large Hadron Collider will crackle back to life, smashing protons together with almost twice the energy achieved in its previous runs. Sifting through the debris with the most complex detectors ever built, researchers will be looking for evidence of anything that doesn't fit within the battle-tested "Standard Model of particle physics," whose final prediction, the Higgs boson, was confirmed just before the machine went on hiatus. Although it is likely that the revamped machine is still far too weak to see strings themselves, it could provide clues pointing in the direction of string theory.

Many researchers have pinned their hopes on finding a new class of so-called "supersymmetric" particles that emerge from string theory's highly ordered mathematical equations. Other collider signals could show hints of extraspatial dimensions, or even evidence of microscopic black holes, a possibility that arises from string theory's exotic treatment of gravity on tiny distance scales.

Though none of these predictions can properly be called a smoking gun—various nonstringy theories have incorporated them too—a positive identification would be on par with the discovery of the Higgs particle and would, to put it mildly, set the world of physics on fire. The scales would tilt toward string theory.

But what happens in the event—likely, according to some—that the collider yields no remotely stringy signatures?

Experimental evidence is the final arbiter of right and wrong, but a theory's value is also assessed by the depth of influence it has on allied fields. By this measure, string theory is off the charts. Decades of analysis filling thousands of articles have had a dramatic impact on a broad swath of research cutting across physics and mathematics. Take black holes, for example. String theory has resolved a vexing puzzle by identifying the microscopic carriers of their internal disorder, a feature discovered in the 1970s by Stephen Hawking.

Looking back, I'm gratified at how far we've come but disappointed that a connection to experiment continues to elude us. Whereas my own research has migrated from highly mathematical forays into extra-dimensional arcana to more applied studies of string theory's cosmological insights, I now hold only modest hope that the theory will confront data during my lifetime.

Even so, string theory's pull remains strong. Its ability to seamlessly meld general relativity and quantum mechanics remains a primary achievement, but the allure goes deeper still. Within its majestic mathematical structure, a diligent researcher would find all of the best ideas physicists have carefully developed over the past few hundred years. It's hard to believe such depth of insight is accidental.

I like to think that Einstein would look at string theory's journey and smile, enjoying the theory's remarkable geometrical features while feeling kinship with fellow travelers on the long and winding road toward unification. All the same, science is powerfully self-correcting. Should decades drift by without experimental support, I imagine that string theory will be absorbed by other areas of science and mathematics and slowly shed a unique identity. In the interim, vigorous research and a large dose of patience are surely warranted. If experimental confirmation of string theory is in the offing, future generations will look back on our era as transformative, a time when science had the fortitude to nurture a remarkable and challenging theory, resulting in one of the most profound steps toward understanding reality.

The Pioneering Role of the Sierpinski Gasket

Tanya Khovanova, Eric Nie, and Alok Puranik

The famous Sierpinski gasket, shown in Figure 1, appeared in Italian mosaics in the thirteenth century. But it was not named the Sierpinski gasket then; in fact, Wacław Sierpiński was not yet born. He was born in 1882 and described this fractal mathematically in 1915.

The first available description of the Ulam-Warburton (UW) automaton, a fractal we describe in the next section, is in a 1962 article by Ulam [6]. Richard Stanley rediscovered the automaton in 1994 [5], and it has continued to fascinate scientists (e.g., [1, 3, 4]).

The UW automaton and the Sierpinski gasket are fractals, but as we shall see, the connection between these two objects—and the hexagonal analog of the UW automaton—is much deeper.

The Ulam-Warburton Automaton

The UW automaton grows on a square grid. It starts with a single cell in the grid; this is generation one. In each subsequent generation, a new

Figure 1. The Sierpinski gasket.

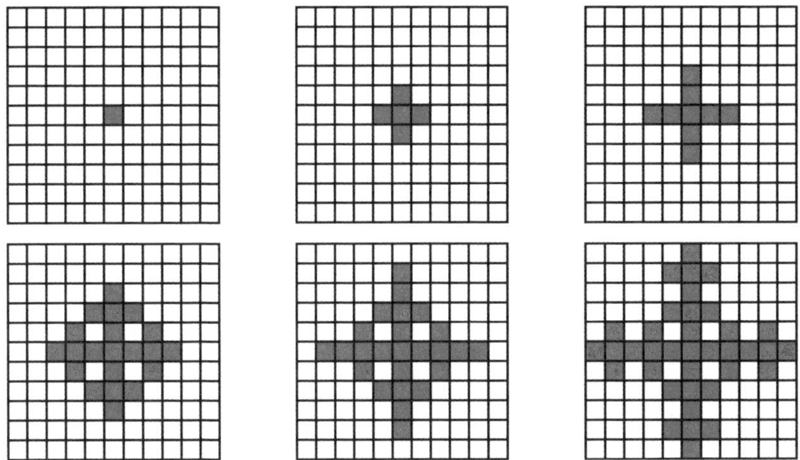

FIGURE 2. The first six generations of the Ulam-Warburton automaton.

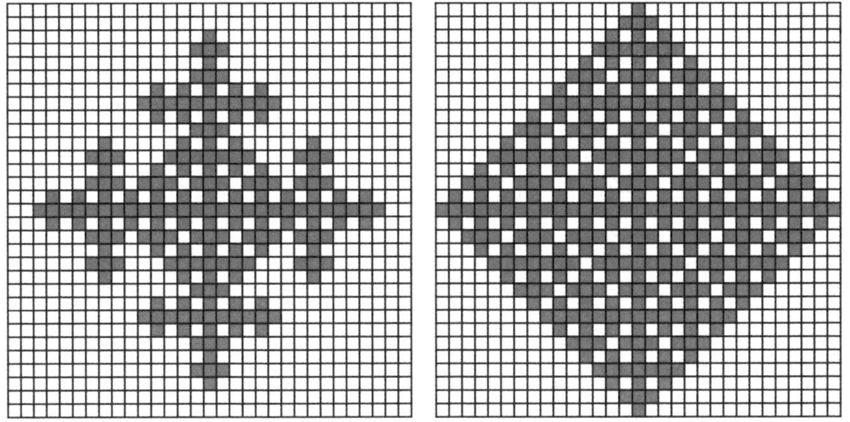

FIGURE 3. The first fourteen and sixteen generations of the Ulam-Warburton automaton.

cell is born if it is adjacent to exactly one live cell. Two cells are adjacent if they share a side. Live cells never die. Figure 2 shows the first six generations.

The population of live cells becomes more elaborate with every generation. Figure 3 shows the set of live cells after fourteen and sixteen generations.

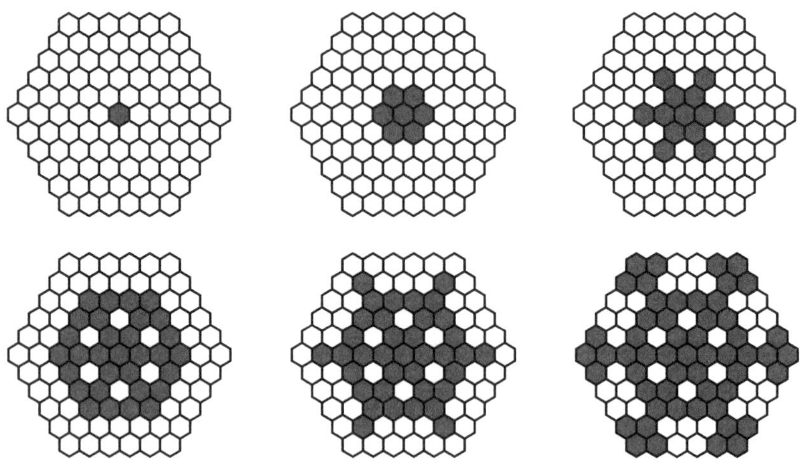

FIGURE 4. First six generations of the Hex-Ulam-Warburton automaton.

The Hex-Ulam-Warburton Automaton

We can generalize this construction to produce an automaton on the hexagonal grid. We follow the same rules: We start with a single cell. A new cell is born if it is adjacent to exactly one live cell, and a live cell never dies. We call this automaton the Hex-Ulam-Warburton (Hex-UW) automaton.

Figure 4 shows the first six generations, and Figure 5 shows the first fourteen, sixteen, and thirty-two generations. To keep things clear, we will call the original automaton the Square-Ulam-Warburton (Square-UW) automaton.

But where is the Sierpinski gasket? The pictures of the UW automata look square and hexagonal, while the gasket is triangular. We will keep the reader in suspense about the full role of the gasket in these two automata, but here is a first glimpse.

The Sierpinski Gasket

We can imagine building the gasket one generation at a time, the same way we built the automata. The top row consists of a single small triangle; it is generation one. We construct the gasket downward, with each row corresponding to a generation.

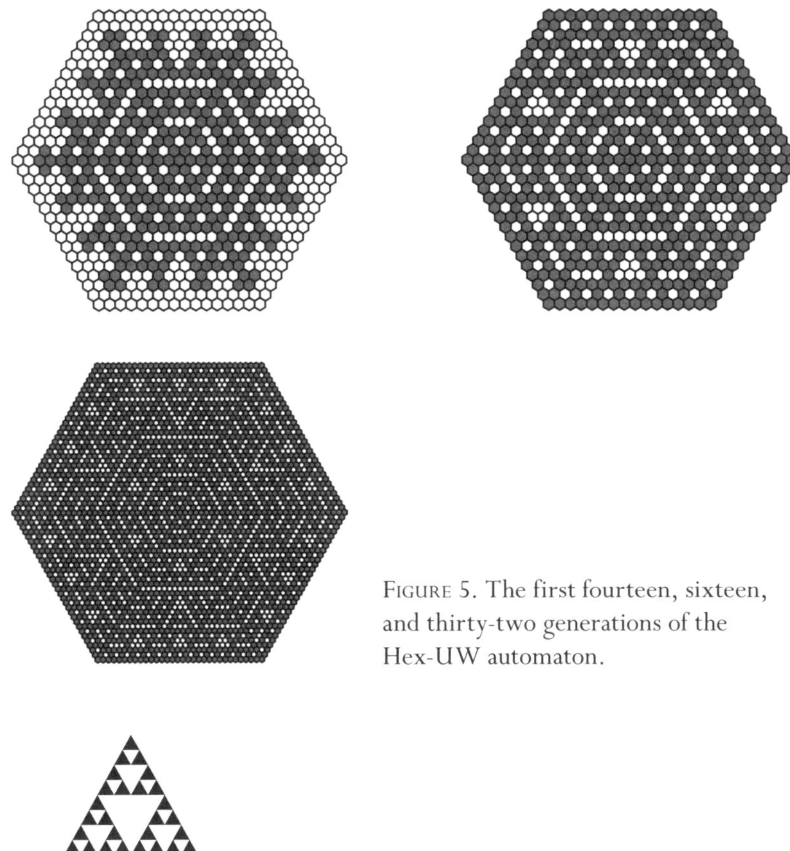

FIGURE 5. The first fourteen, sixteen, and thirty-two generations of the Hex-UW automaton.

FIGURE 6. Sierpinski gasket after fourteen generations.

Figure 6 shows the Sierpinski gasket after fourteen generations. It looks unfinished and asymmetrical. The gasket is a complete triangle only after 2^k generations (Figure 1 shows sixteen generations). Now we see the first similarity: The Square-UW automaton looks like a complete square balanced on its corner, and the Hex-UW automaton looks like a complete hexagon only after 2^k generations.

Divide the square grid into four quadrants originating at the initial cell, so that after 2^k generations we obtain four 45–45–90 triangles. Similarly, divide the hexagonal grid into six parts so that after 2^k

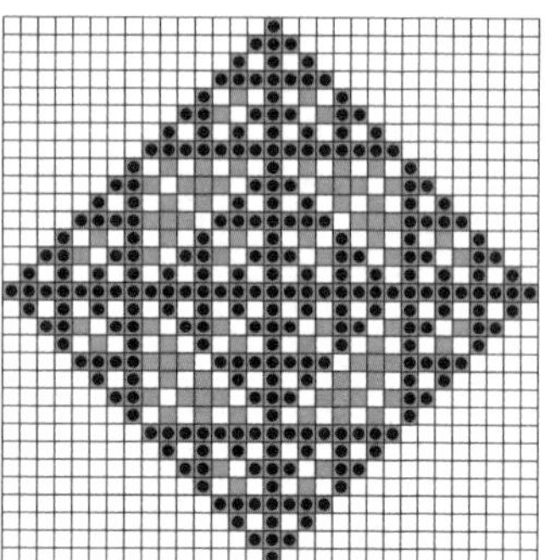

Figure 7. The Sierpinski gasket in the Square-UW automaton.

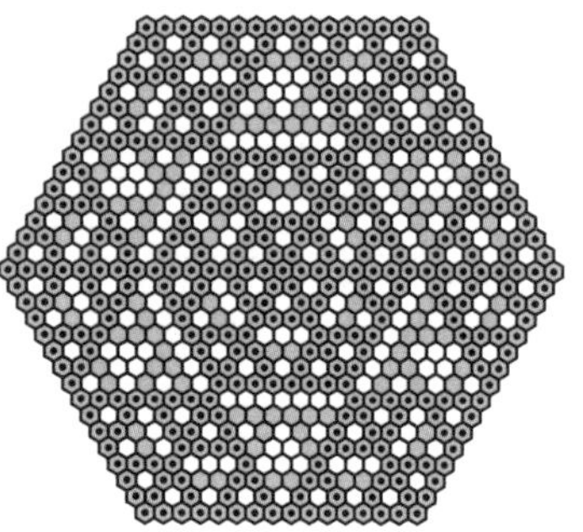

Figure 8. The Sierpinski gasket in the Hex-UW automaton.

generations we obtain six equilateral triangles. We call each section of the automaton at generation *n* a *slice* (think: slice of pizza). Adjacent slices share the cells that lie along the dividing lines.

We can draw the Sierpinski gasket in each slice of the automaton. In Figure 7, we overlay four copies of the Sierpinski gasket, represented by dots, and the Square-UW automaton (in pink)—sixteen generations each. We see that every dot is contained in a pink cell.

Likewise, when we repeat the process in the hexagonal grid, as in Figure 8, the dots fall in the blue cells of the Hex-UW automaton.

Later we prove that the first *n* generations of the Hex-UW and Square-UW automata contain the first *n* generations of the Sierpinski gasket. But that is not all. These cells play a special role in the automata.

Family

We've already defined generations; let's take the family analogy a little further. Call the initial cell in an automaton the *patriarch*. Each living nonpatriarch cell, *a*, was born because it was adjacent to exactly one

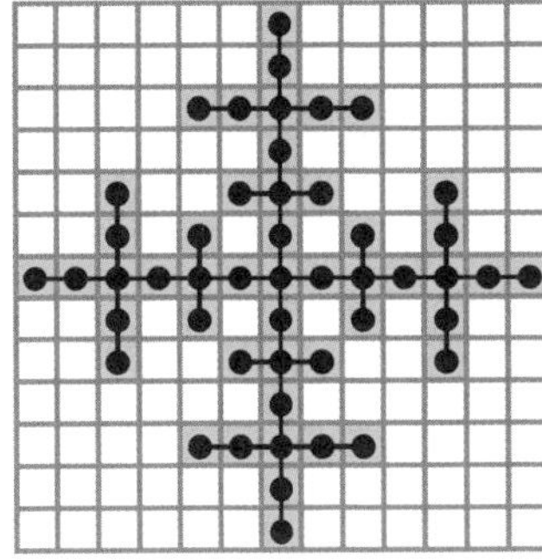

FIGURE 9. Seven generations of the Ulam-Warburton automaton and its family tree.

other cell, b. We say b is the *parent* of a and a is a *child* of b. Notice that by the rules of construction, each child has one parent and a cell born in generation n has a parent born in generation $n - 1$.

Families have family trees. It is traditional to draw a family tree as a graph, in which family members are vertices and parent–child pairs are connected by a segment. Figure 9 shows the family tree for seven generations of the Square-UW automaton superimposed on the automaton. The reader is invited to draw the family tree for several generations of the Hex-UW automaton.

Let the sequence of parents from a cell to the patriarch be its *lineage*. Define an *ancestor* of a cell to be a cell in its lineage.

Each cell has exactly one parent, but how many children can a cell have? The patriarch in the Square-UW automaton has four children, and all other cells have zero, one, or three. A nonpatriarch cannot have more than three children because it has four neighbors and one is the parent.

This automaton exhibits the familiar self-similar fractal behavior. Because we see only zero, one, or three children in these initial stages, we can be sure that future cells will not have two children.

Figure 10 uses color to exhibit the fertility of cells. The cells without children are leaves in a family tree, and they are green in the figure. Parents with three children are blue, and those with one child are red. The patriarch is black. On the border, we colored cells with their eventual fertility; their children will appear in the next generation.

Similarly, the patriarch of the Hex-UW automaton has six children, and no other cell can have more than five children. In reality, the number of children is zero, one, two, or three. In Figure 11, the cells without children are green, the parents with three children are blue, ones

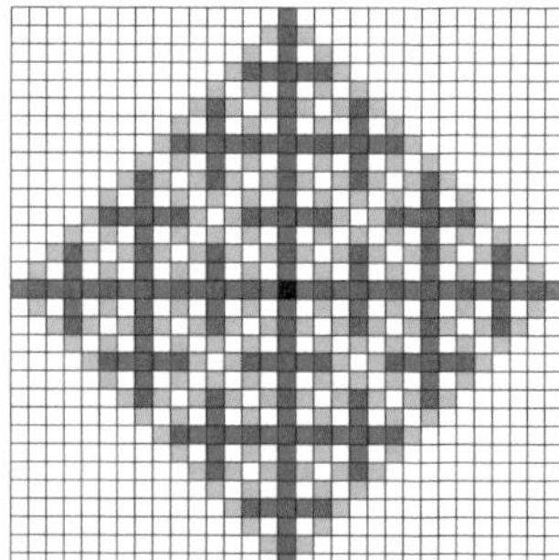

FIGURE 10. Square-UW cells color-coded by the number of children.

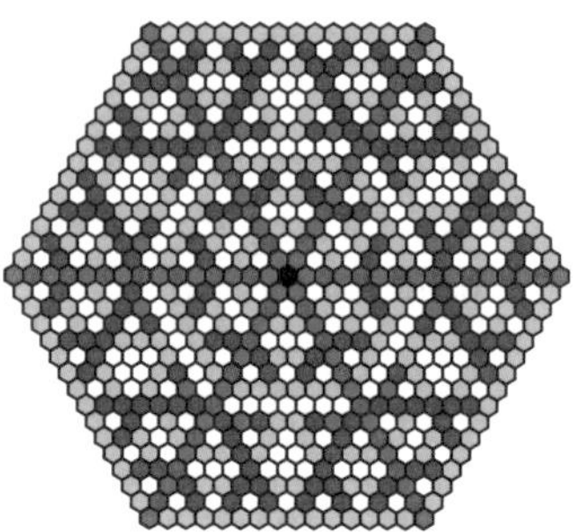

FIGURE 11. Hex-UW cells color-coded by the number of children.

with two children are purple, and those with one child are red. The patriarch is black. Border cells are colored according to their eventual fertility.

Distance

Forget automata for a moment and think about the distance between two cells on a grid. We could define the distance between two cells as the geometrical distance between their centers. But instead, we use a distance function that plays a special role on grids: the *Manhattan distance*.

If you are in Manhattan trying to stroll from one place to another, you cannot cut through buildings; you have to use the streets. And streets in Manhattan form a square grid. In the case of the automaton, the only move that is allowed is from the center of a cell to the center of a neighboring cell. This move is considered to cover distance 1. The Manhattan distance between cells a and b is the length of the shortest path that starts at cell a, ends at cell b, and is restricted as above.

On the square grid, a neighbor of a neighbor might be at Manhattan distance 0 or 2. On the hexagonal grid, a neighbor of a neighbor might be at Manhattan distance 0, 1, or 2.

Now let us return to our automata. Children are at Manhattan distance 1 from their parents. Moreover, the lineage of the cell is a path of length n originating at the patriarch. So a cell born in generation n has a Manhattan distance of at most n from the patriarch.

Pioneers and the Sierpinski Triangle

Define a *pioneer* to be a cell born in generation n with a Manhattan distance of n from the patriarch. This means that the lineage of a pioneer represents the shortest path in the Manhattan sense. Pioneers move outward as quickly as possible and never turn back—all ancestors of a pioneer are pioneers.

The set of all pioneers has a special structure—one that finally shows the connection between the automata and the Sierpinski gasket.

Theorem. *The pioneers in an* n*-generational slice of either automaton coincide with an* n*-generational Sierpinski gasket.*

We will prove this theorem by induction on the generation. For generation one, the patriarch is a pioneer, and it is the top cell of the Sierpinski gasket.

Now assume that the pioneers of generation n correspond to the small triangles in the nth row of the gasket; that is, they belong to generation n of the gasket. Note that all grid elements in row j are a Manhattan distance $j - 1$ from the patriarch. Consider a grid cell a in row $n + 1$ of the gasket.

It is well known that if we overlay the Sierpinski gasket and Pascal's triangle, the gasket corresponds to the odd entries in Pascal's triangle (Figure 12). An entry in Pascal's triangle is odd when the two numbers above it have different parities. Thus, exactly one of the neighbor cells of a in row n is in the gasket, or equivalently, by the inductive hypothesis, exactly one of the neighbor cells in row n is a pioneer. But this means that only this one neighbor of a is alive in generation $n - 1$, and thus a is born in generation n. Hence, it is a pioneer. A similar argument shows that every pioneer in row $n + 1$ is in the gasket. This completes the proof.

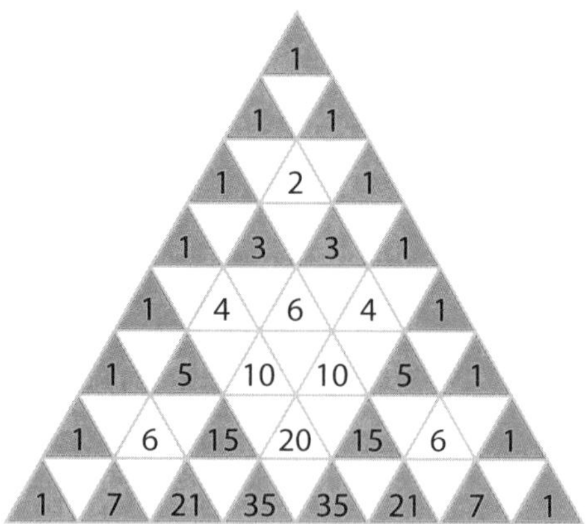

FIGURE 12. The Sierpinski gasket in Pascal's triangle.

It is fitting that the Sierpinski gasket—a pioneer in the history of fractals—also plays a pioneering role in the creation of the UW automata.

We are grateful to Richard Stanley for suggesting this project. We are also grateful to the MIT-PRIMES program for sponsoring this research. After both the project and the paper were complete, we found existing work by Gravner and Griffeath [2] covering many of the same topics.

Further Reading

1. D. Applegate, O. E. Pol, and N. J. A. Sloane, The toothpick sequence and other sequences from cellular automata, arxiv.org/pdf/1004.3036, 2010.
2. J. Gravner and D. Griffeath, Modeling snow crystal growth I: Rigorous results for Packard's digital snowflakes, *Experiment. Math.* **15** no. 4 (2006) 421–444.
3. P. Pnachekha, J. Schneider, G. Xing, A growth process in Z^2, unpublished manuscript.
4. D. Singmaster, On the cellular automaton of Ulam and Warburton, *M500 Magazine of The Open University* **195** (2003) 2–7.
5. R. P. Stanley (proposer) and R. J. Chapman (solver), A tree in the integer lattice, Problem 10360, *Amer. Math. Monthly* **101** (1994) 76; **105** (1998) 769–771.
6. S. M. Ulam, On some mathematical problems connected with patterns of growth of figures, *Mathematical Problems in Biological Sciences* **14** (1962) 215–224.

Fractals as Photographs

Marc Frantz

Can fractals rival fine art photographs? After working toward this goal for more than twenty years, I hope to persuade you that they can. For starters, compare the four photographs in Figure 1 with the four fractal images in Figure 2.

Most people I talk to agree that the fractal images are attractive, intriguing, and photographic in appearance. However, most of those people are my friends, and friends are often kind in their appraisal. So let me explain a little more, and show a little more, and see if I can convince you of the possibility.

Inspired by Masters

My inspiration began in art school, where I spent many hours in the library looking at books of photographs. I loved the black-and-white photography of masters such as Edward Steichen, Minor White, Edward Weston, Paul Caponigro, and Ansel Adams. That's why I chose the four photographs in Figure 1. Despite my enthusiasm, I was too impatient to deal with the techniques of film photography, so I majored in painting instead.

Many years later, when I switched careers from art to mathematics, I realized that I might be able to create such images with tools I *did* enjoy using: the mathematics of fractal geometry, which is built on subjects such as real analysis, topology, and measure theory. I saw this possibility most clearly in Michael Barnsley's book *Fractals Everywhere* (Academic Press, 1988), a text that ostensibly covers material in real analysis and metric space topology, but conveys so much more.

I had grown tired of the typical examples of fractal imagery—Mandelbrot sets and Julia sets, rendered in relentlessly bright colors,

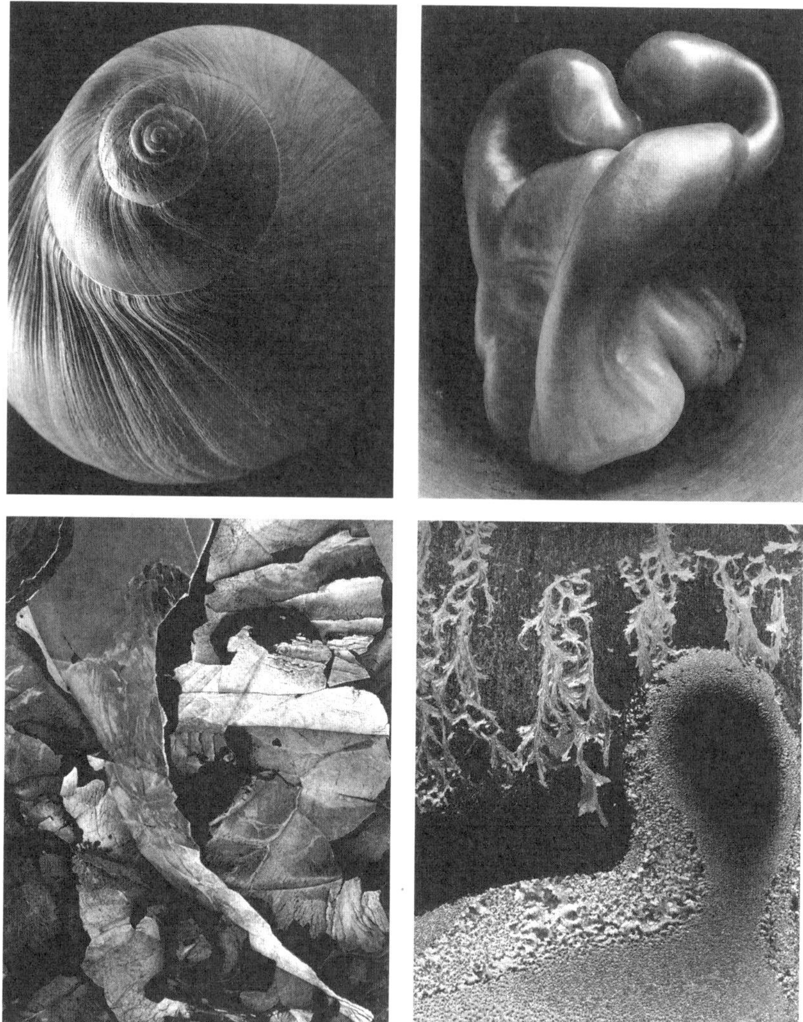

FIGURE 1. Top row: Edward Steichen, *The Spiral Shell*, c. 1921; and Edward Weston, *Pepper*, 1930. Bottom row: Minor White, *Moencopi Strata, Capital Reef, Utah*, 1962, and *Empty Head, Rochester, New York*, 1962.

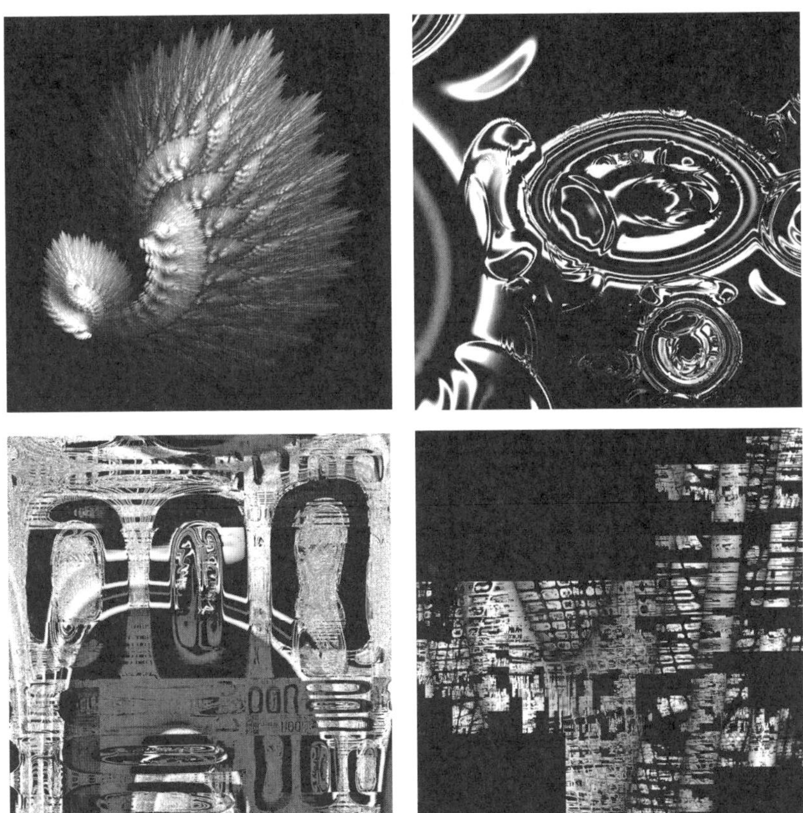

FIGURE 2. Top row: *Courtship Display*, 3-function IFS invariant measure; and *Ripples*, 2-function IFS attractor. Bottom row: *Pow* and *Lift*, 2-function IFS attractors.

using every crayon in the box. Barnsley's images suggested that his techniques might be capable of rendering shapes and textures with beauty, subtlety, and artistic maturity.

Empowered by Mathematics

Here is a brief sketch of how the images in this article were created. We begin with a finite set of functions from the real plane to itself. For simplicity, let's say there are just two functions, f and g. We are going

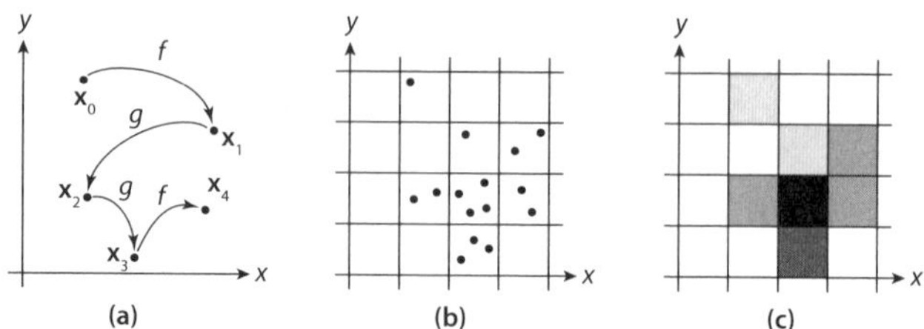

FIGURE 3. An IFS random-iteration orbit in the plane determines a density shading of pixels.

to repeatedly choose a function at random, so we assign nonzero probabilities p_f to f and p_g to g such that $p_f + p_g = 1$.

Pick an arbitrary starting point in the plane, $\mathbf{x}_0$. Suppose that the first four choices come up f, g, g, f. Taking the functions in the chosen order (Figure 3[a]), we let $\mathbf{x}_1 = f(\mathbf{x}_0)$, $\mathbf{x}_2 = g(\mathbf{x}_1)$, ($\mathbf{x}_3 = g(\mathbf{x}_2)$, $\mathbf{x}_4 = f(\mathbf{x}_3)$, and so on.

We keep going until we have generated a large number of points $\mathbf{x}_0, \ldots, \mathbf{x}_n$. (In practice, I generate about a billion points.) We imagine the plane, or some part of it, to be tiled with square pixels. We color each pixel according to how many points lie in it—say black for the highest number, white for the lowest number, and shades of gray in between. This results in a kind of "density shading" shown in Figure 3(c).

The functions f and g, together with their respective probabilities, form an *iterated function system* (IFS) *with probabilities*. The method of generating points is called the *random iteration algorithm*, the set of points $\{\mathbf{x}_0, \ldots, \mathbf{x}_n\}$ is called an *orbit*, and the density shading is said to be a visual representation of a mathematical object called an *IFS invariant measure*. The image *Courtship Display* in Figure 2 is an invariant measure created with three functions.

The other three images in Figure 2—*Ripples, Pow,* and *Lift*—were created using the method illustrated in Figure 4. There we use functions F and G from real three-dimensional space to itself, probabilities p_F and p_G, and an initial point $\mathbf{x}_0$ to create an orbit $\{\mathbf{x}_0, \ldots, \mathbf{x}_n\}$. The projection of this orbit onto the xy-plane is shown in Figure 4(b). We shade darker the points with larger z-values.

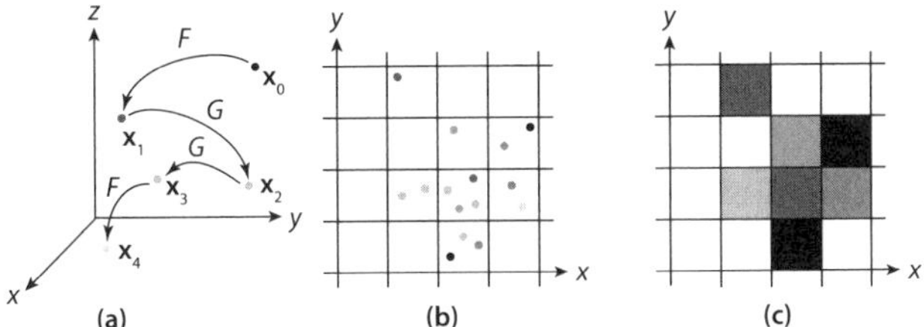

FIGURE 4. An IFS random-iteration orbit in space determines a height shading of pixels.

In Figure 4(c), we shade each pixel to match the highest (darkest) orbit point that projects onto it. Under certain conditions, the orbit approximates an *IFS attractor* in space, and the height shading roughly represents the "roof" of the attractor.

A Small Gallery

Figure 5 exhibits more IFS attractors and invariant measures. Most were made using no more than two functions, although in most cases the functions are more complicated than just affine transformations. After the grayscale image is finalized, I often add a slight touch of sepia or bronze color. That's because black-and-white photographs frequently have a similar colorization to help bring out details.

Small Systems

I love that a carefully chosen IFS with a small number of functions can create a very natural-looking image. (That's an important point; Barnsley showed how to accurately mimic any given photograph with an IFS, provided that one is willing to use a large number of functions.) Even after developing some instinct for what works, however, I ultimately reject about a hundred attempts for each "keeper," and even the keepers require endless tweaking to create a satisfactory composition. I have spent anywhere from a few days to a few years on an image before deciding it was finished. I consider the time thus spent more than justified if the result is a worthy tribute to those great photographers

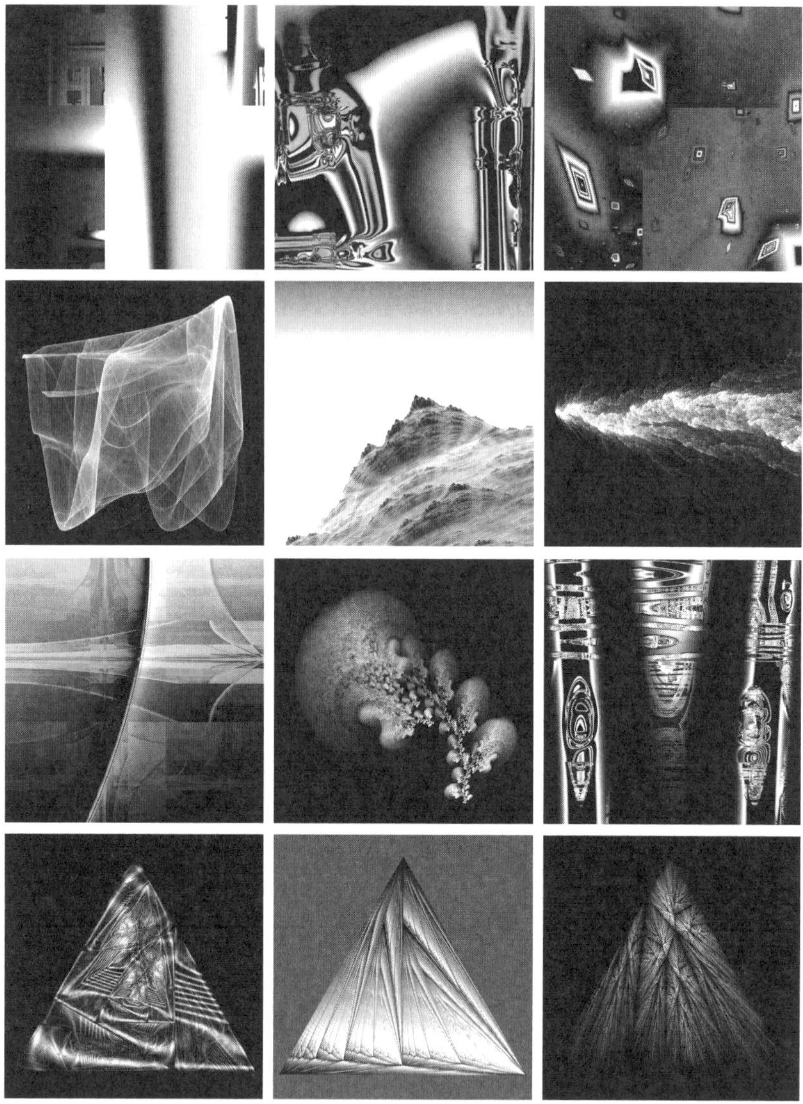

FIGURE 5. Here (DS, *n*) and (HS, *n*) denote a density shading and a height shading, respectively, produced using *n* functions. Row 1: *Room* (HS, 2), *Flesh and Chrome* (HS, 2), and *Mystic Room* (HS, 2). Row 2: *Curtains–Still* (DS, 1), *Visionary Mountains* (DS, 3), and *Meteor* (DS, 2). Row 3: *Reed* (HS, 2), *Liliaceae* (DS, 3), and *Ritual* (HS, 2). Row 4: *Emblem* (HS, 2), *Silk* (DS, 4), and *Broom* (DS, 3).

Math at the Met

Joseph Dauben and Marjorie Senechal

The Metropolitan Museum of Art in New York City—known worldwide and hereinafter as "the Met"—is the largest art museum in the United States, and one of the ten largest in the world. Founded in 1870, it holds two million works of art, from antiquity to today, from all around the world.

We propose here a guided tour of the Met's hidden math. The Met's collections include works of art with surprising mathematical content. These works span the globe, human history, and the Met's curatorial departments. They include numerals, shapes, perspective, astronomy, time, and games. (On a second tour, still in the planning stages, we hope to show you patterns, symmetries, and various mathematical activities.)

Our tour was inspired by the late Dr. David Mininberg, whose fascinating tour "Medicine at the Met" revealed the medical purposes of jars, bowls, and other artifacts from ancient Egypt, Byzantium, the ancient Near East, and Oceania, displayed at the Met for their artistic merit.[1] We dedicate this essay to his memory, with gratitude for his encouragement.

Note: We have organized this tour thematically, not by curatorial department or gallery, or any other scheme to make it easy on your feet. If you take this tour by reading it (in print or online), then you're all set to go. If you take the tour in person, you'll need to plot your path. For help with that, see our notes "If You Go" and "Where to Find It" (among these pages).

Numerals. Although this is a tour of hidden math, let's start with a three-foot high exception. In the wing for Modern and Contemporary Art, the painting *I Saw the Figure 5 in Gold* jumps off the wall at you or, alternatively, rushes away from you (Fig. 1).

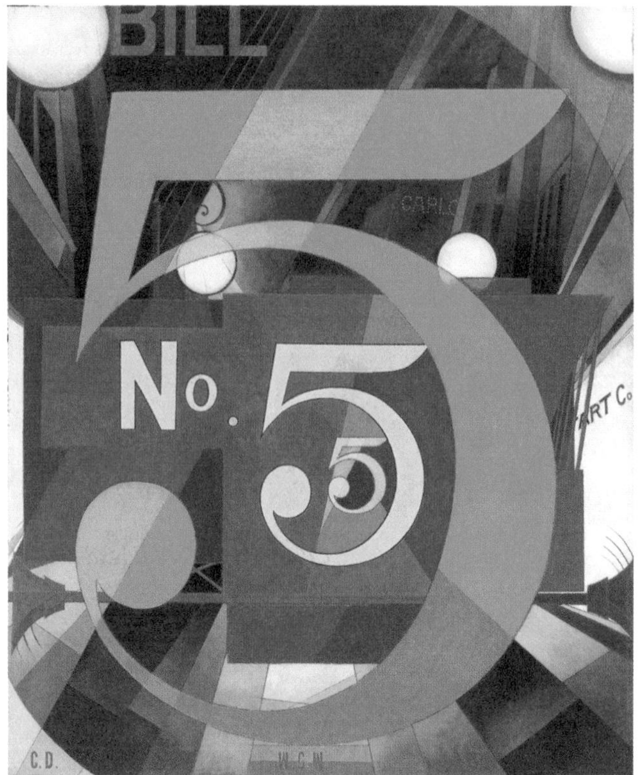

FIGURE 1. *I Saw the Figure 5 in Gold*, Charles Demuth (American, 1883–1935), 1928. Alfred Stieglitz Collection, 1949. Accession no. 49.59.1. http://www.metmuseum.org/.

The painting is a tribute to the artist's friend, the poet William Carlos Williams (look for "Bill," "Carlos," and the poet's initials) and to Williams's poem, "The Great Figure":

> Among the rain/and lights/I saw the figure 5/in gold/on a red/firetruck/moving/tense/unheeded/to gong clangs/siren howls/and wheels rumbling/through the dark city.

Next we go to Drawings and Prints. Here—less obviously—we find numerals in the mysterious, intriguing engraving *Melencolia I*, made by Albrecht Dürer (1471–1528) in 1514 (Fig. 2). Its five hundredth anniversary was heartily celebrated by mathematicians in 2014.[2]

FIGURE 2. Top: Albrecht Dürer, *Melencolia I*, 1514. Engraving. Harris Brisbane Dick Fund, 1943. Accession no. 43.106.1. http://www.metmuseum.org/. Bottom: Detail (the magic square).

Indeed, *Melencolia* is a mathematical feast, with a sphere, a compass, and a mysterious polyhedron. And look closely below the bell: the square with the numbers 1, . . . , 16 is a magic square: the numbers in each row, each column, and each diagonal add to 34. Although magic squares can be traced back some twenty-five hundred years, Dürer is thought to be the first European artist to depict one.

Pondering *Melencolia*, you enter a wonderful world of arithmetic puzzles. What is the smallest magic square—2 × 2? 3 × 3? 4 × 4? How many of each (possible) size exist? Is there a largest size? Are there any magic cubes? Magic hypercubes?

The graceful numerals in this magic square—much like the ones we use today—evolved from the ancient Brahmi numerals of India. Augmented by the tantalizing and multipurpose zero, they diffused westward slowly throughout the centuries, reaching Italy by the medieval era, where they soon replaced the cumbersome Roman numerals then in use.

But another set of symbols then used in the west was not replaced so quickly. Instead, it was used into the Renaissance, coexisting with the "new" Hindu-Arabic numerals, much as typing and word processing coexisted until the touch screen.

These symbols were not written; they were gestures made by flexing finger joints. Rather than simply holding up one, two, three, or more fingers to count from one to ten on two hands, hand-reckoners, as adepts were called, could display all the numbers from one to a hundred with the fingers of just one hand. With two hands, they could tally up to 9,999. And not just tally: they could *reckon*. Hand-reckoners did arithmetic and calculated future dates of holidays such as Easter. This system probably persisted as long as it did because it was the international language of traders. A Middle Eastern camel merchant could strike a bargain with a European wool merchant without exchanging a word of their mutually incomprehensible languages, or putting pen to paper (both difficult to come by). Indeed, traders in the Chicago Board of Trade used a form of hand-reckoning ("open outcry") from 1870 to February 2015.[3]

You can see hand-reckoning at the Met in the Flemish Primitive Adriaen Isenbrant's *Man Weighing Gold* (Fig. 3), in the department of European Paintings, 1250–1800. This early sixteenth-century painting, the Met website tells us, is one of the first portraits depicting a

FIGURE 3. Left: *Man Weighing Gold*, Adriaen Isenbrant (Netherlandish), 1515–1520. The Friedsam Collection, Bequest of Michael Friedsam, 1931. Accession no. 32.100.36. http://www.metmuseum.org/. Right: Detail.

professional activity, although historians aren't sure just which profession is depicted. The man weighing gold may have handled commodities, or he may have been a moneychanger or a banker.

The pan on his left holds a fixed weight; he piles coins on the right-hand pan until they balance. But how does he keep track of the total? Not with an abacus and not with a pen. But notice the crossed fingers of his right hand.

If our interpretation is correct, he is using a variant of the hand positions recorded by his contemporary, Luca Pacioli (Fig. 4), to tally the running sum.

In the department of European Sculpture and Decorative Arts, we find hand-reckoning again, now in an elaborate mirror frame designed by a leading Nuremburg goldsmith of the sixteenth century, Wenzel Jamnitzer (Fig. 5). The frame replicates the frontispiece of his 1568

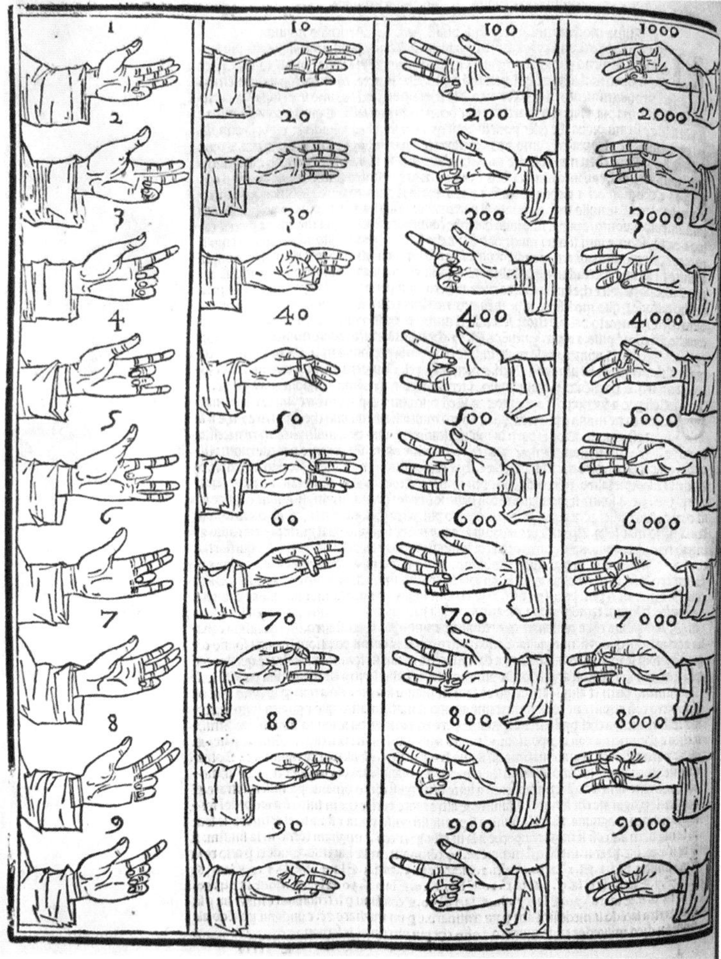

FIGURE 4. Finger positions for the numbers 1 to 10,000. From Luca Pacioli, *Summa de arithmetica, geometria. Proportioni et proportionalita*, Venice, 1494.

book *Perspectiva Corporum Regularium*. Each corner represents one of the medieval university's seven liberal arts; these four, known as the quadrivium, are (clockwise from upper right) geometry, architecture, perspective drawing, and arithmetic. Look closely at "Arithmetic," upper left: the seated female figure is making the sign for the number thirty-six with her left hand and writing Hindu-Arabic numerals with her right.

FIGURE 5. Top: Relief mounted as a mirror frame, Wenzel Jamnitzer, ca. 1568. Gilded silver, ebony, mirror plate, height 11 5/8 × width 9 1/8 in. Gift of J. Pierpont Morgan, 1917. Accession no. 17.190.620. http://www.metmuseum.org/. Left: Upper left detail (from *Perspectiva Corporum Regularium*). "Arithmetica" makes the hand-reckoning symbol for thirty-six with her left hand and writes the Hindu-Arabic numerals with her right.

Next, in the department of Asian Art, we stop before a curious terracotta tile (Fig. 6). This tile, baked in Kashmir in the fifth to sixth centuries, had been placed in an outdoor courtyard. The curious emaciated human figures in relief intrigue art historians, but the incised markings are our interest here.[4]

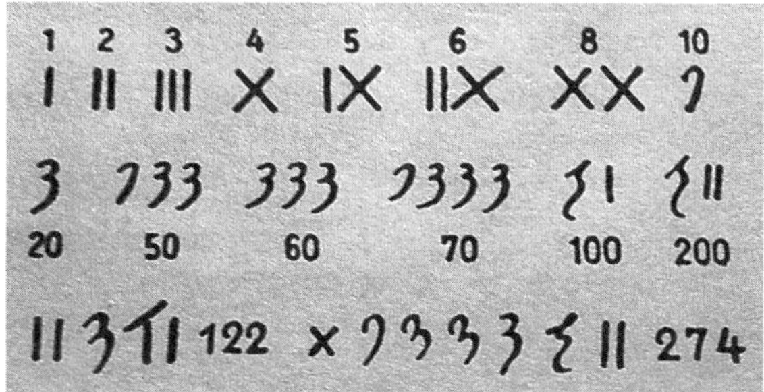

FIGURE 6. Top: Terracotta *Tile with Impressed Figure of Emaciated Ascetics and Couples Behind Balconies*, fifth to sixth centuries, India (ancient kingdom of Kashmir, Harwan). Purchase, Kurt Berliner Gift, 1998. Accession no. 1998.122; http://www.metmuseum.org/. Bottom: The Kharosthi numerals. From K. Menninger, *Number Words and Number Symbols*.

The markings on this tile are numerals, but not ones we easily recognize (except perhaps | | | for 3). Nor are these the Brahmi predecessors of today's Hindu-Arabic numerals. These numerals are written in the ancient Kharosthi script, a contemporary of the Brahmi. The cane-shaped symbol is a ten; the cross means four. Scribes and merchants in northern

India and neighboring regions (and on the Silk Road) used Kharosthi numerals from about the fifth century BCE to the third century CE.

We also find numerals hidden in the Egyptian Art wing, for example in an elaborate necklace pectoral made of gold, carnelian, lapis lazuli, turquoise, and garnet, to honor the Twelfth Dynasty pharaoh Senwosret II (Fig. 7).

The Met's translation of the hieroglyphic inscription reads, "The god of the rising sun grants life and dominion over all that the sun encircles for one million one hundred thousand years (i.e., eternity) to King Khakheperre (Senwosret II)."

And indeed the Egyptian numerals used at that time to denote a million and one hundred thousand hold center stage: the seated man is the symbol for a million, and one hundred thousand (the tadpole) dangles from his arm (Fig. 7).

Games. Still in the Egyptian wing, we find (Fig. 8) numbers represented by pips on dice from Middle Egypt, from Oxyrhynchus (el-Bahnasa), dating to the Roman period, that is, 30 BCE–330 CE.[5]

These dice look very much like the ones we use today. Not all ancient dice did. Other polyhedra, for example, icosahedra (twenty equilateral triangular facets, grouped together in fives), were also used as dice in games of chance. In its wing for Greek and Roman Art, the Met has several examples of such Roman icosahedra (Fig. 9).

These icosahedra were, the Met suggests, used in connection with oracles. However, because the ancient Greeks used letters of their alphabet as numerals, they may have been used as dice in games as well.

Sheep and goat knucklebones, or molded models thereof (Fig. 10), were also used as dice in certain games.[6] The side on which they landed when thrown determined their "value."

Games with dice were a favorite pastime in ancient Egypt. Some of them are represented in board games in the Met's collections. This example (Fig. 11, left) of the popular board game, Game of Hounds and Jackals, is constructed of ebony and ivory. It is dated ca. 1814–1805 BCE. The game was played with five pins with hounds' and jackals' heads. The board itself (Fig. 11, right) contains fifty-eight holes (twenty-nine on either side of a palm tree incised down the center of the board), with the symbol of a shen ring at the center (a symbol of eternity or infinity; it could also be taken to represent the course of the Sun in the heavens).

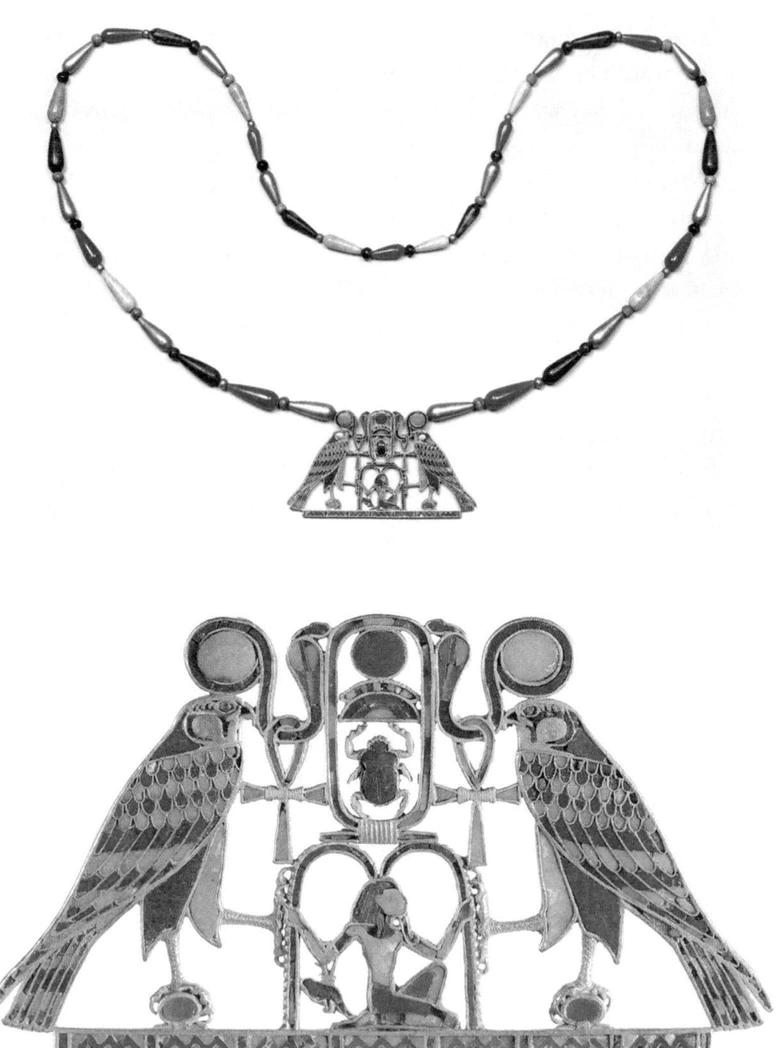

FIGURE 7. Top: Pectoral and necklace of Sithathoryunet with the name of Senwosret II, Middle Kingdom Dynasty, about 1887–1878 BCE. Purchase, Rogers Fund and Henry Walters Gift, 1916. Bottom: The pectoral. Accession no. 16.1.3; http://www.metmuseum.org/.

FIGURE 8. Dice from Oxyrhynchus, made of ivory, dimensions approximately 1 cm (3/8 in.) on each side. Gift of the Egypt Exploration Fund, 1897. Accession no. 97.4.117. http://www.metmuseum.org/.

FIGURE 9. Icosahedra inscribed with letters of the Greek alphabet, Roman, Midimperial, second to third centuries. http://www.metmuseum.org/. Left: Steatite, Fletcher Fund, 1927. Accession no. 27.122.5. Right: Faience, Fletcher Fund, 1937. Accession no. 37.11.3.

FIGURE 10. Eight astragaloi (astragali) (Greek for knucklebones), from the Greek Hellenistic period, made of cast glass, two-part molds, averaging about 1.5 cm in height. The gift of Mr. and Mrs. Jonathan P. Rosen, 1992. Accession no. 1992.266.3–.10. http://www.metmuseum.org/.

Howard Carter and the Earl of Carnarvon, who excavated this board at Thebes in Upper Egypt in 1910, say the game was most likely played with moves of the hounds and jackals determined by the throw of dice or knucklebones (Carter and Carnarvon 1912).[7]

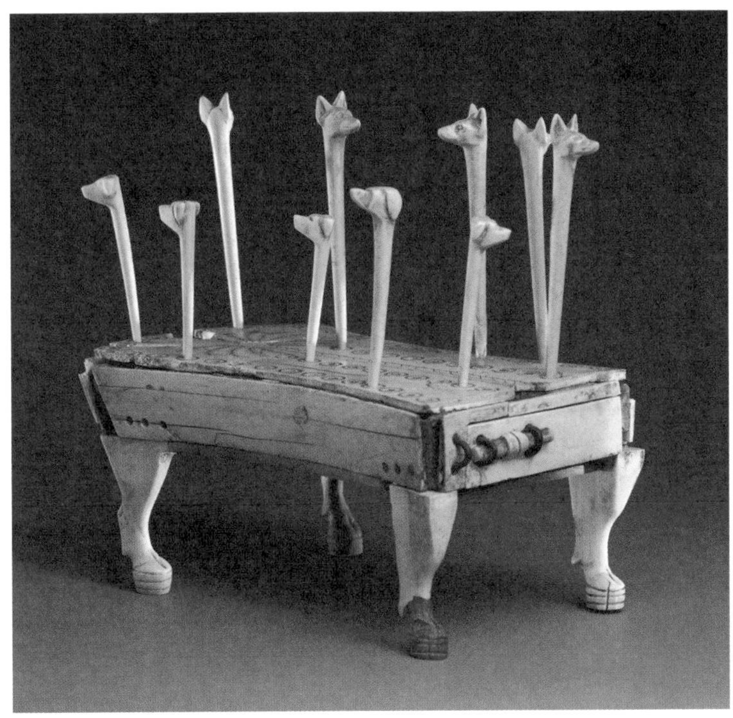

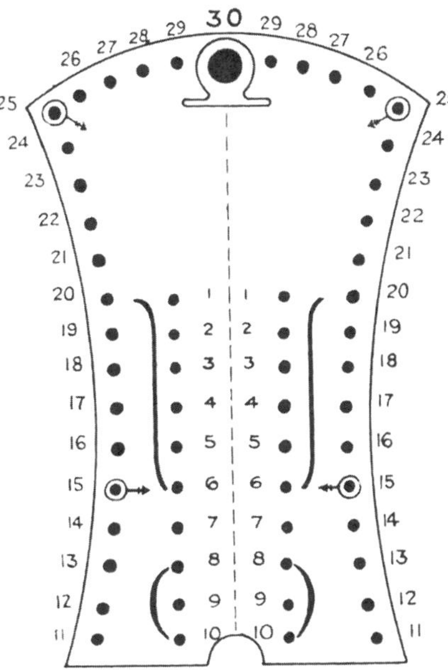

FIGURE 11. Top: Game of Hounds and Jackals, Middle Kingdom, Dynasty 12, reign of Amenemhat IV, from Thebes, made of ebony and ivory, approximate dimensions (height × width × length) with pins: 14 × 10.1 × 15.6 cm. Gift of Edward S. Harkness, 1926. Accession no. 26.7.1287a–k. http://www.metmuseum.org/. Currently on display in Gallery 111. Left: Diagram of the game board (from Carter and Carnarvon 1912).

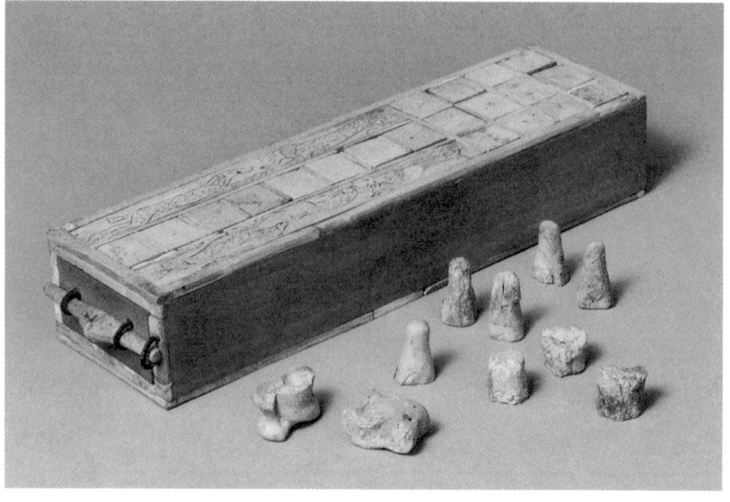

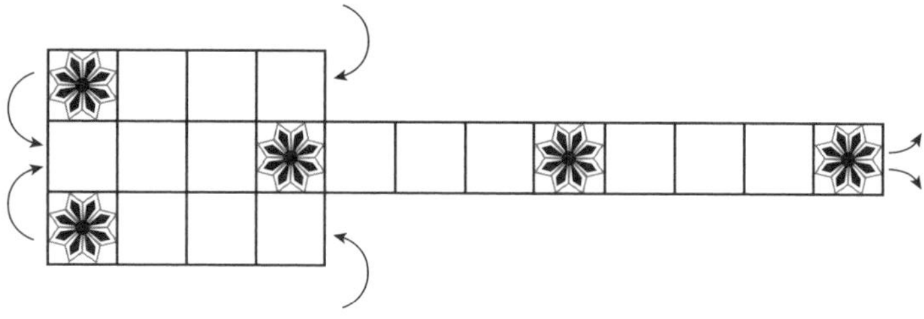

Figure 12. Top: Double-sided game box with playing pieces and a pair of knucklebones. Thebes, Egypt. ca. 1635–1458 BCE. The Metropolitan Museum of Art, New York, Rogers Fund, 1916. Accession no. 16.10.475a. http://www.metmuseum.org/. Bottom: Diagram suggesting the direction of play for the Game of Twenty Squares (Finkel 2007).

Another board game at the Met (Fig. 12), also from Thebes, dates slightly later to 1635–1450 BCE and includes gaming pieces and knucklebones presumably used in the course of the game.

Referred to as the "Game of Twenty Squares" or as the "Royal Game of Ur" (because of the many examples of the game excavated from locations in ancient Sumeria), the game began (Finkel 2007) with two players moving pieces toward the center aisle, as depicted in Figure 12 (right). Every fourth square is marked with a rosette or other geometric symbol. Videos on the Harvard Semitic Museum's website, including a simplified version suitable for children, explain how players

FIGURE 13. *Les Quatres Joueurs de Cartes*, Paul Cézanne, oil on canvas, painted ca. 1890–1892. Bequest of Stephen C. Clark (1960). Accession no.: 61.101.1. http://www.metmuseum. org/.

would move their game pieces depending on the throws of astragaloi or knucklebones.

Card games are represented in the Met's collections, too. One of the best known paintings of these games is Paul Cézanne's *Les Quatres Joueurs de Cartes* (*The Four Card Players*) in the Galleries for 19th and Early 20th Century European Paintings and Sculpture. *Les Quatres Joueurs de Cartes* is one of a series of five paintings that Cézanne devoted to peasants playing cards. His models were local farmhands at the Jas de Bouffan, the family's country home in Provence (Fig. 13). Notice that the legs of the central card player form the Roman numeral V. That's not incidental, says art historian Mary Louise Krumrine (Krumrine 1997): each painting in this series of five includes an instance of the number five in some form.

The Met also includes examples of actual decks of cards, works of art in themselves. One of these is on exhibit from time to time at the branch of the Metropolitan Museum devoted to European medieval art, The Met Cloisters, in Fort Tryon Park (in northern Manhattan). This set of fifty-two Dutch playing cards (Fig. 14) is the only known complete set from the fifteenth century. Instead of the familiar diamonds, spades, clubs, and hearts, the four suits here are represented by familiar items of the hunt: dog collars, hound tethers, game nooses, and hunting horns.

Dimensions.[8] Continuing our search for hidden math, we return to the wing of Modern and Contemporary Art for Josef Albers's *Homage to the Square: With Rays*. This is one in a series of works called "Homage to the Square" that Albers began in 1950. The series grew, the Met's website tells us, into "a body of more than a thousand works executed over a period of twenty-five years, including paintings, drawings, prints, and tapestries." Most depict "several squares, which appear to be overlapping or nested within one another." For Albers, the artist, the series was less about squares than colors, and how they change when we juxtapose them with others.

But there's more math here than meets the eye. The remarkable color effects of the "Homage to the Square" series are discussed ad infinitum on the Internet, but Albers's "mathematically determined format" of the overlapping squares is not. *The New York Times* to the contrary, Albers's squares are not concentric.[9] Can you find a mathematically precise description of the relation among the four squares in Figure 15? And what can you say about the other works in this series on the Met's website?

Moving to the Galleries for Photographs wing and up one dimension, consider *Sugar Cubes*, a photograph by Edward Steichen (1879–1973). In this photograph (Fig. 16), we see a nearly rectangular array of ordinary sugar cubes. It's not the cubes themselves that draw and hold our attention; it's their shadows. Where did Steichen position his lamps to create this plaid effect? Test your answer experimentally, with sugar cubes from your neighborhood grocery store, and an ordinary flashlight.

The sugar cubes and their shadows may remind you of cubes "unfolded" to planar "nets" of squares (Fig. 17).

Look again at the polyhedron in *Melencolia* (Fig. 2). Can you draw a planar net for it, similar to the one shown here for the square? Is your net unique, or can you make it another way? How many different nets can you make for it—how many ways can this polyhedron be cut and "unfolded"? Can every polyhedron be unfolded to a net that lies flat and

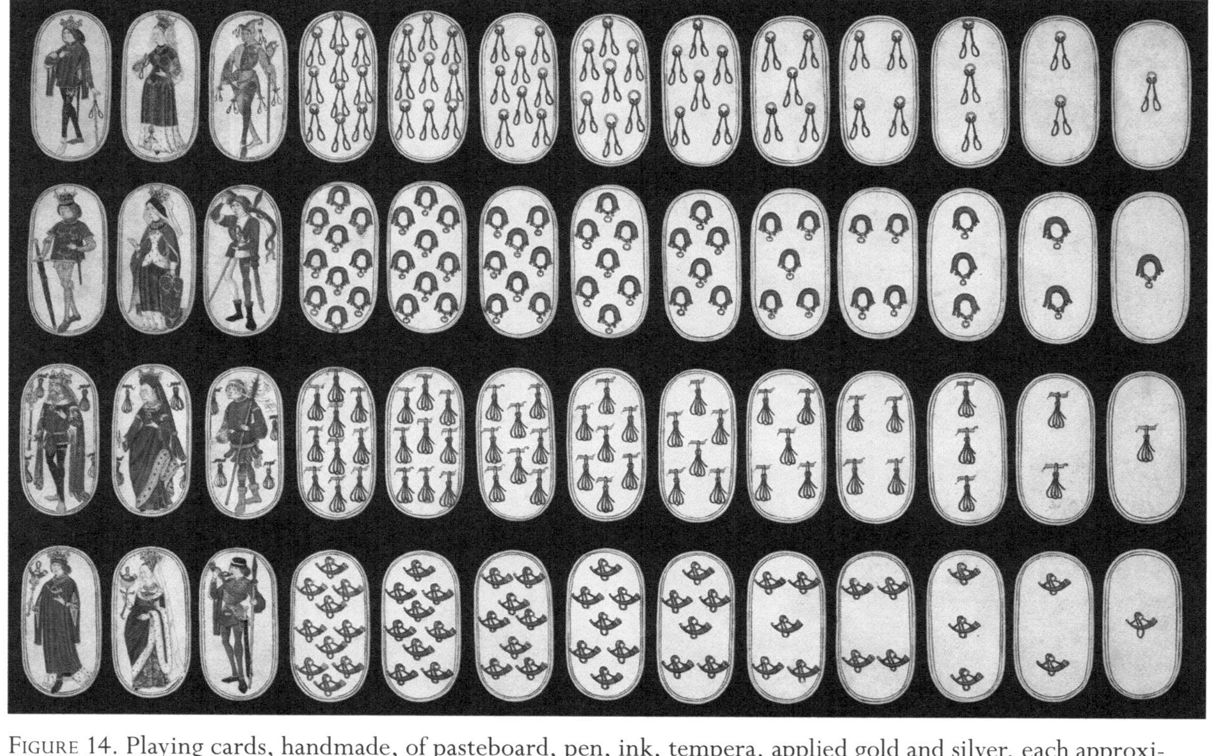

FIGURE 14. Playing cards, handmade, of pasteboard, pen, ink, tempera, applied gold and silver, each approximately 13.7 × 7 cm. Accession no. 1983.515.1–2. http://www.metmuseum.org/.

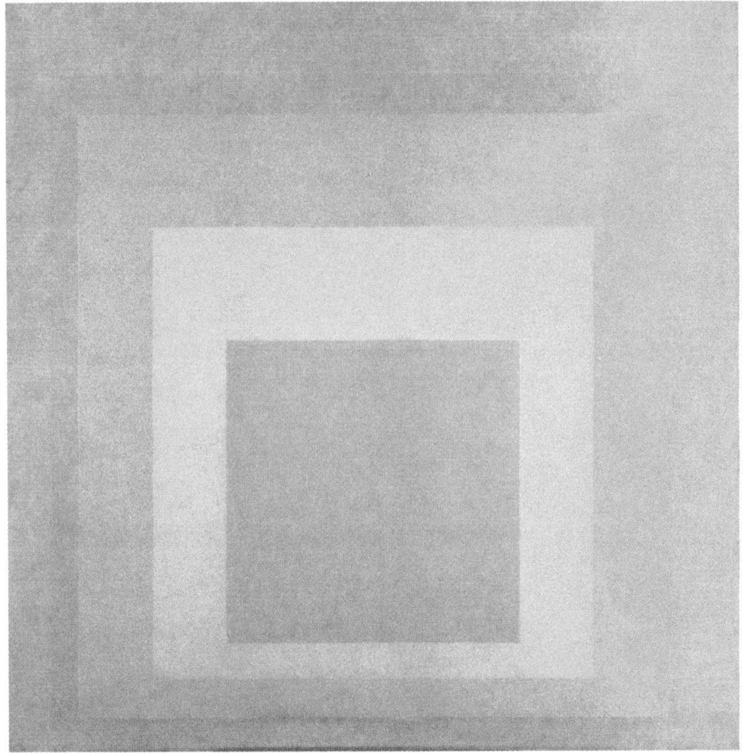

FIGURE 15. *Homage to the Square: With Rays*, Josef Albers, 1959. Oil on Masonite, 48 1/8 × 48 1/8 in., Arthur Hoppock Hearn Fund, 1959. Accession no. 59.160. Copyright © 2015 The Josef and Anni Albers Foundation/Artists Rights Society (ARS), New York. Image copyright © The Metropolitan Museum of Art. Image source: Art Resource, Inc., NY.

doesn't overlap itself? This question, which has applications from biology to industrial design, is an important unsolved geometry problem of our time.

Move up one dimension again, this time from three to four. The eight large cubes forming the cross in Figure 18 are a three-dimensional net of a four-dimensional cube! (Yes, the artist, Salvador Dalí, knew this; hence "hypercubus" in the title.) What would this net look like, in four dimensions, if you could fold it up?

SPACE AND TIME. You'll find timekeeping devices tucked away in works of art throughout the Met (for example, the hourglass on the wall in Fig. 2.)

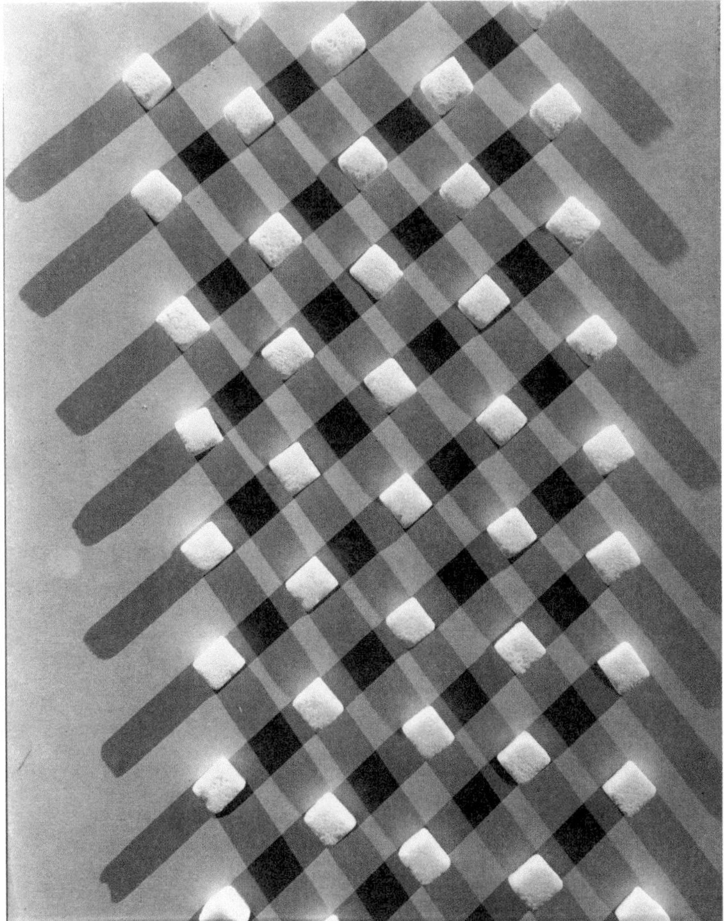

FIGURE 16. *Sugar Cubes: Design for Stehli Silk Corporation*, Edward J. Steichen, 1920s, gelatin silver print, 9 15/16 × 8 in. Ford Motor Company Collection, Gift of Ford Motor Company and John C. Waddell, 1987. Accession no. 1987.1100.217. Copyright © 2015 The Estate of Edward Steichen/ Artists Rights Society (ARS), New York. Image copyright © The Metropolitan Museum of Art. Image source: Art Resource, Inc., NY.

The Met also contains rich collections of astronomical instruments and timekeeping devices.

In the Galleries for Art of the Arab Lands, we find beautifully wrought astrolabes. These ancient devices for locating and predicting the positions of the Sun, Moon, planets, and stars were "the slide-rules

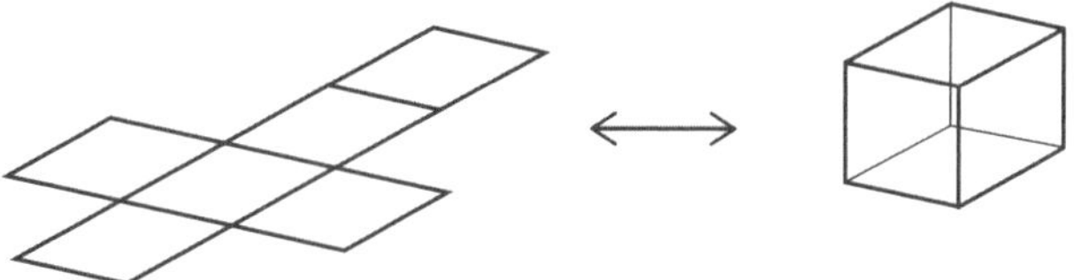

FIGURE 17. The net of six squares on the left folds up to form a cube; conversely, by cutting appropriate edges you can unfold the cube into the net. (Drawn with *Geometer's Sketchpad*.)

of the Middle Ages." (You can learn why on the Internet; for example, search for astrolabes on HowStuffWorks.com.)

The astrolabe of ‘Umar ibn Yusuf ibn ‘Umar ibn ‘Ali ibn Rasul al-Muzaffari, shown in Figure 19 (left), dates to 1291 CE. It was manufactured in Yemen of cast bronze and then was hammered, pierced, chased, and eventually inlaid with silver to produce a handsome instrument measuring a little more than six inches in diameter. Befitting his royal status and before ascending to the throne, Prince ‘Umar ibn Yusuf had been carefully educated, was well versed in mathematics and astronomy, and had even written a treatise on the construction of astrolabes. Another Met astrolabe (Fig. 19, right), manufactured nearly 400 years later in Mashhad, Iran, by Muhammad Zaman al-Munajjim al-As-turlabi, is made of cast and hammered brass and steel and measures less than 7 inches in diameter. Neither of these astrolabes was large enough to make precise astronomical observations, but they were accurate enough to tell time and could be used to determine the positions of planets for astrological purposes.

Returning to the European Decorative Arts, we find an unusual portable sundial that also served to calibrate sundials (Fig. 20). Made of brass and silver, this object was constructed sometime between 1690 and 1708. The inscription in French, "Costé appliqué au mur pur avoir la declinaison des plans," indicates its usefulness for determining the declination of the planets. In addition to marking time or determining longitude, this device included a recessed compass for precise orientation.

As clocks became ever more accurate and ubiquitous, sundials actually increased in importance as a means of setting and regulating them (Chandler and Vincent 1967). "Dialing"—or the art of constructing sundials—was a skill that often went hand in hand with the work of

FIGURE 18. *Crucifixion* (*Corpus Hypercubus*), Salvador Dalí, 1954. Oil on canvas, 76 1/2 × 48 3/4 in. Gift of the Chester Dale Collection, 1955. Accession no. 55.5. Currently on display in Gallery 913. Copyright © 2015 Salvador Dalí, Fundació Gala-Salvador Dalí/Artists Rights Society (ARS), New York. Image copyright © The Metropolitan Museum of Art. Image source: Art Resource, Inc., NY.

FIGURE 19. Left: Astrolabe of Prince ‘Umar ibn Yusuf. Yemen, 1291 CE. Edward C. Moore Collection. Bequest of Edward C. Moore, 1891. Accession no. 91.1.535a–h. Right: Planispheric astrolabe of Muhammad Zaman, Iran, Mashhad, 1654–1655 CE. Accession no. 63.166a–j. http://www.metmuseum.org/.

compass makers and clockmakers. On display at the Met next to the sundial shown in Figure 20 is a miniature sundial with Roman numerals (Accession no. 03.21.60).

THE GUBBIO STUDIOLO. We’ve already covered enough for one tour, you may be thinking. But don’t leave the Met without visiting the Gubbio Studiolo (little studio) of the Grand Duke Federico III da Montefeltro (1422–1482), sold to and reinstalled in the Met in 1939.

The grand duke was not only a successful mercenary general and papal gonfaloniere, he was a humanist admirer of the arts, literature, science, and mathematics. Entering his studiolo, you step centuries back in time (Fig. 21). Pale light suffuses the room from a window as on a crisp autumn morning, and the walls of the little study are inlaid with precisely cut pieces of wood fitted together to give the illusion of cabinets and bookshelves full of items wondrous to the eye and mind.

The Studiolo is a meticulously designed example of the spectacular three-dimensional effects that mathematical perspective has made

FIGURE 20. Combination portable sundial and instrument for calibrating sundials. Brass and silver. Top: As seen from the top down. Bottom: View looking horizontally with the magnetic compass in the rear left; on the right is the gnomon, which would have cast a shadow telling the time at a given latitude. Gift of Mrs. Stephen D. Tucker, 1903. Accession no. 03.21.17. http://www.metmuseum.org/.

FIGURE 21. The Gubbio Studiolo (little studio) of the Grand Duke Federico III da Montefeltro. Designed by the Sienese master Francesco di Giorgio Martini (1439–1501); executed ca. 1478–1482 in the Florentine workshop of Giuliano da Maiano (1432–1490) and Benedetto da Maiano (1442–1497). Walnut, beech, rosewood, oak, and other fruitwoods; purchased through the Rogers Fund, 1939. Accession no. 39.153. http://www.metmuseum.org/.

possible. One of the earliest technical treatises devoted to the subject, *De prospectiva pingendi* (*On Perspective for Painting*, ca. 1475), was written by one of the earliest masters of perspective techniques, Piero della Francesca (ca. 1415–1492). Piero was among the beneficiaries of Montefeltro patronage, and he dedicated *De prospectiva pingendi* to the duke. Another mathematician who received patronage from the Montefeltros was Fra Luca Pacioli (1447–1517), a page of whose famous *Summa de arithmetica, geometria. Proportioni et proportionalita* we saw in Figure 4. This book, too, is dedicated to Guidobaldo da Montefeltro.

On your visit to the Studiolo (with children, or not), see how many items you can find in the room's many panels that are related to mathematics. Some are obvious, such as the armillary sphere and quadrant, an

FIGURES 22 and 23. Details of Figure 21.

architect's plumb bob and carpenter's square, a compass, and a pair of dividers. Others are perhaps less so, such as the hourglass for measuring time, and the musical instruments that reflect the duke's interest in the mathematics of music as one of the liberal arts.[10] The largest musical instrument depicted in the Gubbio paneling is a portative organ, just to the right of the door, and the different lengths of the organ's pipes are clear reminders of the link between the measured proportions and musical harmonies the instrument produces when played.

Note too the checkerboard-patterned torus "on" the "table" in Figure 22! Although of mathematical interest, this is not a mathematical instrument. It's a Renaissance mazzocchio, a hat especially popular in Florence in the late fifteenth century.

Piero found the mazzocchio particularly well suited for didactic exercises in perspective drawing. The proper rendering of the mazzocchio using mathematical perspective is a notable exercise in his *De prospectiva pingendi*, where in Book 3 he uses it to explain the mathematical method of proportionally foreshortening column bases and capitals whereby they may be realized in terms of the principles of perspective (Raggio 1999). In the Gubbio Studiolo, the representation of the mazzocchio was both a test of the mathematical skill of the designer and a mark of the craftsman's ability to fool the eye and to trick

Figure 24. The Met's front entrance, April 2015. Photograph by Marjorie Senechal.

the mind. Before leaving the Studiolo, take one last look and imagine the Duke da Montefeltro here, in a moment of silence, perhaps pondering one of the mathematical manuscripts Piero had written and had dedicated to him. Two of the musical instruments provoke an odd question: why are some of the strings of the lute and harp broken? Not for lack of attention or maintenance; note that below the harp is a finely wrought tuning mechanism, to keep it in perfect mathematical tune. The duke would have been aware of the physical and philosophical link between the mathematics and music that is represented throughout the Studiolo, the link that Plato emphasized in the *Phaedo*, recounting the final hours of the life of Socrates, wherein the strings of the lyre represent harmony, invisible, incorporeal, divine. Yet the lyre is a physical object, and thus the strings are also material, visible, of this world. The broken lute strings, it has been suggested, may be a physical reference to the death of the duke and his wife, the Studiolo itself not having been completed until shortly after the duke's death. In any case, one cannot leave the Gubbio Studiolo without a sense of the peace and harmony of the mathematics that the duke himself must surely have felt here.

Where to Find It

Figure	Title	Accession Number	Gallery*
Modern and Contemporary Art			
1	*I Saw the Figure 5 in Gold*	49.59.1	902
15	*Homage to the Square: with Rays*	59.160	—
18	*Crucifixion (Corpus Hypercubus)*	55.5	913
19th and Early 20th Century European Paintings and Sculpture			
11	*Les Quatres Joueurs de Cartes*	61.101.1	825
Drawings and Prints			
2	*Melencolia I*	43.106.1	—
European Paintings, 1250–1800			
3	*Man Weighing Gold*	32.100.36	644
Art of the Arab Lands, Turkey, Iran, Central Asia, and Later South Asia			
19	Astrolabe of Prince ‘Umar ibn Yusuf and Planispheric astrolabe of Muhammad Zaman	91.1.535a–h, 63.166a–j	453
European Sculpture and Decorative Arts			
5	Relief mounted as a mirror frame	17.190.620	520
20	Combination Portable Sundial and Instrument for Calibrating Sundials	03.21.17	532
21–23	*The Gubbio Studiolo (little studio) of the Grand Duke Federico III da Montefeltro*	39.153 .	501
Asian Art			
6	Tile with Impressed Figure of Emaciated Ascetics and Couples Behind Balconies	1998.122	—

Where to Find It (*continued*)

Figure	Title	Accession Number	Gallery*
Photographs			
16	*Sugar Cubes: Design for Stehli Silk Corporation*	1987.1100.217	—
Egyptian Art			
7	Pectoral and Necklace of Sithathoryunet with the Name of Senwosret II	16.1.3	111
8	Dice from Oxyrhynchus	97.4.117.	138
11	Game of Hounds and Jackals	26.7.1287a–k	111
12	Double-sided game box with playing pieces and a pair of knucklebones	16.10.475a.	114
Greek and Roman Art			
9	Icosahedra inscribed with letters of the Greek alphabet	37.11.3	171
10	Eight astragaloi	1992.266.3	171
Medieval Art and Architecture			
14	Playing cards	1983.515.1–2	13**

*Verify all gallery numbers (search the Met's database at http://www.metmuseum.org by accession number).

- The works that were not on display when this table was printed are listed as gallery number "—"; check the database to see whether they've been returned to view.
- Some works may have been moved to another gallery, or retired temporarily to storage.

**This gallery is not in the Met on Fifth Avenue, but at The Met Cloisters, the Met's collection of medieval art and architecture. http://www.metmuseum.org/visit/visit-the-cloisters.

If You Go

The Met's address is 1000 Fifth Avenue (at 82nd Street), New York, NY 10028 (Fig. 24). The museum is open seven days a week. Check http://www.metmuseum.org for opening hours and other information.

Before you visit the Met, check the table "Where to Find It" to verify the gallery numbers provided there, as the artwork is moved from time to time. To do this, go to the Met's website (http://www.metmuseum.org) and search the database by accession numbers.

Map your route carefully; the museum is vast. You can download a map of the Met from the website or pick one up at the information desk in the museum's entrance hall. *Note:* The last stop on this tour, the Gubbio Studiolo of the Grand Duke Federico III da Montefeltro, is permanently installed in Gallery 501, but nevertheless it is not easy to find. Entering the museum from Fifth Avenue, continue straight ahead either to the right or to the left of the grand staircase leading to the European portrait galleries on the second floor. Instead of taking the staircase, follow the first floor corridors of medieval objects on either side until reaching the Spanish courtyard, whereupon, turning right, the main corridor leading to the hall of medieval armor takes you into a small vestibule, and immediately on entering this, turn right to find the doorway leading into the Studiolo.

Notes

1. You can still visit Dr. Mininberg's 2005 exhibition on ancient Egyptian medicine at http://www.metmuseum.org/en/exhibitions/listings/2005/medicine-in-ancient-egypt/.

2. See http://engineering.nyu.edu/events/2014/05/17/500-years-melancholia-mathematics.

3. http://www.wsj.com/articles/end-of-an-era-as-cme-to-close-almost-all-floor-trading-for-futures-1423100335.

4. The markings have nothing to do with the human figures, apparently; the Met's website suggests they may indicate where in the courtyard the tile was to be placed.

5. Oxyrhynchus is best known to mathematicians for the discovery there of fragments of the most ancient version of Euclid's *Elements* on papyrus, including diagrams, dating to about 100 CE, and for the vast trove of astronomical and astrological papyri recovered from the site during the past century or so, including texts, tables, ephemerides, almanacs, and horoscopes (Jones 1999).

6. Knucklebones were also used in games much like jacks.

7. "Presuming the 'Shen' sign . . . to be the goal, we find on either side twenty-nine holes, or including the goal, thirty aside. Among these holes, on either side, two are marked *nefer*, 'good'; and four others are linked together by curved lines [Fig. 14]. Assuming that the holes marked 'good' incur a gain, it would appear that the others, connected by lines, incur a loss. Taking this for granted, and that the play terminates at the goal 'Shen,' the game seems then to commence at the heart of the palm—the only place where five playing pieces aside could be placed without clashing with the obstacles (i.e., holes incurring gain or loss). . . . Now the moves themselves could easily have been denoted by the chance cast of knuckle-bones or dice, both being known to the ancient Egyptians at an early period; and if so we have before us a simple, but exciting, game of chance" (Carter and Carnarvon 1912).

8. Reproduction, including downloading of the three ARS-licensed works in this section, is prohibited by copyright laws and international conventions without the express written permission of Artists Rights Society (ARS), New York.

9. http://www.nytimes.com/2015/06/05/arts/design/review-robert-irwin-shows-a-calming-installation-at-diabeacon.html?ref=arts&_r=0.

10. The cittern with its nine strings, bridge, and pegbox "brings to mind the theory of music's harmonic proportions. During the 15th century the concept of musical harmony and the principles of linear perspective and architectural proportions came to be seen as expressions of the same mathematical truth" (Raggio 1999).

Works and Websites Cited

Howard Carter and the Earl of Carnarvon. *Five Years' Explorations at Thebes, A Record of Work Done 1907–1911*. London: Oxford University Press, 1912.

Irving Finkel. *Ancient Board Games in Perspective*. London: British Museum Press, 2007.

Wenzel Jamnitzer. *Perspectiva Corporum Regularium*. Facsimile reproduction. Graz, Austria: Akademische Druck u. Verlagsanstalt, 1973.

Mary Louise Krumrine. "Les 'Joueurs de cartes' de Cézanne: Un jeu de la vie," *Cézanne aujourd'hui*. Paris, 1997.

Karl Menninger. *Number Words and Number Symbols: A Cultural History of Numbers*. New York: Dover Publications, 1992.

Olga Raggio. *The Gubbio Studiolo and Its Conservation. I. Federico da Montefeltro's Palace at Gubbio and Its Studiolo*. New York: Metropolitan Museum of Art, 1999.

Harvard Semitic Museum. "Throwing Bones: The Game of 20 Squares," Video presentation created for the exhibition *Houses of Ancient Israel; Domestic, Royal, Divine*. Cambridge, MA: Harvard Semitic Museum, 2006: http://semiticmuseum.fas.harvard.edu/games.

http://electronics.howstuffworks.com/gadgets/clocks-watches/astrolabe. htm.

http://engineering.nyu.edu/events/2014/05/17/500-years-melancholia-mathematics.

http://archaeologicalmuseum.jhu.edu/the-collection/object-stories/archaeology-of-daily-life/childhood/knucklebones/.

Further Reading

Numerals

Keith Devlin. *The Man of Numbers: Fibonacci's Arithmetic Revolution*. New York: Walker & Co., 2011.

Seymour Block and Santiago Tavares. *Before Sudoku: the World of Magic Squares*. New York: Oxford University Press, 2009.

David Eugene Smith. *History of Mathematics*. Boston, MA: Ginn and Co., 1923–1925, vol. 2, "Finger Reckoning": 196–202.

Burma P. Williams and Richard S. Williams. "Finger Numbers in the Greco-Roman World and the Early Middle Ages," *Isis* 86 (4) (1995): 587–608.

Games

Nina Maria Athanassoglou-Kallmyer. *Cézanne and Provence: The Painter in His Culture*. Chicago: University of Chicago Press, 2003.

Bill Chen and Jerod Ankenman. *The Mathematics of Poker*. Pittsburgh: ConJelCo, 2006.

Persi Diaconis and Joseph B. Keller. "Fair Dice," *The American Mathematical Monthly*, 96 (4) (1989): 337–39.

Anne-Elizabeth Dunn-Vaturi. "Twenty Squares: An Ancient Board Game," Exhibition Blog: *Assyria to Iberia at the Dawn of the Classical Age*. Metropolitan Museum of Art, posted Tuesday, December 9, 2014.

http://www.metmuseum.org/blogs/listing?auth=Dunn-Vaturi%2c+Anne-Elizabeth&st=lauthor.

Alexander Jones. *Astronomical Papyri from Oxyrhynchus*. Vols. I and II. Philadelphia: American Philosophical Society, 1999.

Edward W. Packel. *The Mathematics of Games and Gambling*. Washington, DC: Mathematical Association of America, 1981; 2nd ed. 2006.

Edward Thorp. *The Mathematics of Gambling*. Secaucus, NJ: Lyle Stuart, 1985.

Dimension

Edwin Abbott. *Flatland: A Romance of Many Dimensions*. London: Seeley and Co., 1884. See in particular the annotated version by Ian Stewart: *The Annotated Flatland*. Toronto, Canada: Perseus Publishing, 2002.

Joseph O'Rourke. *How to Fold It: The Mathematics of Linkages, Origami, and Polyhedra*. Cambridge: Cambridge University Press, 2011.

Space and Time

Bruce Chandler and Clare Vincent. "A Sure Reckoning: Sundials of the 17th and 18th Centuries," *The Metropolitan Museum of Art Bulletin* 26 (4) (1967): 154–69.

Lennox-Boyd, Mark. *Sundials: History, Art, People, Science*. London: Frances Lincoln Ltd., 2006, esp. pp. 48–85.

Vincent, Clare. "Magnificent Timekeepers: An Exhibition of Northern European Clocks in New York Collections," *The Metropolitan Museum of Art Bulletin* 30 (4) (1972): 154–65.

The Studiolo

Luciano Cheles. *The Studiolo of Urbino: An Iconographic Investigation*. University Park: Pennsylvania State University Press, 1986.

Olga Raggio. "The Liberal Arts Studiolo from the Ducal Palace at Gubbio," *The Metropolitan Museum of Art Bulletin* 53 (4) (Spring, 1996): 5–35.

Pasquale Rotondi. *The Ducal Palace of Urbino: Its Architecture and Decoration*. London: Alex Tiranti, 1969.

Antoine M. Wilmering. *The Gubbio Studiolo and Its Conservation. II. Italian Renaissance Intarsia and the Conservation of the Gubbio Studiolo*. New York: Metropolitan Museum of Art, 1999.

Common Sense about the Common Core

ALAN H. SCHOENFELD

Is the Common Core the best thing since sliced bread or the work of the devil? Is it brand new or a rehash of old ideas? Is it anything more than a brand name, or is there substance? Can it work, given the implementation challenges in our political and school systems? Opinions about the Common Core are everywhere, but the op-eds I've seen are often short on facts and equally short on common sense. A mathematician by training, I've worked for almost forty years as an education researcher, curriculum materials developer, test developer, standards writer, and teacher. What follows is a Q&A based on that experience. I focus on the Common Core State Standards for Mathematics, known as CCSSM, but the issues apply to all standards (descriptions of what students should know and be able to do).

What's the CCSSM About?

Take a look for yourself—the Common Core documents are available at http://www.Corestandards.org/. If you read the first eight pages of CCSSM and then sample the rest, you'll get a good sense of what's intended. In brief, CCSSM focuses on two deeply intertwined aspects of mathematics: the content people need to know, and the knowhow that makes for its successful use, called mathematical practices. You can think of the content as a set of tools—the things you do mathematics with. The practices emphasize problem solving, reasoning mathematically, and applying mathematical knowledge to solve real-world problems. Without the practices, the tools in the content part of the CCSSM don't do much for you. It's like being taught to use a saw, hammer, screwdriver, lathe, and other woodworking tools without having any sense of what it means to make furniture.

At heart, the CCSSM are about thinking mathematically. Here are two visions of a third-grade class, both taken from real classrooms. In one, students are practicing addition and subtraction, getting help where needed to make sure they get the right answers. In another, the students have noticed that every time they add two odd numbers, the sum is even. A student asks, "Will it always be true?" Another says "but the odd numbers go on forever; we can't test them all." Pretty smart for a third grader! But later, a student notices that every odd number is made up of a bunch of pairs, with one left over. When you put two odd numbers together, you have all the pairs you had before, and the two leftovers make another pair—so the sum is even. And this will always be the case, no matter which odd numbers you start with. Now that's mathematical thinking—and it's what the core should be about. Of course, kids should do their sums correctly, and, they should be able to think with the mathematics.

It's important to understand what the Common Core is *not*. Most importantly, the Common Core is not a curriculum. CCSSM provides an outline of the mathematics that students should learn—an outline endorsed by forty-three states. Equally important, the Common Core does not prescribe a particular teaching style: effective teachers can have very different styles. To date—and despite what you read or hear—the desired reality of the Common Core has not made its way into even a small minority of U.S. classrooms. What happens in classrooms will depend on the curricula that are developed and adopted, on the high-stakes tests that shape instruction (for better or worse), on the capacity of teachers to create classrooms that really teach "to the Core," and on the coherence or incoherence of the whole effort.

What Do Powerful Classrooms Look Like?

CCSSM describes what kids should be able to do mathematically, including problem solving, producing and critiquing mathematical arguments, and more. Students won't get good at these things unless they have an opportunity to practice them in the classroom and get feedback on how they're doing. (Imagine a sports coach who lectured the team on how to play and then told the team to practice a lot before the big match. You wouldn't bet on that coach's success.) So, classrooms that produce students who are powerful mathematical thinkers must provide meaningful opportunities for students to *do* mathematics. Just as there are many successful (and different) coaches and coaching styles, there are many ways to run a successful classroom. At the same time, there's consistent evidence that classrooms that produce powerful mathematical thinkers have these five properties:[1]

- *High-quality content and practices.* Students have the opportunity to grapple with powerful ideas in meaningful ways, developing and refining skills, understandings, perseverance, and other productive "habits of mind" as they do.
- *Meaningful, carefully structured challenge.* Solving complex problems takes perseverance; students should neither be spoon-fed nor lost. In powerful classrooms, students are supported in "productive struggle," which helps them build their mathematical muscles.
- *Equitable opportunity.* We've all seen the classroom where the teacher moves things along by calling on the few kids who "get it," leaving the rest in the dust. It shouldn't be that way. In the kind of classroom that lives up to the standards, all students are productively engaged in the mathematics.
- *Students as sense makers.* In powerful classrooms students have the opportunity to "talk math," to exchange ideas, to work collaboratively, and to build on each other's ideas (just as in productive workplaces). In contrast to classrooms where students come to learn that they're not "math people," students in these classes come to see themselves as mathematical sense makers.
- *A focus on building and refining student thinking.* In powerful classrooms, the teachers know the mathematical terrain and how students come to understand that content so well that they can anticipate common difficulties, look for them, and challenge the students in ways that help them make progress, without simply spoon-feeding them.

We call this kind of powerful teaching "Teaching for Robust Understanding" (see http://ats.berkeley.edu/tools.html). Our goal should be to provide such learning experiences for all students. It's very hard to do this well—which is why the issue of supporting teachers' professional growth is crucially important. There are no quick fixes. We should be thinking in terms of consistent, gradual improvement.

What's New in the CCSSM?

The ideas behind CCSSM are not new. We've known for some time that students need a well-rounded diet of skills, conceptual understanding, and problem solving—rich mathematics content *and* the opportunities to develop strong mathematical practices.[2] The "standards movement" began in 1989, when the National Council of Teachers of Mathematics issued its *Curriculum and Evaluation Standards*. NCTM's (2000) *Principles and Standards for School Mathematics* represented an updating of the 1989 standards, based on what had been learned, and the fact that technology had changed so much over the 1990s. CCSSM can be seen as the next step in a progression.

So what's different? First, the organization is new. CCSSM offers grade-by-grade standards for grades K through 8, rather than the "grade band" standards of its predecessors. It represents a particular set of "trajectories" through subject matter, being very specific about what content should be addressed. Second and critically important, the Common Core has been adopted by the vast majority of states. Before the Common Core, each of the fifty states had its own standards and tests. Some of these were world class, with a focus on thinking mathematically; some were focused on low-level skills and rote memorization. Some states compared favorably with the best countries in the world, and some scored near the bottom of the international heap. Mathematics education across the United States was totally incoherent; where you lived determined whether you got a decent education or not. That's no way to prepare students across the United States for college and careers, or the nation's workforce for the challenges of the decades to come. And it's inequitable when your ZIP code determines whether or not you have access to a good education. If CCSSM are implemented with fidelity in the states that adopted them, we'll have something like nationwide consistency and opportunity instead of the crazy-quilt patchwork that we've had.

What's Wrong with CCSSM?

I can find lots of things to complain about—everyone can. Can you think of a class you took that was so perfect that you wouldn't change a thing? With fewer than a hundred pages to outline all of school mathematics, the authors made a series of choices. Those choices can be defended, but so could other choices. However, if schools and classrooms across the United States make strides toward implementing the vision of the Common Core described above, we'll make real progress.

What *is* wrong is our political system, and the fact that teachers and schools are not being provided adequate preparation and resources to implement the Common Core. This lack of support can destroy the vision, because real change is needed. Teaching the same old way, called "demonstrate and practice," just doesn't cut it. (How much of the math that you memorized in school do you remember and actually use as part of your tool kit?) The math we want kids to get their heads around is deeper and richer. Kids need to work hard to make sense of it; and to provide powerful learning environments, teachers need to learn how to support students in grappling with much more challenging mathematics. This isn't a matter of giving teachers a few days of "training" for teaching the Common Core; it's a matter of taking teaching seriously and providing teachers with the kinds of sustained help they need to be able to create classrooms that produce students who are powerful mathematical thinkers. The *real* reason some nations consistently score well on international tests (pick your favorite: Finland, Japan, Singapore, . . .) is that those nations take teaching seriously, providing ongoing support and professional development for teachers. When teachers have a deep understanding of the mathematics and are supported in building the kinds of rich classroom environments described above, the students who emerge from those classrooms are powerful mathematical thinkers.

What Do "Common Core Curricula" Look Like?

I could say, "Who knows?" It bears repeating that the Common Core is not a curriculum. What might be called Common Core curricula—widely accessible curricula intended to be consistent with the Common Core—don't really exist yet, although publishers are rushing to get them out. When those curricula do emerge, we'll have to see how faithful they are to the vision of problem solving, reasoning, and sense making described here.

One thing is for sure: the vast majority of materials currently labeled "Common Core" don't come close to that standard. Here's a case in point: A student recently brought home a homework assignment with "Common Core Mathematics" prominently stamped at the top of the page. The bottom of the page said, "Copyright 1998." That's more than a decade before the CCSSM were written. Remember when supermarkets plastered the word "natural" on everything, because it seemed to promise healthy food? That's what's being done today with phony "Common Core" labels. To find out whether something is consistent with the values of the Common Core, you have to look at it closely and ask, "Are kids being asked to use their brains? Are they learning solid mathematics, engaging in problem solving, asked to reason, using the math to model real-world problems? In short, are they learning to become mathematical sense makers?" If not, the "Common Core" label is just plain baloney.

Now, there *are* materials that support real mathematical engagement. For one set of such materials, look at the Mathematics Assessment Project's "Classroom Challenges," at http://map.mathshell.org/materials/index.php. However, such materials do not a curriculum make—and again, materials without support are not enough. What really counts is how the mathematics comes alive (or doesn't) in the classroom.

What about Testing?

Do you know the phrase "What you test is what you get"? When the stakes are high, teachers will—for their and their students' survival!—teach to the test. If the tests require thinking, problem solving, and reasoning, then teaching to the test can be a good thing. But if a high-stakes test doesn't reflect the kinds of mathematical thinking you want kids to learn, you're in for trouble. I worked on the specs for one of the big testing consortia, to some good effect—the exams will produce separate scores for content, reasoning, problem solving, and modeling—but I'm not very hopeful at this point. To really test for mathematical sense making, we need to offer extended "essay questions" that provide opportunities for students to grapple with complex mathematical situations, demonstrating what they know in the process. Unfortunately, it appears that test makers' desire for cheap, easy-to-grade, and legally bulletproof tests may undermine the best of intentions. It takes time to grade essay questions, and time is money. The two main tests being developed to align with the CCSSM[3] barely scratch the surface of what we can do. That's an issue of political will (read: it costs money and will shake people up), and the people footing the bill for the tests don't seem to have it.

The best use of testing is to reveal what individual students know, to help them learn more. That is, the most important consumers of high-quality tests should be teachers and students, who can learn from them. It *is* possible to build tests that are tied to standards and provide such information; there are plenty of examples at all grade levels. In addition, scores from such tests can be used to tell schools, districts, and states where they're doing well and where they need to get better. It's a misuse of testing when test scores are used primarily to penalize "underperforming" students and schools, rather than to help them improve. (Moreover, high-stakes testing leads to cheating. How many testing scandals do we need to make the point?) Finally, it's just plain immoral to penalize students when they fail to meet standards they were never prepared for. Holding students accountable for test scores without providing meaningful opportunities to learn is abusive.

What's Needed to Fix Things?

There's no shortage of "solutions." To mention one suggestion that's been bandied about, why not just adopt the curricular materials from high-performing countries? That would be nice, if it would work—but it won't. If conditions were the same in different countries—that is, if teachers here were provided the same levels of preparation, support, and ongoing opportunities for learning as in high-performing countries, then this approach could make sense. But the United States is not Singapore (or Finland, or Japan), and what works in those countries won't work in the United States until teachers in the United States are supported in the ways teachers in those countries are. Singaporean teachers are deeply versed in their curricula and have been prepared to get the most out of the problems in their texts. Japanese teachers are expected to take a decade to evolve into full-fledged professionals, and their work week contains regularly scheduled opportunities for continuous on-the-job training with experienced colleagues. Finnish teachers are carefully selected, have extensive preparation, and are given significant amounts of classroom autonomy.

In short, if importing good curricula would solve the problem, the problem would have been solved by now. It's been tried, and it failed. Of course, good curricular materials make life better—*if* they're in a context where they can be used well. The same is true of any quick fix you can think of, for example, the use of technology. Yes, the use of technology can make a big, positive difference—*if* it's used in thoughtful ways to enhance students' experience of the discipline. I started using computers for math instruction in 1981. With computers, you can gather and analyze real data instead of using the "cooked" data in a textbook; you can play with and analyze graphs, because the computer can produce graphs easily; and so on. But in those cases, the technology is being used in the service of mathematical reasoning and problem solving. You can get much deeper into the math if you use the technology well, but the presence of technology in the classroom doesn't guarantee anything. In particular, putting a curriculum on tablets is like putting a book on an e-reader: it may be lighter to carry, but it's the same words. The serious question is, how can the technology be used to deepen students' sense making, problem solving, and reasoning?

The best way to make effective use of technology is to make sure that the teachers who use it in their classrooms are well prepared to use it effectively. Fancy technology isn't going to make much of a difference in a world where half of the new teacher force each year will drop out within the next five years (within three years in urban school districts)—a world in which there are more teachers in their first year of teaching than at any other level of experience. In professions with a stable professional core, the number of newcomers is a much smaller percentage of the total population: there are more established professionals to mentor the newcomers, and a much smaller dropout rate. The best educational investment, as the highest performing nations make clear, is in the professionalization of teachers—so that they can make powerful instruction live in the classroom. In nations where teachers are given consistent growth opportunities, the teachers continue to develop over time. And they stay in the profession.

Living up to the vision of the Common Core requires focus and coherence. Curricula and technology need to be aligned with the vision and implemented in ways true to the spirit of sense making described here—including equitable access to the mathematics for all students. Administrators need to understand what counts, and support it. Testing needs to focus on providing useful information to teachers and students. Most important, we need to provide steady support for the teaching profession, so that teachers can make that vision live in their classrooms. We owe this to our kids.

Notes

1. The quickest path to documentation is through the website ats.berkeley.edu. The front page shows the big ideas; click on the "Tools" tab to see evidence about, and tools for, productive thinking.

2. There's a massive amount of research behind this statement. For one early summary, see Schoenfeld, A. H. (2002, January/February). Making mathematics work for all children: Issues of standards, testing, and equity. *Educational Researcher*, 31(1), 13–25.

3. See the websites of the Partnership for Assessment of Readiness for College and Careers, PARCC, at http://www.parcconline.org/, and the Smarter Balanced Assessment Consortium, SBAC, at http://www.smarterbalanced.org/.

Explaining Your Math: Unnecessary at Best, Encumbering at Worst

KATHARINE BEALS AND BARRY GARELICK

At a middle school in California, the state testing in math was underway via the Smarter Balanced Assessment Consortium (SBAC) exam. A girl pointed to the problem on the computer screen and asked, "What do I do?" The proctor read the instructions for the problem and told the student, "You need to explain how you got your answer."

The girl threw her arms up in frustration and said, "Why can't I just do the problem, enter the answer, and be done with it?"

The answer to her question comes down to what the education establishment believes "understanding" to be, and how to measure it. K–12 mathematics instruction involves equal parts procedural skills and understanding. What "understanding" in mathematics means, however, has long been a topic of debate. One distinction popular with today's math-reform advocates is between "knowing" and "doing." A student, reformers argue, might be able to "do" a problem (i.e., solve it mathematically) without understanding the concepts behind the problem-solving procedure. Perhaps he or she has simply memorized the method without understanding it and is performing the steps by "rote."

The Common Core math standards, adopted in forty-two states and the District of Columbia and reflected in Common Core-aligned tests like the SBAC and the Partnership for Assessment of Readiness for College and Careers (PARCC), take understanding to a whole new level.

"Students who lack understanding of a topic may rely on procedures too heavily," states the Common Core website. "But what does mathematical understanding look like?" And how can teachers assess it? "One way is to ask the student to justify, in a way that is appropriate to the student's mathematical maturity, why a particular mathematical statement is true, or where a mathematical rule comes from."

The underlying assumption here is that if a student understands something, he or she can explain it—and that deficient explanation signals deficient understanding. But this raises yet another question: What constitutes a satisfactory explanation?

While the Common Core leaves this unspecified, current practices are suggestive. Consider a problem that asks how many total pencils there are if five people have three pencils each. In the eyes of some educators, explaining why the answer is 15 by stating, simply, that $5 \times 3 = 15$ is not satisfactory. To show that they truly understand why 5×3 is 15, and why this computation provides the answer to the given word problem, students must do more. For example, they might draw a picture illustrating five groups of three pencils. (And in some instances, as was the case recently in a third-grade classroom, a student would be considered to not understand if he or she drew three groups of five pencils.)

Consider now a problem given in a pre-algebra course that involves percentages: "A coat has been reduced by 20 percent to sell for \$160. What was the original price of the coat?"

A student may show the solution as follows:

$$x = \text{original cost of coat in dollars}$$
$$100\% - 20\% = 80\%$$
$$0.8x = \$160$$
$$x = \$200$$

Clearly, the student knows the mathematical procedure necessary to solve the problem. In fact, for years students were told not to explain their answers, but to show their work, and if presented in a clear and organized manner, the math contained in this work was considered to be its own explanation. But the above demonstration might, through the prism of the Common Core standards, be considered an inadequate explanation. That is, inspired by what the standards say about understanding, one could ask, "Does the student know why the subtraction operation is done to obtain the 80 percent used in the equation or is he or she doing it as a mechanical procedure—i.e., without understanding?"

In a middle school observed by one of us, the school's goal was to increase student proficiency in solving math problems by requiring students to explain how they solved them. This was not required for all problems given; rather, they were expected to do this for two or three problems in class per week, which took up to 10 percent of total weekly class time. They were instructed on how to write explanations for their math solutions using a model called "Need, Know, Do." In the problem example given above, the "Need" would be "What was the original price of the coat?" The "Know" would be the information provided in the problem statement, here the price of the discounted coat and the discount rate. The "Do" is the process of solving the problem.

Students were instructed to use "flow maps" and diagrams to describe the thinking and steps used to solve the problem, after which they were to write a narrative summary of what was described in the flow maps and elsewhere. They were told that the "Do" (as well as the flow maps) explains what they did to solve the problem and that the narrative summary provides the why. Many students, though, had difficulty differentiating the "Do" section from the final narrative. But in order for their explanation to qualify as "high level," they couldn't simply state "100% − 20% = 80%"; they had to explain what that means. For example, they might say, "The discount rate subtracted from 100 percent gives the amount that I pay."

An example of a student's written explanation for this problem is shown in Figure 1:

A coat has been reduced by 20% to sell for $160. What was the original price of the coat?

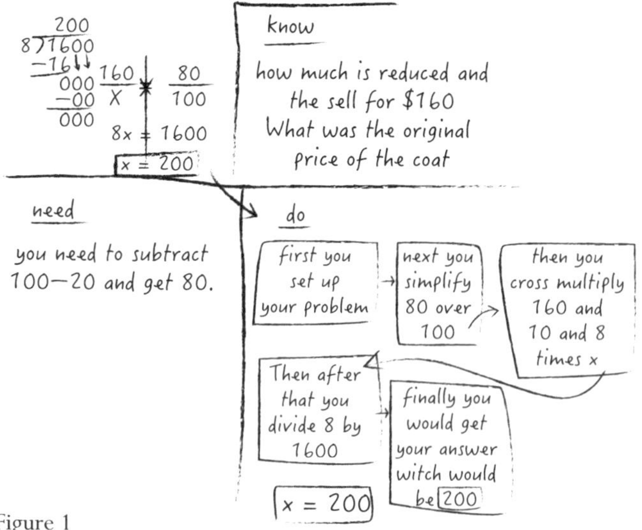

Figure 1

For problems at this level, the amount of work required for explanation turns a straightforward problem into a long managerial task that is concerned more with pedagogy than with content. While drawing diagrams or pictures may help some students learn how to solve problems, for others it is unnecessary and tedious. As the above example shows, the explanations may not offer the "why" of a particular procedure.

Under the rubric used at the middle school where this problem was given, explanations are ranked as "high," "middle," or "low." This particular explanation would probably fall in the "middle" category since it is unlikely that the statement "You need to subtract 100 − 20 to get 80" would be deemed a "purposeful, mathematically grounded written explanation."

The "Need" and "Know" steps in the above process are not new and were advocated by George Polya in the 1950s in his classic book *How to Solve It*. The "Need" and "Know" aspect of the explanatory technique at the middle school observed is a sensible one. But Polya's book was about solving problems, not explaining or justifying how they were done. At the middle school, problem solving and explanation were intertwined, in the belief that the process of explanation leads to the solving of the problem. This conflation of problem solving and explanation arises from a complex history of educational theories. One theory holds that being aware of one's thinking process—called "metacognition"—is part and parcel to problem solving. Other theories that feed the conflation predate the Common Core standards and originated during the Progressive era in the early part of the twentieth century when "conceptual understanding" began to be viewed as a path to, and thus more important than, procedural fluency.

Despite the goal of solving a problem and explaining it in one fell swoop, in many cases observed at the middle school, students solved the problem first and then added the explanation in the required format and rubric. It was not evident that the process of explanation enhanced problem-solving ability. In fact, in talking with students at the school, many of the students shared that they found the process tedious and said they would rather just "do the math" without having to write about it.

In general, there is no more evidence of "understanding" in the explained solution, even with pictures, than there would be in mathematical solutions presented in a clear and organized way. How do we know, for example, that a student isn't simply repeating an explanation provided by the teacher or the textbook, thus exhibiting mere "rote learning" rather than "true understanding" of a problem-solving procedure?

Math learning is a progression from concrete to abstract. The advantage to the abstract is that the various mathematical operations can be performed without the cumbersome attachments of concrete entities—entities like dollars, percentages, groupings of pencils. Once a particular word problem has been translated into a mathematical representation, the entirety of its mathematically relevant content is condensed onto abstract symbols, freeing working memory and unleashing the power of pure mathematics. That is, information and procedures that have been learned become automatic, and this learning frees up working memory. With working memory less burdened, the student can focus on solving the problem at hand. Thus, requiring explanations beyond the mathematics itself distracts and diverts students away from the convenience and power of abstraction. Mandatory demonstrations of "mathematical understanding," in other words, can impede the "doing" of actual mathematics.

Advocates for math reform are reluctant to accept that delays in understanding are normal and do not signal a failure of the teaching method. Students learn to do, they learn to apply what they've mastered, they learn to do more, they begin to see why, and eventually the light comes on. Furthermore, math reformers often fail to understand that conceptual understanding works in tandem with procedural fluency. Doing a procedure devoid of any understanding of what is being done is actually hard to accomplish with elementary math because the very learning of procedures is, itself, informative of meaning, and the repetitious use of them conveys understanding to the user.

Explaining the solution to a problem comes when students can draw on a strong foundation of content relevant to the topic currently being learned. As students find their feet and establish a larger repertoire of mastered knowledge and methods, they can become more articulate in explanations. Children in elementary and middle school who are asked to engage in critical thinking about abstract ideas, more often than not, respond emotionally and intuitively, not logically and with "understanding." It is as if the purveyors of these practices are saying, "If we can just get them to do things that look like what we imagine a mathematician does, then they will be real mathematicians." That may be behaviorally interesting, but it is not mathematical development, and it leaves them behind in the development of their fundamental skills.

The idea that students who do not demonstrate their strategies in words and pictures or by multiple methods don't understand the underlying concepts is particularly problematic for certain vulnerable types of students. Consider students whose verbal skills lag far behind their mathematical skills—non-native English speakers or students with specific language delays or language disorders, for example. These groups include children who can easily do math in their heads and solve complex problems, but often are unable to explain—whether orally or in written words—how they arrived at their answers.

Most exemplary are children on the autism spectrum. As the autism researcher Tony Attwood has observed, mathematics has special appeal to individuals with autism: It is, often, the school subject that best matches their cognitive strengths. Indeed, writing about Asperger's Syndrome (a high-functioning subtype of autism), Attwood in his 2007 book *The Complete Guide to Asperger's Syndrome* notes that "the personalities of some of the great mathematicians include many of the characteristics of Asperger's syndrome."

And yet, Attwood adds, many children on the autism spectrum, even those who are mathematically gifted, struggle when asked to explain their answers. "The child can provide the correct answer to a mathematical problem," he observes, "but not easily translate into speech the mental processes used to solve the problem." Back in 1944, Hans Asperger, the Austrian pediatrician who first studied the condition that now bears his name, famously cited one of his patients as saying, "I can't do this orally, only headily."

Writing from Australia decades later, a few years before the Common Core took hold in America, Attwood added that it can "mystify teachers and lead to problems with tests when the person with Asperger's syndrome is unable to explain his or her methods on the test or exam paper." Here in Common Core America, this inability has morphed into an unprecedented liability.

Is it really the case that the non-linguistically inclined student who progresses through math with correct but unexplained answers—from multidigit arithmetic through to multivariable calculus—doesn't understand the underlying math? Or that the mathematician with the

Asperger's personality, doing things headily but not orally, is advancing the frontiers of his or her field in a zombielike stupor?

Or is it possible that the ability to explain one's answers verbally, while sometimes a sufficient criterion for proving understanding, is not, in fact, a necessary one? And, to the extent that it isn't a necessary criterion, should verbal explanation be the way to gauge comprehension?

Measuring understanding, or learning in general, isn't easy. What testing does is measure "markers" or by-products of learning and understanding. Explaining answers is but one possible marker.

Another, quite simply, are the answers themselves. If a student can consistently solve a variety of problems, that student likely has some level of mathematical understanding. Teachers can assess this more deeply by looking at the solutions and any work shown and asking some spontaneous follow-up questions tailored to the child's verbal abilities. But it's far from clear whether a general requirement to accompany all solutions with verbal explanations provides a more accurate measurement of mathematical understanding than the answers themselves and any work the student has produced along the way. At best, verbal explanations beyond "showing the work" may be superfluous; at worst, they shortchange certain students and encumber the mathematics for everyone.

As Alfred North Whitehead famously put it about a century before the Common Core standards took hold:

> It is a profoundly erroneous truism . . . that we should cultivate the habit of thinking of what we are doing. The precise opposite is the case. Civilization advances by extending the number of important operations which we can perform without thinking about them.

Teaching Applied Mathematics

David Acheson, Peter R. Turner,
Gilbert Strang, and Rachel Levy

How can we inspire the next generation of students about applied mathematics? The four contributors to this group of articles, who have all thought deeply about this question, were asked to give their personal views.

I. David Acheson: What's the Big Picture?

Let *A* and *B* be two teachers of applied mathematics (at any level) and suppose that, generally speaking, *A* is a much better teacher than *B*.

Why is *A*'s teaching so much better? Even without any further information, can we at least hazard a guess?

I wonder, for instance, if you might be prepared to bet that *A* is more trained in "communication skills"? Or perhaps *A* knows more mathematics than *B* or is nearer to the cutting edge of research? Then again, maybe *A* just has a more lively personality?

All these things can be advantageous, of course, but I would not actually bet on any of them.

In fact, in the absence of any further information, there is only one thing that I would be prepared to bet good money on. I would be prepared to bet that *A*'s teaching is so much better—so inspirational, at best—mainly because *A wants* it to be that way, for reasons that we will probably never learn and that *A* may not even know.

This is only an opinion, of course, but it comes from thinking back to my own inspirational teachers when I was young. Some were notable for their scholarship, some for their eccentricity, but—so far as I can see—they only really had one thing in common: they had a great story to tell, *and they really wanted to tell it.*

"Removing Some of the Rubbish . . ."

It is simple common sense with applied mathematics teaching—and possibly with mathematics teaching of any kind—to start with the basics and work up. In other words, "Don't try to run before you can walk."

But I believe it is a terrible mistake not to also bear in mind a very different piece of advice: namely, "If you have no idea where you are going, do not be too surprised if you never get there."

This is, I suspect, what the author John Ward meant a long time ago, in his *Plain and Easie Introduction to the Mathematicks* (1729), when he wrote:

> Tis Honour enough for me to be accounted as one of the under Labourers in Clearing the Ground a little, and Removing some of the Rubbish that lies in the way to Knowledge.

In any event, I believe that a major difficulty with mathematics, at all levels, is that people can easily get bogged down in things of little consequence instead of engaging with things that really matter.

And what could help them, more than anything else, is some kind of "big picture."

My own big picture of mathematics starts with *wonderful theorems*, by which I mean major results, usually with considerable generality and often an element of surprise. Secondly, *beautiful proofs*, i.e., concise deductive arguments, possibly containing a truly "light bulb" moment when all suddenly becomes clear. And finally, *great applications*, particularly to physics, and hence to our understanding of how the world really works.

I would argue, in fact, that mathematics is at its very best when you get *all three things at once.* That, in my view, is when you should really open the champagne.

More controversially, perhaps, I believe that we can, and should, offer some such big picture to virtually anyone, including very young children and the rest of the general public.

Nonetheless, the majority of my teaching experience has been with university students, and that is where I would like to turn next.

Lectures and Classes

One way of bringing a student lecture to life is through a picture or video, but best of all, perhaps, is a live experiment, and my own subject, fluid dynamics, lends itself particularly well to this.

As an example, take two glass plates and put a blob of dishwashing liquid on one of them. (I dye the blob red, with food coloring, for dramatic effect.) Now press down with the other plate, so that the narrowing gap causes the blob to spread out in a nice, symmetric fashion, with a more or less circular boundary.

But if we now gradually pull the plates apart again, the *reverse* motion is hopelessly unstable; tiny ripples appear in the boundary, for no apparent reason, and grow rapidly into long viscous fingers (Figure 1).

If performed on an overhead projector or visualizer, this experiment can often make an audience gasp with astonishment.

But my real point here is a little more subtle. For why do a demonstration like this only in an advanced course on fluid mechanics, along with all the associated theory? Why not stimulate interest by first showing it much earlier, perhaps even in an elementary course on particle dynamics, as soon as the whole idea of stability and instability first arises?

Another way of helping people see the "big picture" is through the history of the subject, provided that the history in question has some real scholarship and depth to it.

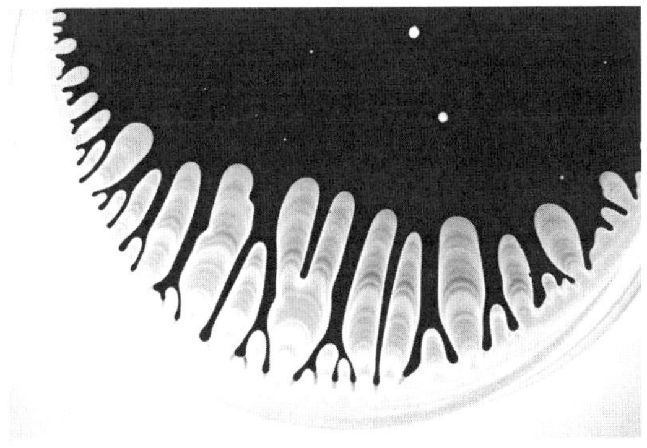

Figure 1. Viscous fingering.

In a first course on particle dynamics, for example, my experience is that students find it genuinely interesting to actually see, with their own eyes, that their textbook treatment of planetary motion is spectacularly different from the one in Newton's *Principia* and that it was not until about sixty years later, in the subsequent works of Euler and others, that dynamics came to be done in more or less the way it is done today.

To take a more lightweight example, my own research in fluid mechanics once gave a new twist to a hundred-year-old problem in vortex motion, first studied by Augustus Love in 1894. And whenever I present this (as a short diversion) in student lectures, I am convinced that it is enlivened by snippets from Love's original paper, to say nothing of an early photograph of Love and his striking Victorian moustache (which was apparently much admired at the time).

But there is another, possibly more unusual, way in which it is possible to bring student lectures to life.

Imagine, if you will, that you have just arrived at what you perceive to be the high point of the lecture, where the next line in the mathematical argument is very clever or inventive in some way. (I would even include here taking the curl of the momentum equation in fluid mechanics, to eliminate the pressure.)

For many years now, whenever this happens I tend to ask the audience whether they have any idea what the next, clever step might be.

Now, conventional wisdom is, I think, that if you can do this sort of thing at all, you can do it only with very small audiences. In my experience, however, even with audiences of 200 or more, once they realize that no one can possibly be *expected* to know the answer and that they are being invited—just for a moment—to more or less put themselves in the shoes of some genius like Newton or Euler, the suggestions will start coming *if you hold your nerve.*

Like so many things of this kind, it all depends on just how much you want to do it.

Books

A well-known publisher once said:

> Everybody has a book inside them. And it should usually stay there.

However true that may be, it could be argued that the sheer impact and reach of a sufficiently original book can completely dwarf what its author might ever hope to achieve through direct, face-to-face teaching, and I am a great optimist about the future of books in the teaching of applied mathematics.

And while, as far as I can tell, it takes considerable imagination and skill to write either an outstanding textbook or a successful popular mathematics book, I have long wondered if a real breakthrough in the future may instead come from some thoroughly original approach that combines the best elements of both.

Public Engagement

One of the most striking developments in recent years has been the rapid increase in popular mathematics lectures for either school students or the wider public.

I count myself fortunate to have been involved in several mathematical shows of this kind, mainly for teenagers. They are often held in mainstream city center theaters, with all the paraphernalia of stage lighting, sound technicians, etc., and the pressure to be entertaining as well as informative is therefore intense. So, to illustrate applied mathematics, I often use the formula for the frequency of a vibrating string, thereby smuggling in a practical demonstration of harmonics (and a self-composed tune!) on my electric guitar.

But so-called community lectures (which are usually held in the evening) can be even more rewarding because the age range at them can be enormous: from grandparents to very young children indeed. All you can really assume on these occasions is that each family group includes at least one person who is good at sums.

It was at one of these events, at a school in North London, that I was midway through a "proof by pizza" (for the sum of an infinite series) when I happened to notice a particular little boy, age about ten, in the audience. A split-second after delivering the punch line of my proof—at the moment when a deep idea suddenly becomes almost obvious—I practically saw the "light bulb" go on in his head, and he got so excited that he fell off his chair.

And, in a sense, that fleeting moment says it all.

For mathematics at its best, at any level, lifts the human spirit, by showing us that the world—whether the world of the mind or the actual physical world in which we live—is an even more weird and wonderful place than we thought.

II. Peter R. Turner: Computation, Modeling, and Projects

Introduction

This article presents a personal philosophy for teaching applied, and particularly computational, mathematics at the undergraduate level. It is largely drawn from my own experience over more than forty years, mostly at three institutions in the United Kingdom and the United States.

That experience has been enhanced by my activities on the Society for Industrial and Applied Mathematics (SIAM) Education Committee, including four years as vice president for education. During this time, I have gained awareness of broader aspects of the role of applied mathematics education at the undergraduate level. It is important here to note that I am not a mathematics education specialist but a mathematician who is interested in education.

Important among these broader aspects was the February 2012 report from the President's Council of Advisors on Science and Technology entitled *Engage to Excel: Producing One Million Additional College Graduates with Degrees in Science, Technology, Engineering, and Mathematics*, which emphasized the role of a good and relevant applied mathematics education within the framework of STEM (science, technology, engineering, and mathematics) education. The national emphasis in the United States on STEM has been a hallmark of recent education policy development.

One of the key points raised was the "math gap." This is a term used to highlight the difficulty in transition from high school to undergraduate study in STEM disciplines—a problem that is exacerbated by what the colleges perceive as a lack of mathematical preparation in high school. There is a gap between colleges' expectations and reality. The fundamental thesis is that this gap can be addressed through a stronger founding in mathematical modeling of real-world situations, and the solution, analysis, and validation of these models using computational and theoretical applied mathematics.

The issue of college mathematics, or broader STEM, readiness has been an area of interest for many people—some at a local and highly detailed level and others at a broader big-picture level, such as the studies carried out by the Mathematical Association of America. The rest of this article addresses a few ideas about how applied and computational mathematics might improve the situation. My basic thesis is that the use of projects that in turn require some modeling and (computational) problem solving enhances almost all (not just applied) mathematics teaching *in all mathematics classes.* For example, calculus can be taught with applied projects replacing endless drills once basic skills are acquired.

Use of Projects in Numerical Methods Classes

Well over a decade ago, the basic structure of my undergraduate numerical methods/scientific computing courses changed to being entirely based on projects. The topical syllabus remained essentially unaltered, covering the fundamentals of nonlinear equations, linear systems, polynomial and spline interpolation, quadrature and numerical solution of ordinary differential equations, with each major theme being approached through an extended project.

The scheme was modified to incorporate more homework assignments to avoid the issue of procrastination. The homework assignments included some of the theoretical background and some of the preliminary steps in addressing the projects. The motivation for the changes was an (oversimplistic but illustrative) model that can explain why good students often found their first computational course difficult. To a faculty member, the class had the beauty of bringing together much of the students' prior experience in calculus, linear algebra, differential equations, and perhaps modeling, too—together with drawing on their programming skills, or even learning some algorithmic programming for the first time.

This same set of properties was the primary source of difficulty for the students. Suddenly, and for the first time, students were required to synthesize methods and solutions from multiple courses. Furthermore, their mathematical and programming experience had typically been totally disjoint up to that point. Thus we had an audience who may have been "good B students" in both their mathematical and programming ability, but simplistically multiplying these independent

0.85 probabilities (the middle of the common B grade range), we had a success rate of only 61%. In other words, these good students were struggling to get a D in numerical methods.

More importantly, the effect on students' attitudes to the course was affected by a failure to see the forest (or the wood) for the trees. The perception was of a required course they had to endure rather than an exciting culmination of all that had gone before. The use of projects was broadly successful in countering this.

The particular choice of projects is one that is important to the success of the course but also one that can be tailored to the individual instructor and audience. The initial set of topics I chose (and which have been modified successively over the years, both by me and by others who have taken over the immediate teaching responsibility) is described briefly here and illustrates the linkage to the main syllabus topics.

The Length of a Telephone Cable

A cable above even ground, and with physical parameters in the model simplified to a specified sag, introduces iterative solution of a single nonlinear equation. The full project referred to a multiloop cable above undulating ground with a profile determined by geographic data, connected with a simple cubic spline. The problem for each loop is then a nonlinear system of two equations, and the solutions over different pieces have to be matched to ensure continuity of the cable.

Rats in a Maze

Based on simple psychology experiments on rats' learning abilities, this was an open-ended project that introduced iterative solution of linear systems. Even a simple rectangular, say 6×5, maze results in a 30×30 linear system. This is an eye-opener for many students, who rarely see systems much larger than 3×3 in introductory linear algebra. The basic problem is to compute the probability of a rat successfully finding food at some set of exits of the maze from an arbitrary starting point. These results provide the baseline against which to measure the rats' success at learning the maze by comparing actual performance with the simulated random decisions.

Students are then required to modify the maze in ways of their choosing: adding diagonal passages, removing certain links, adding a second level, and modifying the decision model from purely random to some bias (perhaps to go straight ahead) are all variations that students came up with and solved. One even tried to apply some artificial intelligence to simulate the rat learning.

Reproduce a Picture

Although the concept of splines had been mentioned in the telephone cable problem, this project was the real introduction to interpolation. The objective was simply to reproduce a chosen picture or line drawing using interpolation. Polynomial interpolation was explored and usually quickly discarded for all but the simplest of shapes. Splines and other functions were introduced. A more modern treatment would probably extend this to using subdivision surfaces.

The Gamma Function

When I started this project-based course, most students were concurrently enrolled in an applied statistics course—hence the choice of computing the gamma function as the quadrature-based project. One benefit is that this necessarily requires modification of conventional quadrature routines to handle both singularities and an infinite range of integration. The basic idea was simply to find appropriate bounds for the infinite tail and a region close to the singularity, and then to compute the major contribution from the resulting bounded integral. Using the recurrence to reduce the need to compute for all values of the argument α improved computational efficiency but necessitated introducing careful, though fairly simple, error analysis to control the required accuracy.

Human Cannonball

A shooting problem for a projectile with nonlinear air resistance was the vehicle for the introduction of numerical solution of differential equations. The setting was finding the appropriate launch angle in order to hit a specified target (described as an escape window to escape from the course).

The particular list of projects above is certainly not intended to be prescriptive. Many improvements have been made, while other changes have, of course, proved less successful! The list here is only intended to illustrate the feasibility of such an approach and advance my thesis that project-based learning can significantly enhance the success of introductory scientific computing courses.

Modeling across the Curriculum

The emphases on projects, computation, and modeling have combined more recently in a general "modeling across the curriculum" philosophy. This started to take shape in the work of a SIAM Education Committee working group, which led to a 2011 *SIAM Review* paper on undergraduate computational science and engineering (CSE) programs. The report emphasized that curriculum design needs to fit local conditions, but it also stressed student experiences such as internships. Other key points in the paper concerned the role that undergraduate CSE education plays in regard to both industrial expectations for graduates and feeding the educational pipeline.

Undergraduate CSE programs can take many forms. A few are full-fledged undergraduate majors in Computational *X*, where *X* could be physics, biology, or finance, for example. More common is some form of minor that accompanies an undergraduate major in either (applied) mathematics or some field of science or engineering. The latter model seems better suited to ensuring some depth in a core discipline while maintaining the breadth that such a minor introduces to the program.

Undergraduate education in applied and computational mathematics feeds the K-12 (preschool through completion of high school) education system, industrial appointments, and, of course, graduate schools in all areas of science and engineering as well as in applied mathematics itself.

One of the main obstacles here is that teacher education in mathematics in the United States is often very light in applied content (including statistics), and good programs of continuing education and professional development for teachers are therefore a necessary precursor to any real change. With colleagues, I have been involved in a very successful design-based summer activity for middle and high school students, including some professional development for their teachers. Students, and their teachers, are introduced to the mathematics and physics of designing a physical roller coaster and to modeling software to simulate their design. This work has been successful in bringing appropriately adapted real-world applications and relevance to mathematics education. An increasing proportion of these students, mostly from economically disadvantaged backgrounds or other underrepresented groups, have subsequently entered STEM college programs, demonstrating the benefit of exposure to such applied content. The benefit is realized not just in their mathematics but also in the science that accompanies it.

Final Thoughts

The main point I am making is that students learn better when they perceive their studies as being relevant to their lives and future careers. In the case of mathematics, this provides a strong motivation to increase the applied and computational content at all stages of a student's development.

Early emphasis on problem solving leads to more advanced projects and full-scale modeling experiences as the students' abilities and background knowledge develop. This essay addresses some of those issues at a "big-picture" level rather than in detail because the details have to be right for the combination of institutional philosophy, instructor, and students where they are to be applied.

Understanding applied mathematics is inherently difficult because of the combined demands of the theoretical basis, the modeling and understanding of the application field, and the computational abilities that are needed to solve the problems. Determining that a "solution" really addresses the original issue, and if necessary refining it and solving again, are important aspects that only add to the inherent difficulty of the subject.

In my opinion, these difficulties oblige the educational community to address them throughout the curriculum. For example, I believe that the call for more emphasis on modeling and applications in the K-12 curriculum needs to be heard and that this move should be implemented soon. This must continue into the undergraduate program to help to address the "math gap" identified in the President's Council of Advisors on Science and Technology report. In summary, "early and often" is perhaps not sufficient to achieve the desired improvements. I advocate, instead, "early and always" for modeling and applications throughout the STEM curriculum.

Further Reading

President's Council of Advisors on Science and Technology. 2012. "Engage to Excel: Producing One Million Additional College Graduates with Degrees in Science, Technology, Engineering, and Mathematics." Office of Science and Technology Policy Report (February). Available at www.whitehouse.gov/sites/default/files/microsites/ostp/pcast-engage-to-excel-final_2–25–12.pdf.

Schalk, P. D., D. P. Wick, P. R. Turner, and M. W. Ramsdell. 2009. "IMPACT: Integrated mathematics and physics assessment for college transition." In Frontiers in Education Conference, San Antonio, TX, October 2009. IEEE Press.

———. 2011. "Predictive assessment of student performance for early strategic guidance." In Frontiers in Education Conference, Rapid City, SD, October 2011. IEEE Press.

SIAM Working Group on CSE Undergraduate Education. 2011. "Undergraduate computational science and engineering education." *SIAM Review* 53:561–74.

Turner, P. R. 2001. "Teaching scientific computing through projects." *Journal of Engineering Education* 90:79–83.

———. 2008. "A predictor-corrector process with refinement for first-year calculus transition support." *Primus* 18:370–93.

Turner, P. R., and K. R. Fowler. 2010. "A holistic approach to applied mathematics education for middle and high schools." In Proceedings of the Education Interface between Mathematics and Industry Conference, EIMI 2010, Lisbon, Portugal, pp. 521–31. International Commission on Mathematical Instruction/International Council of Industrial and Applied Mathematics.

III. Gilbert Strang: What to Teach and How?

In expressing these brief but strongly held thoughts about teaching, I would like to distinguish between two separate questions.

- What should we teach?
- How should we teach it?

What to Teach?

In my experience, a good decision on what to teach tells students that you care about them: you are thinking about them and their needs. This *sincerity of effort*—what you are contributing and what you want for them—is the most important message you could send to students. We all respond favorably to sincerity.

The teacher may think that he or she has no freedom. In calculus that may be partly true (and partly untrue). In applied mathematics, the syllabus is seldom so rigid. Our subject is extremely large! There is no hope of presenting or comprehending all the important directions, so there is an opening for independent thought. And a teacher who thinks independently will pass on an all-important message to students: that they can begin to work by themselves and think for themselves.

I will compare and contrast my thoughts about linear algebra and about computational science. Linear algebra in 1970 was in a very abstract and unsatisfactory state as a basic undergraduate course. Its enormous importance in practice was quite unappreciated by many of the standard textbooks. *Finite-Dimensional Vector Spaces* by Paul Halmos, for example, was written carefully and concisely, but it was written for mathematics students—stronger ones or weaker ones—and not for the much larger number who needed to *use* the subject.

Outside the classroom, every year brought new applications of linear algebra. Matrices became part of the core language of science and engineering and economics. To solve differential equations or to study large networks, linearity is the first step. At the same time, a flood of data began to arrive from much better sensors, and it often came in matrix form. The challenge was (and still is) to interpret those data and extract what is useful.

Along with these big external changes came, inevitably, a new set of options for our teaching. Examples became more important. Instead of inventing subspaces to study, the subspaces came from the matrices themselves: the null space and the column space of A and A^{T}. The ideas of basis and dimension and orthogonality apply to those concrete subspaces. The abstraction of a linear transformation need not come first!

Instead of starting with the abstract case, understanding emerged from the examples themselves. This is how most mathematicians think and learn. Why should the minds of our students not work in the same way?

There is certainly a danger that the new approach could also become too rigid. There seems to be general acceptance of an overall syllabus by textbook authors and textbook choosers. Algebraic ideas like independence, basis, and dimension combine with algorithms like LU and Gram-Schmidt. Eigenvalues are introduced for a purpose: to decouple the variables and make the matrix diagonal. Computing A and e^{At} then becomes a one-dimensional problem. Factorizations of A express the essential facts: $A = LU$, $A = QR$, $A = S\Lambda S^{-1}$, $A = U\Sigma V^{\mathrm{T}}$. Matrices that are important in practice (reflections, rotations, differences) provide genuine examples.

This subject is still partly driven by what happens outside the classroom. Further changes in the syllabus will come. In all of this renaissance, I want to emphasize that the beautiful ideas of this subject are not suppressed! The very opposite, in fact; they become better understood and appreciated.

Now for the comparison and contrast with applied mathematics and computational science as a whole. This is a vast subject, and our classroom time is quite limited. I do not see rapid convergence to one fully developed core curriculum. My own course did converge over a twenty-five-year period to focus on key ideas, and those ideas went into a textbook. But more examples keep emerging, and new codes. Courses on applied and engineering mathematics are still in a (healthy) state of flux.

How to Teach?

The heading here is a question and not a statement! Teaching is far too difficult, and success is too uncertain and ill-conditioned, to give an algorithm for success. I will give suggestions below, not rules.

A key point: the subject is to be uncovered, not covered.

It is natural to prepare for a class by deciding on a plan. Start with a question that it is important to answer. What is the inverse of a matrix and which matrices have inverses?

The requirements $A^{-1}A = I$ and $AA^{-1} = I$ are straightforward. But those are only letters! Examples are needed right away. Write down

$$\begin{bmatrix} 0 & 1 \\ 1 & 0 \end{bmatrix}, \quad \begin{bmatrix} 1 & 1 \\ 1 & 1 \end{bmatrix}, \quad \begin{bmatrix} 1 & 1 \\ 0 & 1 \end{bmatrix}, \quad \begin{bmatrix} a & b \\ c & d \end{bmatrix}.$$

Invert the first and third of these and describe A^{-1} in words. Show that the second matrix is not invertible. (The best way to do this is to see a specific vector x in the null space.) The fourth case above can reduce to a test for parallel rows or parallel columns. Determinants can be mentioned, as can nonzero pivots. Multiple ways of recognizing invertibility or its opposite are highly valuable.

The important point is that in working with simple examples you are giving the students a chance. The key is to build their confidence as active users of mathematics. The teacher has to be saying, in many ways but not in so many words, *you can do it*.

In an applied mathematics course, a correspondingly simple question might be the following.

What are the solutions to $d^2y/dx^2 = \lambda y(x)$?

Here we are looking for eigenfunctions. The matrix in $Ay = \lambda y$ is replaced by the second derivative. No boundary conditions have been imposed so far.

Now add the conditions $y(0) = 0$ and $y(1) = 0$. That should leave only the eigenfunctions $y = \sin k\pi x$ with their eigenvalues $\lambda = -k^2\pi^2$. The goal is to make the idea of an "eigenfunction" familiar. The answer was already known, and it illuminates the question.

I would like to emphasize the importance of *the teacher's voice*. All of us watch for verbal signals from a speaker: "This is exciting." "This is very ordinary." "Pay attention to this!" Boredom or enthusiasm comes through so clearly. We are virtually announcing low expectations or raised expectations. If we are not interested ourselves, that message overrides our words. And fortunately, if our own curiosity is aroused about where a particular example leads, students understand what applied mathematics mostly is: *following an example to the end*.

Instead of consulting published references, I recommend a severe critique of lecture videos. You can find the author's own courses on the OpenCourseWare site (ocw.mit.edu): 18.06 (linear algebra) and 18.085 (computational science). The version 18.06SC of the former involves brief lectures on problem solving by six teaching assistants. What is it that makes each of them succeed or fail?

If only we knew more about teaching, we could define success.

IV. Rachel Levy: Industrial Mathematics Inspires Mathematical Modeling Tasks

Motivation

A common refrain from high school mathematics students goes something like this: "We have to learn this stuff for twelve years! Nobody ever even uses it!" As a mathematics professor and former middle and high school mathematics teacher, I am sadly unsurprised. Much of the mathematics written in textbooks, even when it is framed as practical, can strike students as overly academic and divorced from reality. Real problems require skills that most students never have the opportunity to practice. In this article, I outline a set of skills that we can incorporate into assignments—initially one at a time, and eventually all together—to prepare students for the types of problems they will face in the real world, especially in science, technology, engineering, and mathematics fields. While I have compiled the list below with undergraduate students in mind, students would ideally have opportunities to practice simple versions of these skills throughout their mathematics education.

I used to think modeling meant story problems. And why not? After all, story problems come from real situations. But they do not ring true because they are too neat. Traditional story problems provide the task, the methodology, and the exact information needed to complete the task. Real problems are messy. Someone actually cares whether we solve them or not. We do not usually know how to solve them a priori. To prepare students to solve ill-defined, messy problems, we need to increase the cognitive demand we place on students by incorporating genuine and engaging modeling problems into our curricula. By cognitive demand, I mean the complexity of the tasks and the degree of decision making required.

What types of mathematical modeling tasks can we provide? Of course, as modelers we might start with a model of modeling itself. Modeling can be viewed as an iterative process in which assumptions are made and then questioned, models are proposed and then refined, and data from real situations help to test the validity of the model through multiple versions of the solution. Models generally balance simplicity with realism, just as computations must often balance efficiency with accuracy. Every step of the modeling process may not be necessary and the process may not always proceed in the same order, but the spirit is one of iteration and balance.

With this iterative process in mind, we can begin to design tasks that enable students to practice modeling in meaningful ways. My ideas about mathematical modeling stem from my experience in modeling camps, British-style industrial mathematics study groups, and the Harvey Mudd College Mathematics Clinic. In these intensive experiences, groups of faculty and students gather to solve problems posed by companies, governmental organizations, and sometimes individuals. Rather than create a hierarchical taxonomy, I will suggest elements that can be incorporated into modeling tasks to increase their cognitive demand and make problems richer and more realistic.

The Modeling Tasks

The order of the tasks below roughly follows one iteration of the modeling process. However, individual elements can be incorporated into assignments, to provide practice with various challenges from real-world mathematics.

Tackle realistic problems. Problems that can be described as "real," of course, are ones that have not yet been solved. The task of identifying interesting problems with an appropriate level of challenge that are still tractable could be an interesting assignment for students, or it could fall to the faculty. Many businesses have problems that they need to solve or processes they would like to better understand.

Interact with a vested party. Ideally, the modeling problem will have been posed by a party who cares about the outcome. When students interact with the problem sponsors (e.g., from a company) and end users of a project, they can practice professional communication and negotiation. If the students communicate with the sponsor on a regular basis, the solution will be more likely to satisfy the sponsor. The sponsor can provide realistic constraints—such as time, money, hardware, and software limitations—as well as information about company/sponsor culture and policy. Communication with the sponsor can motivate the students, since the problem they must solve matters to someone outside the classroom.

Define the problem statement. When a problem is first proposed in writing, the ideas put down on paper may not reflect exactly what the sponsor wants. Sometimes the sponsor will have omitted critical information or constraints that necessitate a specific approach. The suggested method of solution may not be the best approach for the proposed problem. After researching background on the problem, students can propose reasonable goals and their preferred approach. This step often involves negotiation among the students about how to proceed as well as with the sponsor about the approach and final deliverables.

Disregard extraneous information. Modeling projects often begin with a literature search, in which students seek relevant approaches, data, or parameter values. The variety and scope of digitally available information can both facilitate and overwhelm problem-solving efforts. Sophisticated search strategies, such as those taught by librarians, can help students navigate the information overload. Students need to be given an opportunity to decide which of the available information about a problem should be used to develop their solution. Few textbook problems contain extraneous information, whereas with a real problem any available information (data, mathematical techniques, approaches) can be brought to bear. Students must determine which ideas and information are most salient to the problem at hand.

Cope with messy data. Textbooks (as well as many models) assume that data are normally distributed or that they follow some other distinguishable distribution, pattern, or trend. Real data often do not fall into a nice category and furthermore contain outliers, faulty entries, and missing values that need to be identified. In addition, the governing principles underlying the data may not be known.

Define and justify assumptions. Models simplify reality via assumptions. Even when students are presented with a model, they can discern and question the model's assumptions. They can also predict what might happen if the various assumptions are relaxed. When students do have the opportunity to make modeling assumptions, they can learn to identify ways to simplify problems and to balance generality with specificity.

Choose an approach. When we present mathematical models in lectures, we often have a particular approach in mind and actively steer the discussion in that direction. For example, when I introduce the mass-spring system in a differential equations class, I know which equation I want the students to use to model the system. Ideally, students will tackle modeling problems for which a variety of approaches could succeed. When that is not possible, students who debate the motivations for a particular approach can begin to see modeling as a set of choices based on underlying principles rather than an application of absolute and obvious laws.

Combine mathematical, statistical, and computational skills. In academic classes, students generally learn to apply a specific set of skills designated by the current topic, readings, and problem set. In many texts, the title of the section will indicate which technique to use. Real problems leave more to the imagination. In addition, for most industrial problems, mathematics students must bring their programming skills to bear, including collaborative coding, version control, commenting, testing, validating, and debugging. Projects sometimes also require proficiency in and use of a particular piece of software, even if the students have a different preferred platform or solution.

Validate and test results. Once students have run their models, how can they conduct a sanity check on their results? If they have used data to create a model or algorithm, did they reserve some data for testing purposes? Under what circumstances should the results be most accurate? What metrics best define a good solution? Can the model run in a reasonable amount of time? Does the solution meet the requirements of the sponsor? Students must report the accuracy and reliability of their data, assumptions, and model as well as estimate the error introduced by the parameter values and data and methods they have used.

Iterate to refine the solution. Iteration can begin at many stages, as rethinking becomes necessary. Preliminary results may uncover unnecessary assumptions or new model requirements. While iteration is always a potentially useful step in the process, it is less likely to be feasible for short-term projects.

Draw conclusions. Once students have designed, refined, and run their model, they must decide what conclusions they can draw from the results. The conclusions will of course depend on the assumptions and choices that were made in the previous parts of the modeling process, and on the results.

Communicate results to both general and technical audiences. Deliverables can provide opportunities for students to practice mathematical writing, software documentation, peer review, and public speaking, as well as visual communication of key ideas and relationships. People interested in the results of their model will likely have various levels of understanding of the problem and its solution. Students will need to be able to communicate effectively with managers and officers as well as with technical teams. Regular communication with a sponsor, midproject oral updates, and periodic written reports can all provide opportunities to practice these skills before deliverables are due.

Practical Matters

Typical textbook story problems rarely require any of the skills discussed above, even though many of the tasks may be required in jobs that involve mathematical modeling. Modeling camps, study groups, and industrial mathematics workshops provide opportunities to practice combinations of the skills, but not every student will have those opportunities or be prepared to fully participate. As a way to provide practice, one or more of the skills can be incorporated into a mathematics course through class activities, projects, and homework problems. In this way, students can benefit from practicing each skill in isolation before attempting to combine them.

A simple place to start might be to incorporate some extraneous information into a problem set. Another possibility would be to give students a raw data set and ask them how they would approach the information to draw a conclusion. Governmental organizations, such as the U.S. Environmental Protection Agency, make data and models available online. A third option could require students to analyze a problem and suggest a course of action. The Mathematical Contest in Modeling provides ten years of online example problems and winning papers.

Long-term projects, such as the year-long Mathematics Clinic at Harvey Mudd College, require students to practice all of these skills. With long-term projects, students have the opportunity to work in teams and practice project management. With some guidance, students can learn to prepare reasonable time lines, plan for failure and other contingencies, and assign tasks so that each team member contributes to the solution.

As we design modeling activities for students, we should consider how much iteration is feasible. Ideally, we would discuss the rationale for iteration with students even when they do not have time to implement the iteration themselves. As we teach mathematical modeling and provide mentoring, we also need to decide how much scaffolding we will provide to lead their modeling efforts in a desirable direction. Cognitive research in mathematics education supports the idea that in the early stages of learning, worked examples can promote skill acquisition, but problem solving is superior during later stages. While we may want to provide students with a framework for approaching particular problems, scaffolding can unintentionally reduce the cognitive demand of the task. Students need to struggle in order to develop strategies for real, messy problems. Therefore, as students advance, we must gradually take away the scaffolding and make sure that the problems are tackled by the students rather than being demonstrated to them.

As we teach mathematical modeling, the tasks outlined above can provide students with opportunities to develop individual skills they can apply to real problems. When they can combine the skills, work well individually or in teams, and produce relevant results for real unsolved problems, they will be ready to make valuable contributions to modeling problems in science, technology, engineering, and mathematics fields. We can provide these challenges through problems with high cognitive demand, inspired by industrial mathematics.

Further Reading

Borrelli, R. 2010. "The doctor is in." *Notices of the American Mathematical Society* 57(9):1127–29. (This article describes the Harvey Mudd College Mathematics Clinic.)

Renkl, A., and R. Atkinson. 2003. "Structuring the transition from example study to problem solving in cognitive skill acquisition: A cognitive load perspective." *Educational Psychologist* 38:15–22.

Soon, W., L. T. Lioe, and B. McInnes. 2011. "Understanding the difficulties faced by engineering undergraduates in learning mathematical modelling." *International Journal of Mathematical Education in Science and Technology* 42: 1023–39.

Stein, M. K., M. S. Smith, M. A. Henningsen, and E. A. Silver. 2000. *Implementing Standards-Based Mathematics Instruction: A Casebook for Professional Development*. New York: Teachers College Press.

Williams, J. A. S., and R. C. Reid. 2010. "Developing problem solving and communication skills through memo assignments in a management science course." *Journal of Education for Business* 85:323–29.

Circular Reasoning: Who First Proved That C Divided by d Is a Constant?

David Richeson

Ask any mathematician: Who first proved the fundamental theorem of algebra? He or she will answer with confidence: Carl Friedrich Gauss. Who proved that $\sum_{n=1}^{\infty}(1/n^2) = \pi^2/6$? Leonhard Euler. The nondenumerability of the real numbers? Georg Cantor. Fermat's last theorem? Andrew Wiles. But ask the question in the title of this article, and the response will likely be, "Isn't that in Euclid's *Elements*?" or "It is obvious; all circles are similar," or, most frequently, "I don't know."

The expression $C = \pi d$, which gives the relationship between the circumference and the diameter of a circle, is one of the few formulas known to almost all children and adults, regardless of how long they have been out of school. It is the basis of our definition of π. Yet it has no name attached to it.

It is not in *Elements*. Euclid (fl. c. 300 BCE) gave Eudoxus of Cnidus's (c. 400–c. 347 BCE) proof of, essentially, the existence of the area constant for circles (A/r^2), which we know is also π. But he did not mention the invariance of C/d or anything equivalent to it. The argument that "all circles are similar" is intriguing and has some merit, but it does not easily lead to a rigorous proof. And none of the standard history of mathematics textbooks answer the question.

Our answer is necessarily hesitant: probably Archimedes of Syracuse (287–212 BCE). Clearly, Archimedes was aware of the result; it is implicit in and can be easily proved from his mathematics. But as far as we know, he never stated it explicitly. The treatise that would most likely contain the theorem, *Measurement of a Circle* [**1**, pp. 91–98] has not come down to us fully intact, and we have some evidence that the full version contained the theorem. In the fragment that we possess, Archimedes gave his famous bounds on this constant, $223/71 < C/d < 22/7$. He also proved that $A = Cr/2$, which, when combined with the theorem in *Elements*, can be used to prove that C/d is constant. Moreover, in another work, *On the Sphere and the Cylinder I* [**18**, pp. 31–184], Archimedes gave two postulates which, when added to those in *Elements*, enable this theorem to be proved.

We must point out that throughout this article we will use modern terminology, notation, and mathematical concepts unless it is important to understand the original approach. The phrase "the value A/r^2 is independent of the circle and it is equal to the circumference constant $C/d = \pi$" would, for many reasons, be completely foreign to the Greeks. For example, they would express these ideas through ratios and proportions, and the terms in these ratios, A, r^2, C, and d, would be geometric objects, not numbers. And, of course, the use of π to represent this constant was many years in the future. (The first use is attributed to William Jones (1675–1749) in 1706 [**12**].)

Is There Something to Prove?

One common response to the question in the title is: "It is obvious; all circles are similar." For example, Katz writes that "It may be obvious that the circumference is proportional to the diameter . . ." [**13**, p. 20] Indeed, if we scale a figure by a factor of k, then all lengths increase by this same factor. Thus, the ratios of corresponding diameters and circumferences remain constant. The "it's obvious" argument probably explains why the circumference constant was rediscovered by so many cultures.

But this response is mathematically unsatisfying; it is not a simple process to turn it into a rigorous proof, and Euclid did not attempt to do so. He defined similarity for polygons (Def. VI.1: "Similar rectilinear figures are such as have their angles severally equal and the sides about the equal angles proportional" [**11**, vol. 2, p. 188]) but not for circles or other curves. It is not clear how he would do so. He could have said that two circles are similar provided $C_1 : d_1 :: C_2 : d_2$, but of course that is what we want to prove.

Today, we define similarity in terms of functions; a function f from the Euclidean plane to itself is a *similarity transformation* if there is a positive number k such that for any points x and y, $d(f(x), f(y)) = k \cdot d(x, y)$.

Two geometric shapes are *similar* if one is the image of the other under such a mapping. If AB and CD are two segments in a geometric figure and $A'B'$ and $C'D'$ are the corresponding segments in a similar figure, then, assuming that the length of a line segment is the distance between its endpoints,

$$\frac{A'B'}{C'D'} = \frac{d(A',B')}{d(C',D')} = \frac{d(f(A),f(B))}{d(f(C),f(D))} = \frac{k \cdot d(A,B)}{k \cdot d(C,D)} = \frac{d(A,B)}{d(C,D)} = \frac{AB}{CD}.$$

Despite the fact that it is easy to show that two circles are similar, we still cannot say that C/d is constant. To do so, we need a definition for the length of the circumference, but it is not clear how to extend the definition for the length of a segment to the length of a curve. This brings us to a key question: What *is* the length of a curve? How do we define arc length?

The history of arc length is long and fascinating, but a thorough discussion would send us too far afield. (See Gilbert Traub's Ph.D. dissertation [**26**] for a detailed history.) In this article, we are interested in circles, not general curves, and in fact, we are not interested in lengths but relative lengths (is a given arc longer or shorter than a given line segment?) and ratios of lengths ($C : d$).

The Early History of Pi

Books and articles on the history of mathematics are quick to point out that ancient civilizations knew about the circle constant π—or circle constants, we should say, for π plays a dual role as the area constant and the circumference constant. A great deal has been written about whether the Egyptians, the Babylonians, the Chinese, the Indians, and the writers of the Bible knew about the constants, about what value they used for them, and about whether they knew that the two constants were the same. (For a history of π, see, for example, [**2, 3, 4, 6, 23, 24**].)

Proposition XII.2 of Euclid's *Elements* implies the existence of the area constant π: "Circles are to one another as the squares on their diameters" [**11**, vol. 3, p. 371]. Symbolically, if we have circles with areas A_1 and A_2 and diameters d_1 and d_2, respectively, then $A_1/A_2 = d_1^2/d_2^2$. In other words, A/d^2 is the same value for every circle. Thus, so is A/r^2. This theorem may have been proved first by Hippocrates of Chios

(c. 470–c. 410 BCE), but Euclid's proof—a classic application of the method of exhaustion—is due to Eudoxus.

One expects to see a proposition in Euclid's *Elements* such as "The circumferences of circles are to one another as their diameters." But on this topic, Euclid was silent. In fact, Euclid, whose *Elements* contained Eudoxus's theory of magnitudes and ratios, only compared two magnitudes (lengths, areas, angles, etc.) if they are of the same type. In Definition V.4, he wrote that "magnitudes are said to have a ratio to one another which can, when multiplied, exceed one another" [**11**, vol. 2, p. 114]. Thus, it must be possible to put together several copies of one magnitude to obtain a magnitude larger than the other and vice versa. This definition was flexible enough for him to find ratios of incommensurable magnitudes (such as the ratio of the diagonal of a square to its side). But it did not allow him to find the ratio of an angle to a line segment, an area to a volume, or—and this is the key fact for us—the arc of a circle to the arc of a circle with a different radius or the arc of a circle to a line segment. Thus, Euclid would not consider $C_1 : C_2$ or $C : d$. (See Proposition VI.33 and the discussion that follows it in [**11**, vol. 2, pp. 273–76].)

In [**24**], Seidenberg argues that even though the theorem is not in *Elements*, it is likely that Euclid and his predecessors knew it; it was inherited knowledge. For instance, Seidenberg gives several examples showing that Aristotle (384–322 BCE) knew this relationship. Aristotle wrote on the circular motion of the stars, "it is not at all strange, nay it is inevitable, that the speeds of the circles should be in the proportion of their sizes." He argues that when Aristotle refers to the size of a circle, he means the diameter, so the circumference is proportional to the diameter. Thus, we may suspect that Eudoxus, Euclid, and the rest knew the theorem but were unable to prove it.

Euclid's Common Notion 5

The reason that the theorem on circular areas was proved before the theorem on circumferences is that arc length is inherently more complicated than area (just ask any first-year calculus student!). Gottfried Leibniz (1646–1716) wrote that "areas are more easily dealt with than curves, because they can be cut up and resolved in more ways" [**16**, p. 69]. In particular, Euclid could apply his fifth common notion—"the

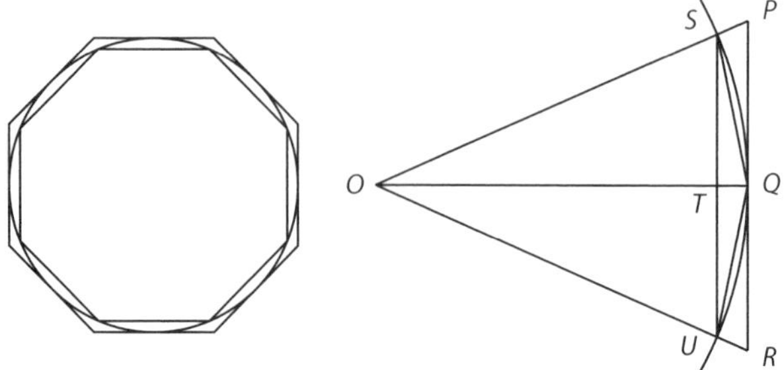

FIGURE 1. A circle with an inscribed and a circumscribed n-gon.

whole is greater than the part" [**11**, vol. 1, p. 155]—to areas but not to curves. For example, suppose a triangle is inscribed in a circle. The filled triangle is a subset of the circular disk. By Euclid's common notion, the area of the triangle is less than the area of the circle. However, while it is intuitively clear that the circumference of the circle is greater than the perimeter of the triangle, Euclid's common notion does not guarantee this and it is not true in general. An inscribed star-shaped polygon, or worse, an inscribed region with a fractal boundary, like a Koch snowflake, has smaller area than the circle but larger, perhaps infinite, perimeter.

As an exercise, let us try to prove that the perimeter of an inscribed regular n-gon is less than the circumference of the circle, which is less than the perimeter of a circumscribed regular n-gon (Figure 1). Suppose the circle has radius r, circumference C, and area A. For simplicity, we may restrict our attention to a sector of the circle in which SU and PR are sides of the inscribed and circumscribed polygons, respectively. We would like to prove that $SU < \mathrm{Arc}(SQU) < PR$.

We know that

$$\text{quadrilateral } OSQU = \Delta OSQ \cup \Delta OQU \subset \text{sector } OSQU \subset \Delta OPR.$$

By Euclid's fifth common notion,

$$\mathrm{Area}(\Delta OSQ) + \mathrm{Area}\,(\Delta OQU) < \mathrm{Area}\,(\text{sector } OSQU) < \mathrm{Area}\,(\Delta OPR),$$

and hence

$$\frac{1}{2}OQ \cdot ST + \frac{1}{2}OQ \cdot TU < \mathrm{Area}\,(\text{sector } OSQU) < \frac{1}{2}OQ \cdot PR.$$

We know that $OQ = r$ and $ST + TU = SU$, so

$$\frac{1}{2}SU \cdot r < \text{Area (sector } OSQU) < \frac{1}{2}PR \cdot r,$$

and

$$SU < \frac{2 \text{ Area (sector } OSQU)}{r} < PR.$$

Because

$$\frac{\text{Area (sector } OSQU)}{A} = \frac{\text{Arc }(SQU)}{C},$$

it follows that

$$SU < \frac{2A}{Cr}\text{Arc }(SQU) < PR.$$

In order to conclude that $SU <$ Arc(SQU) $< PR$, we need the fact that $2A/Cr = 1$. But this is precisely Archimedes's theorem ($A = rC/2$), and as we shall see, his proof used the fact that

Perimeter(inscribed polygon) $< C <$ Perimeter(circumscribed polygon).

The argument is circular (pun intended)!

Incidentally, this conversation is intimately related to that of how to prove that $\sin \theta < \theta < \tan \theta$, a key ingredient in the proof that $\lim_{\theta \to 0} (\sin \theta)/\theta = 1$. In our diagram, if $\theta = \angle SOQ$, then $SU = 2r \sin \theta$, $PR = 2r \tan \theta$, and—assuming that angle measures arc length—Arc(SQU) $= 2r\theta$. (See **[8**, pp. 216–20], **[10]**, **[22]**, and **[24]** for more on this topic.)

One must suspect that Euclid would have loved nothing more than to include the "circumference theorem" in *Elements*, but he was unable to prove it using his postulates and common notions. It took the genius of Archimedes to realize that the proof required more axioms than those given by Euclid.

Measurement of a Circle

To show that C/d is a constant, we need two of Archimedes's works: *Measurement of a Circle* and *On the Sphere and the Cylinder I*. It is an unstated corollary of a theorem in the first, but the proof of this theorem requires two new geometric axioms that are stated in the second.

The order in which these two treatises were written is unknown. Usually *On the Sphere and the Cylinder* is listed chronologically before *Measurement of a Circle*. Indeed, the Archimedes expert Johan Heiberg (1854–1928) put them in this order, perhaps because that makes the most sense logically—axioms first, theorem second. Yet he put a question mark after the latter of these, not confident in where he placed it. When the influential historian Sir Thomas Heath (1861–1940) reproduced the list, the question mark disappeared. Thus, this order became the standard [**15**, p. 153]. More recently, the eminent historian Wilbur Knorr convincingly argued that they were written in the other order. He gave specific mathematical evidence showing that Archimedes was a more mature mathematician when he wrote the former treatise. He speculated that when writing *Measurement of a Circle* Archimedes accepted the axioms as obvious truths. "The formulation of these principles as explicit axioms in *On the Sphere and the Cylinder* would thus result from Archimedes's own later reflections on the formal requirements of such demonstrations" [**15**, pp. 153–55].

The extant version of *Measurement of a Circle* has only three propositions and is not a faithful copy of Archimedes's original (one of the three results is clearly incorrect as stated). E. J. Dijksterhuis referred to it as "scrappy and rather careless." He pointed out that "it is quite possible that the fragment we possess formed part of a longer work, which is quoted by Pappus under the title *On the Circumference of the Circle*, and that the latter also dealt with the more general question as to the ratio between the length of an arc of a circle and that of its chord" [**9**, p. 222], which would support our thesis. We must wait a little longer to see if Reviel Netz's fresh translation of *Measurement of a Circle* from the recently rediscovered Archimedes palimpsest will yield any new information. (See [**19**] for the fascinating story of the Archimedes palimpsest and [**7**, pp. 3–6], [**9**, pp. 33–49], and [**14**] for details on the various Greek and Arabic sources of this treatise.)

In his first proposition, Archimedes proved that (Figure 2) "The area of any circle is equal to a right-angled triangle in which one of the sides about the right angle is equal to the radius, and the other to the circumference, of the circle" [**1**, p. 91]. That is, $A = Cr/2$.

This theorem has one immediate and important corollary: The problems of squaring the circle and rectifying the circle are equivalent. If it was possible to rectify the circle—that is, use a compass and straightedge

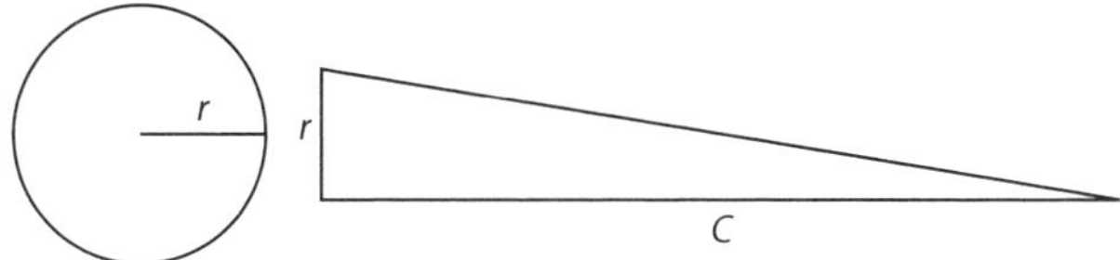

FIGURE 2. The area of a circle with radius r and circumference C equals the area of a right triangle with legs r and C.

to construct a line segment the same length as the circumference in a finite number of steps—then we could construct the triangle in Archimedes's theorem and, from this, a square with the same area. Conversely, if it was possible to square a circle, then we could construct the triangle and thereby rectify the circle. By the work of René Descartes (1596–1650) and Pierre Wantzel (1814–48), we know that for a line segment to be constructible from a unit-length line segment, it must have an algebraic length (but not all algebraic lengths are constructible). Thus, Ferdinand von Lindemann's (1852–1939) 1882 proof that π is a transcendental number **[17, 20]** implied that a circle cannot be rectified. And by Archimedes's theorem, a circle cannot be squared.

Some historians argue that Archimedes's theorem relating the circumference and the area was proved 100 to 200 years earlier because it was used to exhibit the circle squaring properties of the curve now known as the *quadratrix*. But more recent scholarship argues that this use of the quadratrix was not due to Hippias of Elis (born c. 460 BCE), who discovered the curve and used it to trisect angles, or to Dinostratus (fl. c. 350 BCE), as others contend, but to Nicomedes (fl. c. 250 BCE), who was able to benefit from the work of Archimedes and Eudoxus (see [**15**, pp. 80–82, 233] or [**25**] for details).

With Archimedes's result and Euclid's XII.2, we are but a few algebraic steps away from proving that C/d is a constant and that it is the same as the area constant:

$$\frac{C}{d} = \left(\frac{2A}{r}\right)\frac{1}{d} = \frac{A}{r^2} = \pi.$$

Although Archimedes never stated the invariance of C/d explicitly, it is implied by the third proposition, which gives his famous bounds,

$$\frac{223}{71} < \frac{C}{d} < \frac{22}{7}.$$

Many people highlight this pair of inequalities as the key result of this work. One popular textbook calls it "the most important proposition in *Measurement of the Circle*" [**5**, p. 202]. Indeed, it is amazing that Archimedes overcame the clumsy Greek numerical system to obtain these bounds and that he devised an algorithm that was used for centuries by digit-hunters searching for increasingly more accurate bounds for π (approximating the circle with inscribed and circumscribed polygons of many sides). But we argue that the first theorem is the more important one, for it shows the relationship between the circumference and the area, and from this it follows that C/d is constant and that this circumference constant equals the area constant.

We now sketch Archimedes's proof that $A = Cr/2$. Suppose we begin with a circle with radius r, area A, and circumference C. Let T be a right triangle with legs of length r and C. For the sake of contradiction, suppose Area$(T) \neq A$. There are two cases, either Area$(T) < A$ or Area$(T) > A$.

Suppose Area$(T) < A$, that is, $A -$ Area$(T) > 0$. By using, although not explicitly citing, Euclid's Proposition XII.16, Archimedes observed that by beginning with an inscribed square and by repeatedly doubling the number of sides, we can obtain an inscribed regular polygon P_{in} such that $A -$ Area$(P_{\text{in}}) < A -$ Area(T). So Area$(P_{\text{in}}) >$ Area(T). Let r' be the length of the apothem of P_{in}. Then $r' < r$ (Figure 3). It follows that

$$\text{Area}\,(P_{\text{in}}) = \frac{1}{2}r' \cdot \text{Perimeter}\,(P_{\text{in}}) < \frac{1}{2}rC = \text{Area}\,(T).$$

This is a contradiction.

Suppose Area$(T) > A$. Now Archimedes began with a circumscribed square, and he repeatedly doubled the numbers of sides to obtain a sequence of circumscribed regular 2^n-gons. He proved that each time the

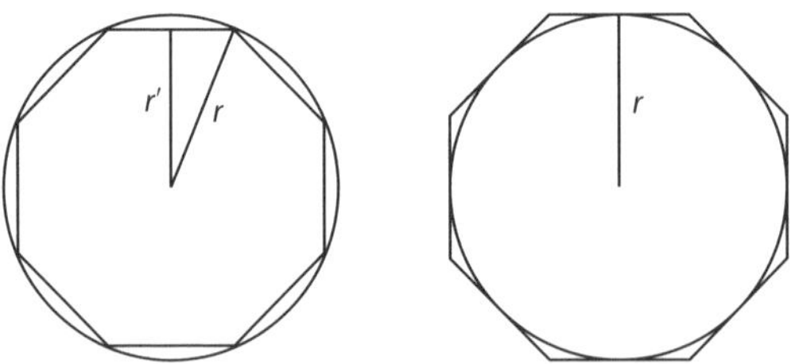

FIGURE 3. The lengths of the apothems of an inscribed (left) and circumscribed (right) polygon are r' and r, respectively.

number of sides is doubled the difference of the areas of the polygon and the circle is more than cut in half. Thus, eventually this process yields a circumscribed regular polygon P_{circ} such that $\text{Area}(P_{\text{circ}}) - A <$ $\text{Area}(T) - A$. So $\text{Area}(P_{\text{circ}}) < \text{Area}(T)$. In this case, the apothem of P_{circ} is the radius of the circle (Figure 3). So

$$\text{Area}\,(P_{\text{circ}}) = \frac{1}{2}r \cdot \text{Perimeter}\,(P_{\text{circ}}) > \frac{1}{2}rC = \text{Area}\,(T),$$

a contradiction. Thus the theorem is proved.

In the proof, Archimedes used $\text{Perimeter}(P_{\text{in}}) < C < \text{Perimeter}(P_{\text{circ}})$ without justification. These inequalities would follow if, given a chord in a circle, AB, the included arc ADB, and the two segments tangent to the circle AC and BC (as in Figure 4), we knew that $AB < \text{Arc}(ADB) <$ $AC + BC$. If we believe Knorr's chronology, it is likely that Archimedes assumed these inequalities when writing *Measurement of a Circle*, only later returning to them with a more rigorous mindset.

One may argue that these inequalities are obvious, especially the first, because a line is the shortest distance between two points. Yet Euclid did not assume this as an axiom. In fact, he proved both inequalities for segments, namely, that in Figure 4, $EF < EH + FH < EG + FG$. Proposition I.20 is the triangle inequality: "In any triangle two sides taken together in any manner are greater than the remaining one" [**11**, vol. 1, p. 286]. From this, we conclude that $EF < EH + FH$. According to Proclus, the Epicureans scoffed at this theorem, saying that it was so intuitive that even an ass knew it was true: If you put food at one vertex of a triangle and an ass at another, then it would certainly walk along the edge between them and not along the other two [**11**, vol. 1, p. 287].

Euclid's next proposition implies that $EG + FG > EH + FH$. "If on one of the sides of a triangle, from its extremities, there be constructed two straight lines meeting within the triangle, the straight lines so

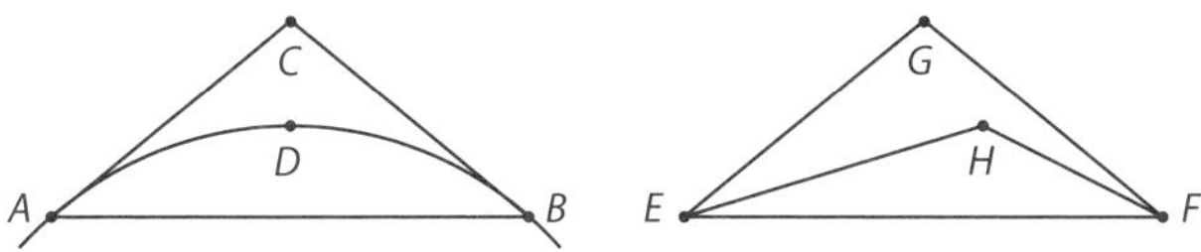

FIGURE 4. In *Measurement of a Circle,* Archimedes assumed that $AB <$ $\text{Arc}(ADB) < AC + BC$, whereas Euclid proved that $EF < EH + FH < EG + FG$ in *Elements*.

constructed will be less than the remaining two sides of the triangle, but will contain a greater angle" [**11**, vol. 1, p. 289].

On the Sphere and the Cylinder I

Archimedes began *On the Sphere and the Cylinder I* by formalizing the notions that he took for granted in *Measurement of a Circle*. He stated that a curve is *concave in the same direction* "if any two points whatever being taken, the straight [lines] between the [two] points either all fall on the same side of the line, or some fall on the same side, and some fall on the line itself, but none on the other side" [**18**, p. 35]. (Note: In these translations, curves are called "lines"; when they are straight they are called "straight lines.") For example, the curve *ABC* in Figure 5 is concave in the same direction (as are *ADC* and *AC*).

Then he stated the following axioms [**18**, p. 36]:

1. That among lines which have the same limits, the straight [line] is the smallest.
2. And, among the other lines (if, being in a plane, they have the same limits): that such [lines] are unequal, when they are both concave in the same direction and either one of them is wholly contained by the other and by the straight [line] having the same limits as itself, or some is contained, and some it has [as] common; and the contained is smaller.

Returning to Figure 5, Archimedes's axiom (1) implies that *AC* is shorter than *ABC* and *ADC*, and (2) implies that *ABC* is shorter than *ADC*. Clearly, these axioms are the analogs of Euclid's propositions for piecewise-linear curves. With these axioms in place, Archimedes was ready to make rigorous statements about lengths of curves. (In fact,

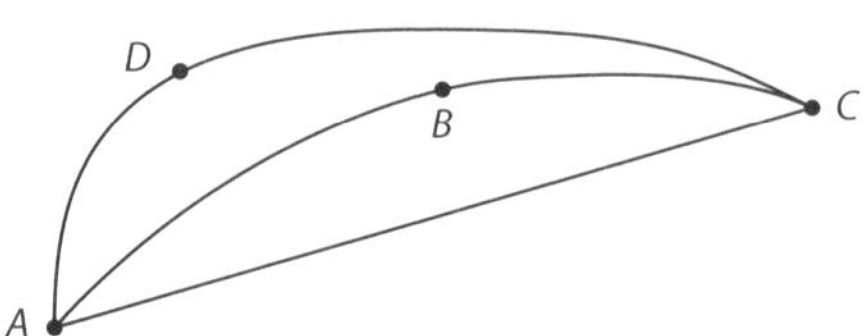

FIGURE 5. Archimedes's axioms imply that *AC* is shorter than *ABC*, which is shorter than *ADC*.

Archimedes should have included a third axiom: finite additivity. He used the fact that if a curve is broken into parts, then the total length equals the sum of the individual lengths. Euclid's *Elements* should also have contained this axiom. To make up for this, Christopher Clavius (1538–1612) added the axiom "the whole is equal to the sum of its parts" to his 1574 version *Elements* [**26**, p. 40], [**11**, p. 323].)

Following the statements of the axioms, Archimedes immediately applied the first one. "Assuming these it is manifest that if a polygon is inscribed inside a circle, the perimeter of the inscribed polygon is smaller than the circumference of the circle; for each of the sides of the polygon is smaller than the circumference of the circle which is cut by it" [**18**, p. 36].

Then he stated his first proposition and used the second axiom to prove it (although we must point out that Archimedes did not prove that the hypotheses of the second axiom are satisfied). "If a polygon is circumscribed around a circle, the perimeter of the circumscribed polygon is greater than the perimeter of the circle" [**18**, p. 41].

These results fill the gaps in the proof from *Measurement of a Circle*. Thus, we may obtain the corollary that C/d is a constant.

Archimedes's Number

We would be remiss if we did not mention Archimedes's other contributions to the study of the number π. We know that π has at least two more lives—it is the volume and surface area constants for spheres.

Just as arc length is trickier than planar area, so too is surface area more subtle than volume. In *Elements* (Prop. XII.18 [**11**, vol. 3, p. 434]), Euclid proved, essentially, the existence of the volume constant for spheres, V/d^3 (that is, $V_1 : V_2 : : d_2^3 : d_2^3$). But, again, it took the genius of Archimedes to prove that S/r^2, where S is the surface area of a sphere, is a constant independent of the sphere (*On the Sphere and the Cylinder I*, Prop. 31: "The surface of every sphere is four times the greatest circle of the [circles] in it" [**18**, p. 144]).

Archimedes then brought all of these results together in one concise, elegant corollary: "And, these being proved, it is obvious that every cylinder having, [as] base, the greatest of the [circles] in the sphere, and a height equal to the diameter of the sphere, is half as large again as the sphere, and its surface with the bases is half as large again as the surface of the sphere" [**18**, p. 150]. In other words, for a sphere inscribed in a cylinder (including its top and bottom disks),

$$\frac{3}{2} = \frac{\text{Volume (cylinder)}}{\text{Volume (sphere)}} = \frac{\text{Surface area (cylinder)}}{\text{Surface area (sphere)}}.$$

Combining this corollary and Archimedes's theorem that $A = Cr/2$, we see that the four circle and sphere constants are intimately related:

$$\pi = \frac{C}{d} = \frac{A}{r^2} = \frac{3}{4}\left(\frac{V}{r^3}\right) = \frac{1}{4}\left(\frac{S}{r^2}\right).$$

Archimedes recognized this work for the grand achievement that it is. Plutarch (45–120 CE) wrote, "Although [Archimedes] made many excellent discoveries, he is said to have asked his kinsmen and friends to place over the grave where he should be buried a cylinder enclosing a sphere, with an inscription giving the proportion by which the containing solid exceeds the contained" [**21**, p. 481].

Because of these results about the nature of π and his very accurate bounds on the value, it seems fitting that we call π *Archimedes's number.*

References

1. Archimedes, *The Works of Archimedes*. Reprint of the 1897 edition and the 1912 supplement, T. L. Heath, ed. Dover, Mineola, NY, 2002.
2. J. Arndt and C. Haenel, *Pi—Unleashed*. Springer, Berlin, 2000.
3. P. Beckmann, *A History of Pi*. St. Martin's, New York, 1971.
4. L. Berggren, J. Borwein, and P. Borwein, *Pi: A Source Book*. Springer, New York, 2004.
5. D. M. Burton, *The History of Mathematics: An Introduction*. McGraw-Hill, Boston, 2007.
6. D. Castellanos, "The ubiquitous π." *Math. Mag.* **61** (1988) 67–98, 148–63, http://dx.doi.org/10.2307/ 2690037, http://dx.doi.org/10.2307/2689713.
7. M. Clagett, *Archimedes in the Middle Ages. Vol. I: The Arabo-Latin Tradition*. Univ. Wisconsin Press, Madison, WI, 1964.
8. M. Comenetz, *Calculus: The Elements*. World Scientific, River Edge, NJ, 2002.
9. E. J. Dijksterhuis, *Archimedes*. Trans. C. Dikshoorn with a contribution by W. R. Knorr. Princeton Univ. Press, Princeton, NJ, 1987.
10. L. Gillman, "π and the limit of (sin α)/α." *Amer. Math. Monthly* **98** (1991) 346–49, http:// dx.doi.org/ 10.2307/2323805.
11. T. L. Heath, *The Thirteen Books of Euclid's Elements*. Three volumes. Univ. Press, Cambridge, U.K., 1908.
12. W. Jones, *Synopsis palmariorum mathesios*. London, 1706.
13. V. J. Katz, *A History of Mathematics: An Introduction*. Addison Wesley Longman, Reading, MA, 1998.
14. W. R. Knorr, "Archimedes' *Dimension of the circle*: A view of the genesis of the extant text." *Arch. Hist. Exact Sci.* **35** (1986) 281–324, http://dx.doi.org/10.1007/BF00357303.

15. ———, *The Ancient Tradition of Geometric Problems*. Dover, Mineola, NY, 1993.
16. G. W. Leibniz, *The Early Mathematical Manuscripts of Leibniz*. Dover, Mineola, NY, 2005.
17. F. Lindemann, "Über die Zahl π." *Math. Ann.* **20** (1882) 213–25.
18. R. Netz, *The Works of Archimedes: The Two Books On the Sphere and the Cylinder*. Cambridge Univ. Press, Cambridge, U.K., 2004.
19. R. Netz and W. Noel, *The Archimedes Codex: How a Medieval Prayer Book Is Revealing the True Genius of Antiquity's Greatest Scientist*. Da Capo, Philadelphia, 2007.
20. I. Niven, "The transcendence of π." *Amer. Math. Monthly* **46** (1939) 469–71, http://dx.doi.org/10. 2307/2302515.
21. Plutarch, *Plutarch's Lives*. Trans. B. Perrin. William Heinemann, London, 1917.
22. F. Richman, "A circular argument." *College Math. J.* **24** (1990) 160–62, http://dx.doi.org/10.2307/ 2686787.
23. H. C. Schepler, "The chronology of PI." *Math. Mag.* **23** (1950) 165–70, 216–28, 279–83, http://dx. doi.org/10.2307/3029000.
24. A. Seidenberg, "On the area of a semi-circle." *Arch. History Exact Sci.* **9** (1972) 171–211, http://dx.doi.org/10.1007/BF0032739000.
25. A. J. E. M. Smeur, "On the value equivalent to π in ancient mathematical texts. A new interpretation." *Arch. History Exact Sci.* **6** (1970) 249–70, http://dx.doi.org/10.1007/BF00417620.
26. G. Traub, *The Development of the Mathematical Analysis of Curve Length from Archimedes to Lebesgue*, New York Univ. thesis, 1984.

A Medieval Mystery: Nicole Oresme's Concept of Curvitas

Isabel M. Serrano and Bogdan D. Suceavă

In a paper published in 1952, J. L. Coolidge (1873–1954) appreciates that the story of curvature is "unsatisfactory" [2], and he points out that "the first writer to give a hint of the definition of curvature was the fourteenth-century writer Nicole Oresme, whose work was called to my attention by Carl Boyer." Then Coolidge comments, "Oresme conceived the curvature of a circle as inversely proportional to the radius; how did he find this out?" The scholarly conditions of the fourteenth century make this discovery phenomenal and the question as to how it was achieved worth researching. In this article, we describe how a fourteenth-century scholar (1) gave a correct definition for curvature of circles and attempted to extend it to general curves, (2) tried to apply curvature to understand the behavior of real-life phenomena, and (3) produced in his research a statement that anticipates the fundamental theorem of curves in the plane.

In various cases, Oresme's work is not cited when the history of curvature is discussed (e.g., [5], [8]), while some authors (e.g., [1], p. 191) make note of his contribution to this concept. Several scholars have even concluded that the medieval sciences contributed very little to the modern scientific revolution. In addressing this perception, Edward Grant [3] writes, "Even if the Middle Ages made few significant contributions to the advancement of the sciences themselves, or none at all, . . . if no noteworthy medieval contributions were made to help shape specific scientific advances in the seventeenth century, in what ways did the Middle Ages contribute to the Scientific Revolution and, more to the point, lay the foundations for it?" We describe in this article that curvature is one of the concepts that was first defined in the Middle Ages. The importance of the idea of curvature is described in many works (e.g., [1], [5], [8], [10]), and we don't feel we should elaborate on this point.

Nicole Oresme was born around 1320 in the village of Allemagne, near Caen, today Fleury-sur-Orne [6]. The first certain fact in his biography is that he was a "bursar" of the College of Navarre from 1348 to October 4, 1356, when he became a master. The College of Navarre, established by Queen Joan I of Navarre in 1305, focused on teaching the arts, philosophy, and theology. Oresme's major was theology [4]. As a student, he had to observe the code at the College of Navarre, where the students were required to speak and write only in Latin; his ability to work in Latin would prove critical in his future work. Oresme studied with, among others, Jean Buridan and Albert of Saxony. It was at this institution where he wrote his most important works, e.g., *De proportionibus proportionum*, which is of particular importance for the history of mathematics, or *Ad pauca respicientes*, of interest for the history of ideas in celestial mechanics. Oresme remained grand master of the college until December 4, 1361, when he was forced to resign [6]. On November 23, 1362, he became a canon of the Rouen Cathedral (a place of major importance in the history of France), and on March 18, 1364, he was promoted to dean of the cathedral. Oresme was in that period the king's confessor and adviser, and some time before 1370 he became one of Charles V's (1364–80) chaplains; at the king's request, he translated from Latin into French Aristotle's *Ethics* (1370) and *Politics*, as well as *Economics*.

As Marshall Clagett points out [7], it is very likely that *De configurationibus* was written in the interval 1351–55. To better depict this historical period, we recall here that during this period, Geoffrey Chaucer, later considered the father of English literature, was still a child in London. These are the same years when Giovanni Boccaccio wrote *The Decameron*, largely completed by 1352. In Florence, Francesco Petrarch, the first to coin the name of the "Dark Ages," was writing in Latin *De vita solitaria*. One of the main historical references for that historical period is Jean Froissart's *Chronicles*, describing the battles from the Hundred Years' War and the Black Death, affecting most of Europe in the interval 1346–53. The cathedral Notre-Dame de Paris was just completed a few years before, in 1345, and dominated the skyline of medieval Paris. In short, this period of time was full of conflict and tragedy, greatly occupying civilian minds and making the main

The Rouen Cathedral where Nicole Oresme served first as canon, then as dean of the cathedral, after 1362. Photo by Vlad Ştefan Barbu.

focus of life survival. The poets and the scientists worked in many cases in isolation for long intervals of time.

In the time frame in which Oresme wrote, the language of functions was not yet used in mathematics. It is impressive that Oresme reached the concept of curvature before the concept of function was established. He had to invent and express his thoughts without several fundamental mathematical concepts to refer to, thus making his explanations on curvature unique. These circumstances justify why Marshall Clagett is correct in discussing a "doctrine" [7] when he described Oresme's original contributions.

Clagett's critical edition including the treatise *De configurationibus* [7] was published in 1968, more than a decade after J. L. Coolidge was hoping to see a more complete history of curvature. Clagett (1916–2005) notes ([7], pp. 50–51) that the first instance of Oresme's comments on representation of quantity by either a line or surface of a body are his remarks to *Questions on the Geometry of Euclid*, more precisely when he discusses questions 10 and 11. Oresme refers to other authors, such as Witelo and Lincoln (i.e., Robert Grosseteste), "who in this manner imagine the intensity of light," and to "Aristotle, who in the fourth [book] of the *Physics* imagines time by means of a line." He also includes "the Commentator [Campanus] in the fifth [book] of this work [the *Elements*] where he holds, in expounding ratios, that everything having the nature of a continuum can be imagined as a line, surface, or body." Clagett points out ([7], p. 51) that the reference to Aristotle's work is "to the effect that every magnitude is continuous, and movement follows magnitude; therefore movement and hence time are continuous, for motion and time seem to be proportional." In short, Oresme's "doctrine" is actually a theory describing how quantities could be described by graphs. The concept was novel at that time, although it was based on Aristotle's earlier work.

In the first part (the first forty chapters) of *De configurationibus*, Oresme sets up the groundwork for the doctrine of configurations; then he applies the doctrine to qualities, focusing on "entities" which are permanent or enduring in time. While discussing these elements, he suggests that his theory could explain numerous physical and psychological phenomena. In the second part of *De configurationibus* (the next forty chapters), Oresme describes how graphical representation can be applied to "entities that are successive"; in particular, he applies the doctrine of "figurations" to motion. He concludes this part

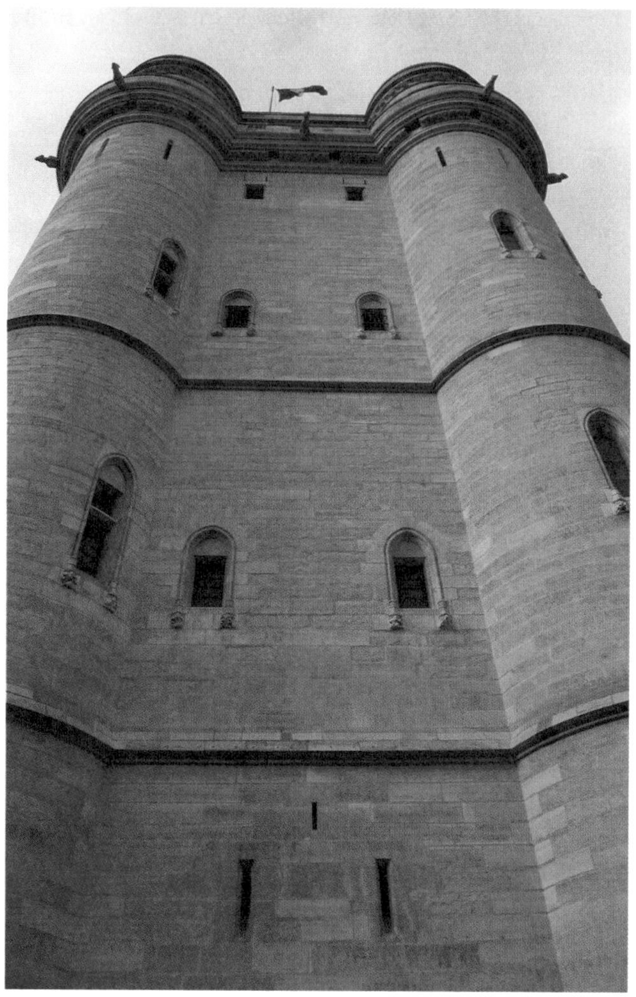

The donjon tower of the Château de Vincennes, in Paris, is 52 m high and represents the tallest medieval fortified structure in Europe. King Philip VI of France started this work about 1337. The work was completed during Charles V's reign. When Nicole Oresme was a scholar in Paris, this donjon was in the process of being erected. Later, during the reign of Charles V, this donjon served as a residence for the royal family. Its buildings are known to have once held the library and personal study of Charles V.

with several examples that could be extended to psychological effects, including the perceptions that are described as magic. Finally, Oresme describes external geometrical figures used to represent qualities and motions. He compares the areas of such figures and concludes that by comparing the areas, one may have a basis for the comparison of different qualities and motions.

To fully describe his theory, Oresme begins his *De configurationibus* with the following clarification: "Every measurable thing except numbers is imagined in the manner of continuous quantity." Then he pursues a discussion of the latitude and longitude of qualities, followed with the presentation of their quantity. He leads into his argument that qualities can be "figured." He spends several chapters discussing suitability of figures and shape of various particular cases. This discussion suggests an early analysis of curves in general position, if we are to refer to the modern concept. One important distinction appears in chapter I.xi, where Oresme examines the differences between uniform and difform qualities. He continues his focus on this topic in I.xiv with a discussion of "simple difform difformity," which is of two kinds: simple and composite. In this chapter he uses "linea curva" for a curve, and "curvitas" to express its curvature. In I.xv he begins describing four kinds of simple difform difformity, which are explained by drawing graphs. There is little doubt that the author builds here an early approach to variable quantities and their corresponding graphical representation. After this extensive discussion, performed without any algebraic notation, Oresme approaches "surface quality." Finally, in chapters I.xix, I.xx, and I.xxi, he introduces curvature. Additionally, in chapters I.xv and I.xvi, Oresme describes graphs that are concave and convex. Due to the context of his analysis, Oresme actually performs the first exploration of the possible connections between curvature and convexity.

A particular case in this doctrine of qualities is represented by curvature (chapter I.xx), endowed with "both extension and intensity." Oresme writes (in M. Clagett's translation, [7], p. 215), "We do not know with what, or with regard to what, the intensity of curvature is measured. For now it appears to me that there are only two [possible] ways [to speak of the measure of curvature]. The first is that the increase in curvature is a function of its departure from straightness, i.e., of its distance from straightness. This is [to be measured] by the quantity of the angle constituted of a straight line and a curve, e.g., an angle of contingence or perhaps another angle also constructed from a straight line and a curve." This intuitive description is consistent with the modern study of signed curvature and its relationship with the change of turning angle with respect to arc length. Even further, Nicole Oresme reaches a more precise description. He writes specifically that the curvature of the circle is the inverse of its radius (in chapter I.xxi, where Oresme cites Aristotle's *On Curved Surfaces*). He delves more into this concept by covering more general curves: "Difform curvature is composed of an infinite number of parts of different nature and unrelatable [to each other]" (I.xx). Thus, his study of curvature is not limited to circles but is extended to more general cases. However, Oresme does not have any precise procedure to compute such a general curvature.

In classical differential geometry, the so-called fundamental theorem of curves states (e.g., [9], p. 29) that if two single-valued continuous functions $\kappa(s)$ and $\tau(s)$, for $s > 0$, are given, then there exists one and only one space curve, determined by its position in space, for which s is the arc length, measured from an appropriate point on the curve, κ is the curvature, and τ is the torsion. This result was obtained in the nineteenth century. It is very surprising to read in Oresme the following reasoning (representative of Oresme's style and his intuition), which leads to a statement quite similar in conclusion to the fundamental theorem of curves ([7], p. 219):

> No intensity of difform curvature can be related to another dissimilar curvature in a ratio of 2 to 1 or [even] in a ratio of $\sqrt{2}$ to 1, i.e., either in a commensurable or incommensurable ratio—or, universally, in any ratio which could be found as existing between line and line. The conclusion is hence evident that intensity of curvature is not to be imagined by lines. Nor is there some curvature which is similar in intensity to some other quality of another species. Nor is curvature to be imagined by some figure. Nor is its intensity to be assimilated to the altitude of a figure, because the altitude of every figure is designated by lines. Finally, it is evident from this that no curvature is uniformly difform, for, by reason of accident, "uniformly difform" exists throughout a whole subject of the same nature and where the ratio of intensity to intensity, or excess of intensity, in the diverse parts is as the ratio of

> distance to distance, and consequently as the ratio of *lines*, as it is evident from the descriptions in chapter eleven, and this [reduction to ratios between lines] can not, as was just said, be suitable for difform curvature. And so it follows finally that every difform curvature is difform in a way different from that in which any other quality of another kind could be, and [so it is difform] with a strange, marvelous, diverse kind of difformity."[1]

Oresme does not work with the distinction between curvature and torsion for skew curves; all of his discussion is about planar curves, and the uniqueness part of the statement is suggested by "strange, marvelous, diverse" in the last sentence from the excerpt cited above. This shows that *curvitas difformis* is special in a unique way.

The generality of Oresme's doctrine resides in the attempt to model various phenomena by this approach. In *De configurationibus*, we encounter his first attempt to apply his doctrine in chapter I.xxiv, where he discusses "On the variety of natural powers dependent on this figuration." He writes ([7], p. 233), "It is manifest from natural philosophy and experience alike that all natural bodies determine in themselves their shapes, as, for example, animals, plants, some stones, and the parts of [all of] these. They also determine in themselves certain qualities which are natural to them. In addition to their shape that these qualities possess from their subject, it is necessary that they be figured with a figuration which they possess from their intensity—to employ the previously described imagery." To mention just one example, in chapter II.xl, titled "On the difformity of joys," Oresme discusses a subjective perception in the same terms as a physical quantity: "One ought to speak in the same way concerning a joy or a pleasure, which I suppose to be a certain quality extended in time and intended in degree."

The question asked by Julian Coolidge is where the idea of curvature comes from. There are many elements to suggest that the definition of curvature for curves is due to Nicole Oresme. One strong argument is that this definition was needed for his doctrine. Oresme developed it to serve his theoretical goals and to understand his configurations. Furthermore, Oresme builds upon Aristotle's conclusions and applies these ideas to a larger array of concepts where his graphs ("configurations") could be used. Some of the concepts he is interested in are today considered part of mathematics, some part of physics, while others approach the realm of psychology (e.g., the study of the question why certain perceptions lead to magic).

If Oresme clearly reached the first recorded definition of curvature for curves, then why do we see a certain hesitation to discuss and refer to his work? Perhaps because after the following generation his influence faded, his work was not continued, and his heritage was less understood. The historical reality of the Hundred Years' War limited the dissemination of Oresme's ideas. Later authors, such as Christiaan Huygens and Isaac Newton, discovered and developed fundamental concepts independently and did not build on Oresme's heritage. When mathematics benefited from the important revolution in sciences after 1600, Oresme's texts were perceived as inherited from a different paradigm. However, by looking back at this work today, we should not imagine that *De configurationibus* is an obscure medieval text that could be described as "religious science." Instead, it should be looked on as the initial approach to introduce curvature in the context of an early theory.

Decorative architecture of the Rouen Cathedral.

Addressing this type of understanding of the medieval books and types of arguments, Edward Grant writes ([3], p. 84), "Theologians had remarkable intellectual freedom and rarely permitted theology to hinder their inquiries into the physical world. If there was any temptation to produce a 'Christian science,' they successfully resisted it. Biblical texts were not employed to 'demonstrate' scientific truths by blind appeal to divine authority. When Nicole Oresme inserted some fifty citations to twenty-three different books of the Bible in his *On the Configurations of Qualities and Motions*, a major scientific treatise of the Middle Ages, he did so only as examples, or for additional support, but in no sense to demonstrate an argument." There is no better answer to address the aforementioned concerns.

Our article does not aim more than to contribute to a long overdue discussion on the first recorded definition of curvature, pursuing J. L. Coolidge's suggestion for a more complete history of this fundamental mathematical idea.

Note

1. The last sentence in the original is this ([7], p. 218): "Et inde sequitur ulterius quod omnis curvitas difformis est difformis aliter quam aliqua alia qualitas alterius generis possit esse et quadam extranea, mirabili, et diversa difformitate."

References

[1] B.-Y. Chen, Riemannian submanifolds, in *Handbook of Differential Geometry*, vol. I, edited by F. J. E. Dillen and L. C. A. Verstraelen, Elsevier Science B.V., Amsterdam, 2000.

[2] J. L. Coolidge, The unsatisfactory story of curvature, *American Mathematical Monthly* **59** (1952), 375–79.

[3] E. Grant, *The Foundations of Modern Science in the Middle Ages: Their Religious, Institutional and Intellectual Contexts*, Cambridge Studies in the History of Science, Cambridge University Press, Cambridge, 1996.

[4] ———, *Science and Religion, 400 B.C. to A.D. 1550: From Aristotle to Copernicus*, Johns Hopkins University Press, Baltimore, 2006.

[5] A. Knoebel, R. Laubenbacher, J. Lodder, and D. Pengelley, *Mathematical Masterpieces. Further Chronicles by the Explorers*, Springer-Verlag, New York, 2007.

[6] Nicole Oresme, *The De Moneta of Nicholas Oresme and English Mint Documents*, translated and edited by Charles Johnson, Ludwig von Mises Institute, Kindle edition by Amazon Digital Services, Inc., 2011.

[7] ———, De configurationibus qualitatum et motuum, in *Nicole Oresme and the Medieval Geometry of Qualities and Motions*, edited with an introduction, English translation and commentary by Marshall Clagett, University of Wisconsin Press, Madison, 1968.

[8] R. Osserman, *Poetry of the Universe: A Mathematical Exploration of the Cosmos*, Anchor Books, New York, 1995.

[9] D. J. Struik, *Lectures on Classical Differential Geometry*, 2nd edition, Dover Publ. Inc., Mineola, NY, 1961.

[10] S. Sternberg, *Curvature in Mathematics and Physics*, Dover Books on Mathematics, Dover Publications, Mineola, NY, 2013.

The Myth of Leibniz's Proof of the Fundamental Theorem of Calculus

VIKTOR BLÅSJÖ

What was Leibniz's take on the fundamental theorem of calculus? He was one of the creators of the field after all, so one is naturally curious. But if you go to the library to find the answer to this question, you will be sold a bait-and-switch. You will be referred to his 1693 article [15], supposedly the only place where Leibniz explicitly stated and proved the fundamental theorem of calculus in print. The passage in question is reproduced in full in English translation in Struik [26, pp. 282–284], Calinger [4, pp. 354–356], and Laubenbacher and Pengelley [13, pp. 133–135]; discussed in full detail in Cooke [5, pp. 470–471], Hahn [11, pp. 125–128], Nitecki [25, pp. 292–293], Bressoud [3, pp. 101–102], and Nauenberg [24]; and cited in Katz [12, p. 529], Edwards [8, p. 260], Volkert [27, p. 104], González-Velasco [9, p. 357], Grattan-Guinness [10, p. 55], and Beyer [1, p. 163], all on the supposition that this is Leibniz's proof of the fundamental theorem of calculus. (Leibniz's complete paper is available in German translations in [23] and [20] and a French translation in [22].) If you study the proof, you will probably recognize it as a rather clunky way of saying $\int_a^b f(x)\,dx = F(b) - F(a)$ (where $F' = f$) in geometrical language. I shall argue that it is not. And this despite the fact that Leibniz clearly writes, "I shall now show that the general problem of quadratures can be reduced to the finding of a curve that has a given law of tangency" [15, p. 390]. Today everybody reads this as shown in the box.

Read through modern eyes in this manner, then, this looks like smoking-gun evidence that Leibniz is announcing his intention to prove the fundamental theorem. So it is not difficult to see how it came to be generally accepted as such in the literature. It is natural that scholars who know the centrality of the fundamental theorem of calculus in the modern conception of the field should go looking for its proof in

> THE STATEMENT OF LEIBNIZ AND ITS MISLEADING TRANSLATION INTO MODERN TERMS
>
> "The general problem of quadratures can be reduced to the finding of a curve that has a given law of tangency."
>
> The evaluation of a general integral $\int_a^b f(x)\,dx$ can be reduced to the finding of a function $F(x)$ that satisfies $F'(x) = f(x)$.

Leibniz, and it is understandable that this passage would then catch their eyes. But I shall argue that this is an anachronistic reading that misses the point of the argument completely. When Leibniz's paper is understood in its historical context, it becomes evident that it is meant to serve a different purpose.

Leibniz's Calculus

Before delving into the forgotten historical context that explains what Leibniz is up to in his paper, we may ask ourselves: if this isn't it, then how *did* Leibniz think about the fundamental theorem of calculus? I believe that, if cornered to argue for this result, Leibniz would have argued essentially as follows.

$$\int_a^b y'dx = \int_a^b \frac{dy}{dx}dx = \int_a^b dy$$

$$= \text{sum of little changes in } y \text{ from } a \text{ to } b$$
$$= \text{net change in } y \text{ from } a \text{ to } b$$
$$= y(b) - y(a).$$

For the other part of the theorem, just note from Figure 1 that if t increases by dt, then the area $\int_a^t y\,dx$ increases by $y(t)dt$, whence

$$\frac{d\int_a^t y(x)\,dx}{dt} = \frac{y(t)dt}{dt} = y(t).$$

This mode of reasoning is very much in line with Leibniz's conceptions of integrals and differentials. Indeed, he would most likely not consider this a "proof" of a "fundamental theorem," but rather a somewhat

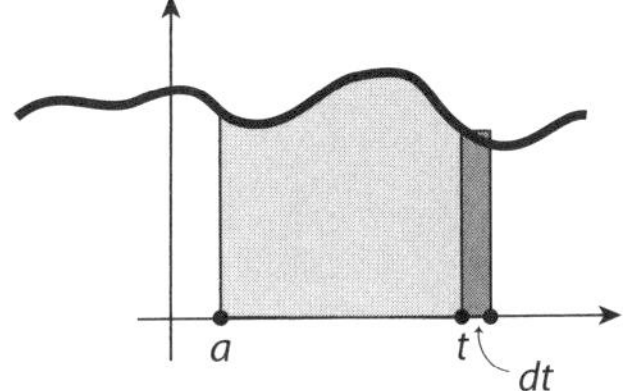

FIGURE 1. The integral $\int_a^t y\,dx$ and its differential.

tedious explication of the meaning of differentiation and integration. That is why he never put it in print. Instead he was satisfied with the casual statement that "as powers and roots in ordinary arithmetic, so for us sums and differences, or $\int$ and d, are reciprocal" [14, p. 297]. The comparison is an apt one not only procedurally but also foundationally: in neither case can there be a question of proof of the reciprocal relationship; rather it is built into the very meaning of the notions involved.

So much for the fundamental theorem, which, however, has nothing to do with the purpose of Leibniz's 1693 paper. To understand what Leibniz *did* intend in this paper, we must first understand its context. In the seventeenth century, Euclid's *Elements* was still the gold standard of mathematical rigor and method. One conspicuous aspect of this work is that Euclid speaks only of figures he can *construct* using ruler and compasses. The scope of Euclid's construction tools was soon found too restrictive, but his emphasis on constructions was retained.

Descartes's Construction Method

The Euclidean requirement of construction as a prerequisite for knowledge was taken very seriously by Descartes, who was to have a great influence on Leibniz. Descartes taught the world coordinate geometry and the identification of curves with equations in his *La Géométrie* of 1637 [6]. In connection with this, he also argued that the scope of mathematics should be extended to include all algebraic curves—to which his new method was especially suited—as opposed to being limited to the lines and circles of Euclid's *Elements* and the handful of more complex curves studied in antiquity. However, Descartes did not present this as a radically new way of doing geometry, different in principle from that of Euclid. Rather he argued at great length that his method was really nothing but the Euclidean program brought to its logical conclusion. In particular, he accepted curves represented by algebraic equations as legitimate mathematical objects only after he had found a way of constructing them in a Euclidean spirit.

Descartes's criterion for an acceptable construction is the following:

> To treat all the curves I mean to introduce here [i.e., all algebraic curves], only one additional assumption [beyond ruler and compasses] is necessary, namely, [that] two or more lines can be moved, one [by] the other, determining by their intersection other curves. This seems to me in no way more difficult [than the classical constructions] [7, p. 43].

The key phrase is "one by the other": Descartes has no objections to assemblages of curves pushing one another in whatever fashion as long as all the motions are ultimately generated by one and only one primitive motion. You can build a curve-tracing machine as intricate as you like, as long as one single point needs to be moved to operate it. This single-motion criterion is the key to Descartes's division of curves into "geometrical" (i.e., exact) and "mechanical" (i.e., not susceptible to mathematical rigor).

Figure 2 shows an example of Descartes's construction method. It can be adapted to generate algebraic curves of higher and higher degree.

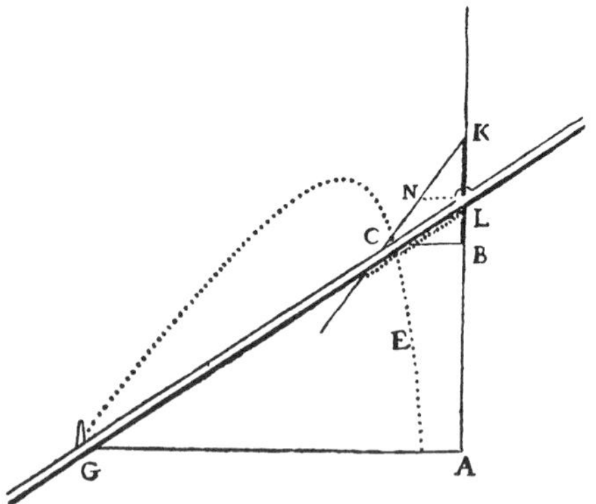

FIGURE 2. Descartes's curve tracing method [6, p. 321]. The triangle *KNL* moves vertically along the axis *ABLK*. Attached to it at *L* is a ruler, which is also constrained by the peg fixed at *G*. Therefore, the ruler makes a mostly rotational motion as the triangle moves upward. The intersection *C* of the ruler and the extension of *KN* defines the traced curve, in this case a hyperbola.

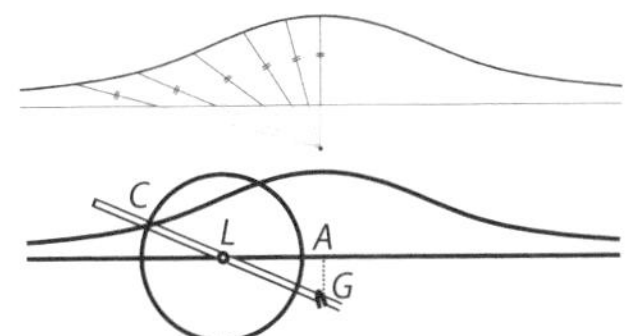

FIGURE 3. Top: The defining property of the conchoid, a famous algebraic curve studied in antiquity. Bottom: Construction of the conchoid using Descartes's method of Figure 2 with a circle in place of the line *KNC*.

For example, it is quite easy to see that replacing the line *KNC* by a circle produces a conchoid (Figure 3). And so it continues: once e.g. the conchoid has been generated it can be taken in place of the starting curve *KNC* to generate an even more complex curve, and so on.

These curve-tracing methods are what made algebraic curves legitimate geometry to Descartes. And they were so not in the sense of incidental or half-hearted attempts at justifying his new mathematics to obstinate colleagues stuck in old ways of thinking. Rather, these considerations formed the basis for his mathematical research from the very beginning. Already in 1619, *before* he had the idea of a correspondence between a curve and an equation, Descartes was concerned with "new compasses, which I consider to be no less certain and geometrical than the usual compasses by which circles are traced" (quoted from [2, p. 232]). The key criterion for these "new compasses," according to Descartes, was that they should trace curves "from one single motion," contrary to the "imaginary" curves traced by "separate motions not subordinate to one another," such as the quadratrix (Figure 4) or the Archimedean spiral (Figure 5). The coordination of motions in both of these constructions involve π, which, since π is transcendental, is nonconstructable (and hence unknowable) by Euclidean and Cartesian standards. As Descartes puts it in the *Géométrie*,

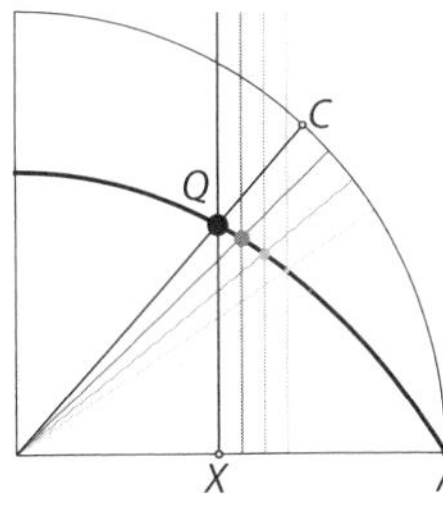

FIGURE 4. The quadratrix. *C* moves along the arc of a circle and *X* along its radius. Both points start at *A* and move at uniform speed in such a way as to reach the vertical axis at the same time, i.e., $\frac{d}{t}AC = \frac{\pi}{2}(\frac{d}{dt}AX)$. The intersection *Q* generates the quadratrix.

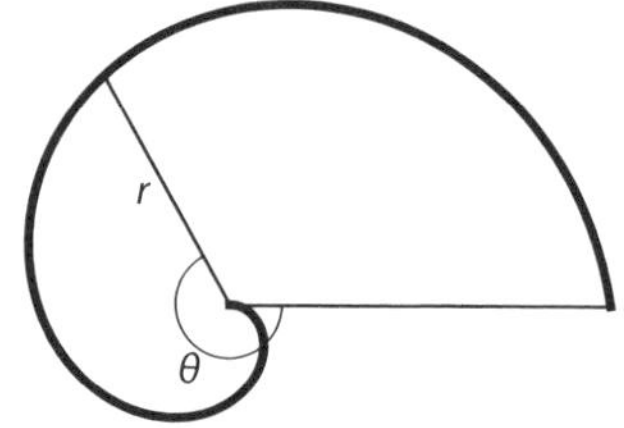

FIGURE 5. The Archimedean spiral $r = \theta$. The radial motion must be exactly coordinated with the rotational motion—a constructively impossible task, according to Descartes.

> The spiral, the quadratrix, and similar curves . . . are not among those curves that I think should be included here, since they must be conceived of as described by two separate movements whose relation does not admit of exact determination, [. . .] since the ratios between straight and curved lines are not known, and I believe cannot be discovered by human minds, and therefore no conclusion based upon such ratios can be accepted as rigorous and exact [7, pp. 44, 91].

By the time he published his *Géométrie*, Descartes had become convinced that his single-motion criterion included all algebraic curves (i.e., curves with polynomial equation of any degree), and nothing else. Convincing his readers of this—and thereby justifying the new algebraic methods in terms of the standards of classical, construction-based geometry—is one of the dominant themes of the *Géométrie*. (This is one of the main points of Bos [2], the definitive study of Descartes's geometry.)

Leibniz's Construction Method

These considerations form the direct background of Leibniz's 1693 article. He believed, as firmly as Descartes, that constructions are the bedrock of geometrical rigor. That is why he offered his own single-motion construction method, which can produce not only any algebraic curve but in fact any curve described by a differential equation of the form $dy/dx = f(x)$, where $f(x)$ can be any previously constructed function, just as the line *KNC* in Descartes's construction can be replaced by any previously constructed curve. This construction is what Leibniz's paper is all about.

Leibniz's construction goes as follows (Figure 6). The plane Π is a horizontal surface, say a table. On it is placed a weight at $C = (x, y)$

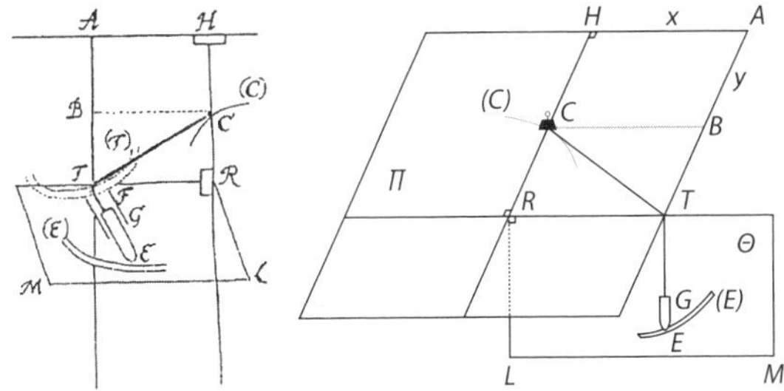

FIGURE 6. Leibniz's tractional-motion device for constructing the solution curve $C(C)$ of any inverse tangent problem. From [15], figure 3 (left), and my reproduction (right).

attached to a string TC. If we move the free end T of the string along the edge ABT of the table, the curve $(C)C$ generated by the moving weight would be the ordinary tractrix. But we shall modify this situation by having part of the string hang over the edge of the table. This end also has a weight attached to it, G, which ensures that it hangs straight down along the vertical plane Θ, until it hits the edge $E(E)$ protruding from this plane. Thus the fixed string length is $CT + TE$, and the length of the part TE hanging below the table is determined by the curve $E(E)$, which catches the weight at a point vertically below T. In fact, the length of TE is a function of the x-coordinate of the weight at C, for as C moves it pushes the "ruler" HR and thereby the vertical plane Θ ahead of it, so that $E(E)$ is effectively the graph of a function with $RT = x$ as input and TE as output. The curve $(C)C$ is traced as T is moved along the edge of the table away from A. The motion of T thus inflicts two separate motions on the plane Θ: one in the y-direction resulting directly from the motion of T, and one along the x-direction resulting from the motion of C.

In this way, we can generate a curve for which the length TC of its tangent is any given function of its x-coordinate. For if we seek a curve $C(C)$ for which $TC = \varphi(x)$, say, then we can always choose the curve $E(E)$ so that TE is the total string length minus $\varphi(x)$, which leaves just the required amount of string for the tangent TC. Thus if we write a for the total string length CTE, the required curve $E(E)$ is simply the graph of the function $a - \varphi(x)$ plotted in the plane Θ with RT as x-axis and RL as y-axis.

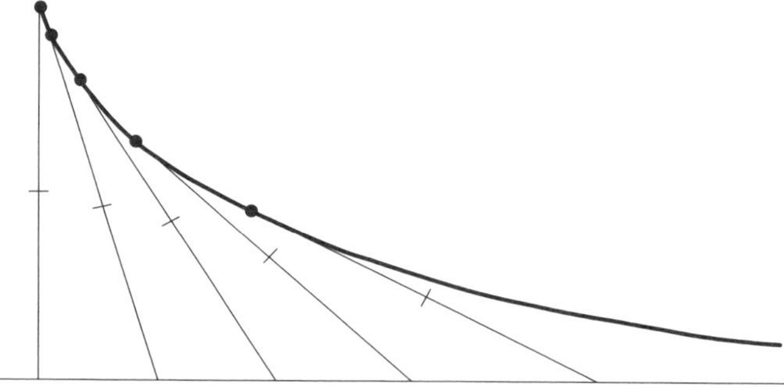

FIGURE 7. The tractrix is the curve traced by a weight dragged along a horizontal surface by a string whose other end moves along a straight line.

Alternatively, we can generate a curve with a given slope $dy/dx = TB/BC$. This reduces to the above problem since $TC = \sqrt{TB^2 + BC^2}$ is a simple algebraic function of TB and BC. Thus if we want to generate the curve $C(C)$ with the given slope $dy/dx = f(x)$, we note that in this case $BC = x$ and $TB = xf(x)$, so that $TC = x\sqrt{f(x)^2 + 1}$. Once we have this expression for TC, we can complete the construction of $E(E)$ as above.

In either case, then, since $\varphi(x)$ or $f(x)$ are given, it takes only "ordinary" Cartesian geometry to construct the required curve $E(E)$ that will enable the curve $C(C)$ with the desired property to be traced. In particular, Leibniz's construction gives the solution to $dy/dx = f(x)$, where $f(x)$ is any previously constructed curve, while assuming nothing more than Cartesian geometry and a single-motion tracing procedure. In this way, he enlarged the domain of constructable curves vastly beyond the algebraic curves admitted by Descartes while still adhering very strictly to Descartes's requirement of single-motion tracing and to the Euclidean-Cartesian construction framework generally.

Construction of Quadratures

Such was the purpose of Leibniz's paper. The confusion regarding the fundamental theorem arises from Leibniz's application to the problem of the construction of quadratures, i.e., the problem of constructing a line segment whose length equals a given area, or integral. This is quite clearly the guiding idea of the whole paper, whose title promises "a general construction of all quadratures by motion." In other words,

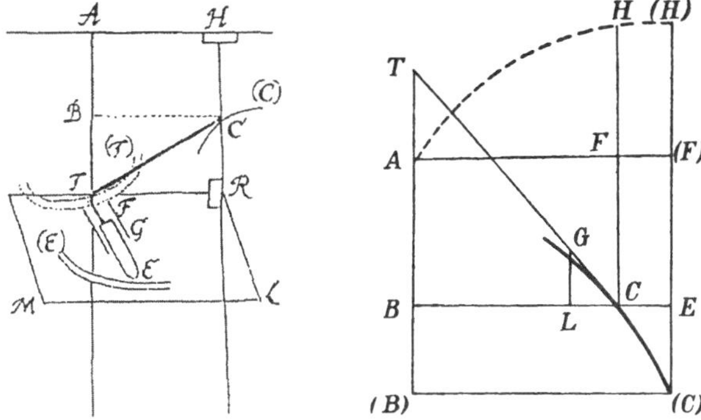

FIGURE 8. Leibniz's reduction of quadratures to rectifications. From [15], Figure 2, and as reproduced in [20, p. 31].

Leibniz wants to clarify that his construction not only solves any differential equation $dy/dx = f(x)$ but also any integral $\int_a^b f(x)dx$. This problem readily reduces to the above as follows (Figure 8). Let $AF = x$ and let $f(x) = FH$ be the function whose integral is to be constructed. As above, construct a curve $C(C)$ such that its slope $dy/dx = TB/BC$ always equals $f(x)$. Then it follows that $FC = y = \int f(x)dx = AFHA$, so the quadrature has been constructed as a line segment, as required.

Since the tractional construction itself is prima facie concerned with constructing curves with given tangent properties, a casual reader of Leibniz's paper might have missed that it can also be used to find a line segment equal to a given integral had Leibniz not taken the trouble to spell out this application specifically and even note it in the title of the paper. This construction of quadratures was a major problem at the time, quite apart from differential equations, so it was certainly worth highlighting.

Leibniz's Statement Reevaluated

It is in the course of this explanation that we encounter Leibniz's sentence quoted above that seemed to be a statement of the fundamental theorem: "I shall now show that the general problem of quadratures can be reduced to the finding of a curve that has a given law of tangency." Now that we understand its context, we see that to Leibniz this is a lemma linking the problem of quadratures to the tractional construction. It is not a fundamental theorem telling you to find an antiderivative F whenever you seek an integral $\int f dx$. Rather it is a specification of how the tractional motion needs to be set up to produce the values of $\int f dx$ as the y-coordinates of the tractional curve $C(C)$.

It is true that Leibniz's argument here concerns the relation between the differential equation $dy/dx = f(x)$ and the integral $\int_a^b f(x)dx$, and as such, to be sure, it is closely related to the fundamental theorem of calculus. But Leibniz's point is a much more specific one, and one very much specifically tailored to the setup of his tractional construction. It would be a big mistake, therefore, to forget about the context of the tractional construction and cut out the few lines relating to the fundamental theorem and study them as if they were meant as a proof of this general theorem. Yet this is precisely the mistake that occurs so often in the historical literature.

Leibniz would certainly consider it madness to apply his construction to an integral $\int f(x)dx$ for which an explicit antiderivative $F(x)$ can be found. Indeed, Leibniz says precisely this in a letter: "One cannot determine by this construction whether the sought quadrature can not also be carried out by common geometry; when this is possible one does not need the extraordinary route" [17, p. 694]. In such cases, he would simply go straight to $F(x)$, as he had done many times in print already before his 1693 paper.

Cases where $F(x)$ is algebraic had long been done and dusted, and logarithmic and trigonometric functions were also becoming common currency at this time. Certainly Leibniz would not spill ink in his 1693 paper on proving the fundamental theorem for use on such trivial cases.

The problem that interested him was integrals such as $\int \sqrt{1+x^4}\,dx$, or the corresponding differential equation $dy = \sqrt{1+x^4}\,dx$. Indeed, whenever Leibniz refers back to his paper, it is certainly never with reference to the fundamental theorem, but rather always as "my general construction of quadratures by traction" [18, p. 127], i.e., as showing that the tractional device "serves to construct all quadratures by an exact and regular motion" [16, p. 665]. Again, Leibniz [19, p. 157] explains that "I wished for the tractional method to be applied to the inversions of tangents [i.e., solving differential equations] rather than to quadratures where we already have [a method, namely finding $F(x)$]."

Conclusion

So, in conclusion, the irony of the story is that what is commonly referred to as Leibniz's proof of the fundamental theorem of calculus is actually his strategy for what to do when the theorem is of no use (in that one cannot find $F(x)$).

References

1. Horst R. Beyer, *Calculus and Analysis: A Combined Approach*, Wiley, 2010.
2. Henk J. M. Bos, *Redefining Geometrical Exactness: Descartes' Transformation of the Early Modern Concept of Construction*, Springer, 2001.
3. David M. Bressoud, "Historical Reflections on Teaching the Fundamental Theorem of Integral Calculus," *American Mathematical Monthly* 118(2) (2011), 99–115.
4. Ronald S. Calinger, ed., *Classics of Mathematics*, Moore, 1982.
5. Roger L. Cooke, *The History of Mathematics: A Brief Course*, 2nd ed., Wiley-Interscience, 2005.
6. René Descartes, *La Géométrie*, Leiden, 1637.
7. René Descartes, *The Geometry of René Descartes*, Dover, 1954.
8. C. H. Edwards Jr., *The Historical Development of the Calculus*, Springer, 1979.
9. Enrique A. González-Velasco, *Journey through Mathematics: Creative Episodes in Its History*, Springer, 2011.
10. Ivor Grattan-Guinness, ed., *Landmark Writings in Western Mathematics 1640–1940*, Elsevier, 2005.
11. Alexander J. Hahn, *Basic Calculus: From Archimedes to Newton to its Role in Science*, Springer, 1998.
12. Victor J. Katz, *A History of Mathematics: An Introduction*, 2nd ed., Addison-Wesley, 1998.
13. Reinhard Laubenbacher and David Pengelley, *Mathematical Expeditions: Chronicles by the Explorers*, Springer, 1999.
14. G. W. Leibniz, "De geometria recondita et analysi indivisibilium atque infinitorum," *Acta Eruditorum* (June 1686), 292–300.
15. G. W. Leibniz, "Supplementum geometriae dimensoriae, seu generalissima omnium Tetragonismorum effectio per motum: Similiterque multiplex constructio lineae ex data tangentium conditione," *Acta Eruditorum* (September 1693), 385–392.
16. G. W. Leibniz, Letter to Huygens, December 1, 1693, in [21, Vol. 5, No. 199].
17. G. W. Leibniz, Letter to R. C. von Bodenhausen, December 20, 1693, in [21, Vol. 5, No. 201].
18. G. W. Leibniz, Letter to Huygens, June 12, 1694, in [21, Vol. 6, No. 45].
19. G. W. Leibniz, Letter to Johann Bernoulli, October 6, 1696, in [21, Vol. 7, No. 39].
20. G. W. Leibniz, *Analysis des Unendlichen*, edited by Gerhard Kowalewski, *Ostwald's Klassiker der exakten Wissenschaften* No. 162, Engelmann, Leipzig, 1908.
21. G. W. Leibniz, *Sämtliche Schriften und Briefe. Reihe III: Mathematischer, naturwissenschaftlicher und technischer Briefwechsel*, Leibniz-Archiv, Hannover, 1976– , leibniz-edition.de.
22. G. W. Leibniz, *La Naissance du Calcul Differentiel: 26 Articles des* Acta Eruditorum, edited by Marc Parmentier, Vrin, 1995.

23. G. W. Leibniz, *Die Mathematischen Zeitschriftenartikel*, edited by H.-J. Heß and M.-L. Babin, George Olms Verlag, 2011.
24. M. Nauenberg, "Barrow, Leibniz and the geometrical proof of the fundamental theorem of the calculus," *Annals of science* 71(3) (2014), 335–354.
25. Zbigniew H. Nitecki, *Calculus Deconstructed: A Second Course in First-Year Calculus*, Mathematical Association of America, 2009.
26. Dirk Struik, *A Source Book in Mathematics, 1200–1800*, Harvard University Press, 1969.
27. Klaus Thomas Volkert, *Geschichte der Analysis*, Wissenschaftsverlag, 1988.

The Spirograph and Mathematical Models from Nineteenth-Century Germany

Amy Shell-Gellasch

What do the Spirograph and German kinematic models constructed in the late nineteenth century have in common? Quite a lot. The Spirograph is a toy that uses toothed rings and disks to draw intricate curves. Kinematic models are designed to draw special curves of interest to mathematicians and engineers.

The Spirograph, produced by Hasbro in the 1960s, is still in production. Unable to find the original set my siblings and I used as children, I recently purchased a new one. I wanted to get it as part of my research into the ten Schilling kinematic models donated to the Smithsonian National Museum of American History by the University of Michigan in 1967.

The German firm of Martin Schilling was one of the most prolific makers of mathematical models; its 1909 catalog lists more than three hundred. These physical and visual representations were popular teaching devices in the late-nineteenth and early-twentieth centuries. You may have seen or used a model of the conic sections in an analytic geometry course. They were used in diverse disciplines such as architecture, engineering, and various fields of applied mathematics.

Sadly for those of us who love to touch our mathematics, mathematical models have largely gone the way of the dodo bird. (Although, with the recent appearance of low-cost 3-D printers, I am hopeful that physical models will once again be a part of our educational tool kit.)

Today we can draw curves using computer algebra systems, graphing calculators, and even cell phones. But not long ago, mathematical curves could not be drawn without a tool. Kinematic models, as the name implies, are devices that move in some way, and these Schilling models show how to generate certain curves mechanically.

Straight-Line Motion

The Schilling models are interesting both for the curves they produce and for the ingenious mechanisms that draw them. For example, how would you create a device to produce a straight line? At the height of the Industrial Revolution, an important engineering challenge was to solve the "problem of parallel motion": to convert rotational motion to linear motion using only rods, joints, and pins.

The Scottish engineer James Watt, who invented improvements to early steam engines, was the first to tackle this prickly problem. His 1784 steam engine patent contains a mechanical linkage that could convert circular movement into *approximately* straight-line movement. Watt's setup allowed the vertical motion of a piston to drive a rod, one end of which moves along a circular arc. Figure 1 shows Watt's three-bar linkage. Points *A, B, C,* and *D* are hinged, with *A* and *D* fixed and *B* and *C* free to move. Point *P* traces out the nearly straight segment of the lemniscate.

In 1864, the French engineer Charles Nicolas Peaucellier created a seven-bar linkage that produced pure linear motion. Figure 2 shows a "Peaucellier cell" or "Peaucellier inversor." In this context, "inverse" is a geometric term that refers to the process of converting, or inverting, one geometric shape into another—the conversion of a circle to a straight line in the case of this model.

In this model, the leftmost point is fixed in place. The radial arm rotates in a circular motion. As it does, the rightmost joint of the linkage moves up and down in a perfectly straight line.

The Smithsonian collection also contains linkages that produce linear motion designed by H. Hart and by A. B. Kempe and J. J. Sylvester.

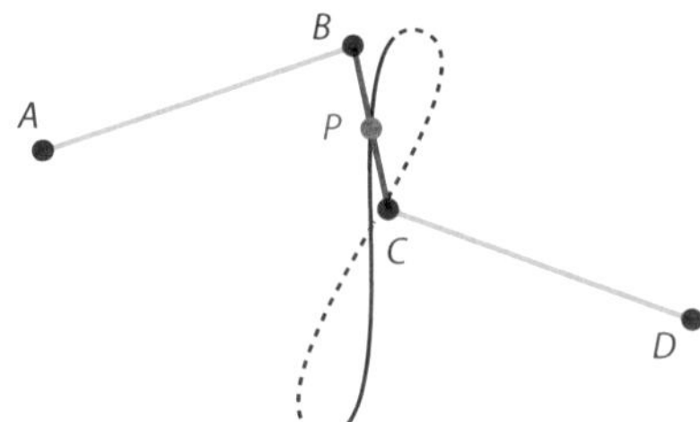

Figure 1. Watt's linkage.

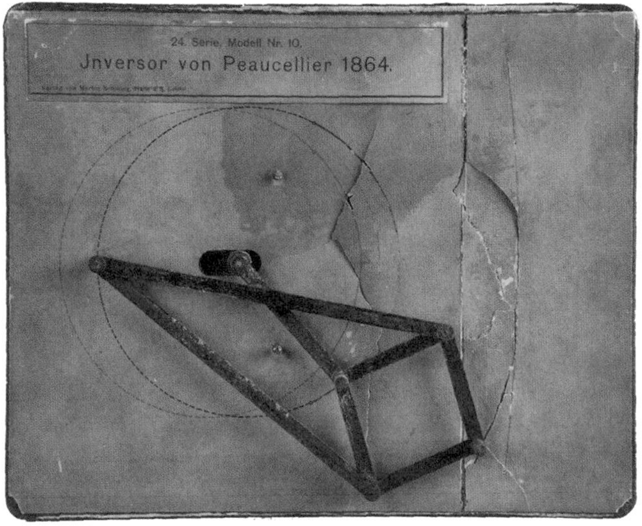

FIGURE 2. Schilling model number ten—Peaucellier's inversor. Smithsonian negative DOR2013-50203

Trochoids

Several of the Schilling models draw curves known as trochoids. Trochoids are formed by tracing a point on either the radius of a circle, the circumference of a circle, or the extension of the radius of a circle as it rolls along another curve.

The cycloids are a famous subfamily of trochoids, in which a circle, called a roulette, rolls along a straight line. The most well known of these curves is the epicycloid, which is often referred to simply as a cycloid—the name given to it by Galileo. Its two sisters are the curate and prolate cycloids. The red curve in Schilling model seven (Figure 3) is a prolate cycloid, the blue curve is the classic cycloid (the epicycloid), and the green curve is a curate cycloid.

The study of the cycloid and its many fascinating properties dates back to the ancient Greeks. The Romans used it to shape the arches of bridges and aqueducts. It has been the plaything of physicists, engineers, architects, and, especially, mathematicians—including many of the greats of the seventeenth century. The eighteenth-century

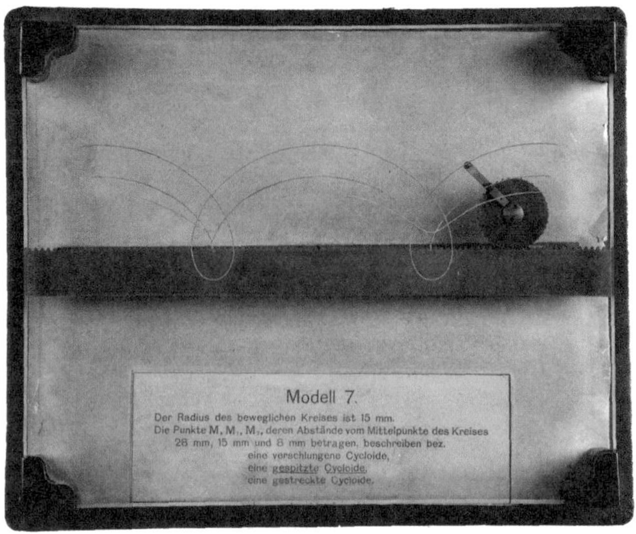

Modell 7.
Der Radius des beweglichen Kreises ist 15 mm.
Die Punkte M, M_1, M_2, deren Abstände vom Mittelpunkte des Kreises 28 mm, 15 mm und 8 mm betragen, beschreiben bez.
eine verschlungene Cycloide,
eine gespitzte Cycloide,
eine gestreckte Cycloide.

FIGURE 3. Schilling model seven shows all three types of cycloids. Smithsonian negative DOR2013-50217

mathematician Jean Montucla called it "the Helen of the geometers" because of the debates over the nature of the cycloid.

The cycloid solves the famous brachistochrone problem, which was posed in 1696 by Swiss mathematician Johann Bernoulli. This problem asks: Given points *A* and *B*, with *B* below *A*, along what path will a ball roll under gravity from *A* to *B* in the shortest time compared with all other paths? The surprising answer: the arc of an inverted cycloid (Figure 4).

The cycloid also solves the tautochrone problem: A ball placed anywhere on a ramp shaped like an inverted cycloid will reach the bottom under gravity in the same amount of time. In other words, two balls placed anywhere on a cycloidal ramp and released simultaneously will reach the bottom at the same instant.

The Dutch physicist Christiaan Huygens used this property of the cycloid to design a better timekeeper. For an accurate pendulum clock, the period of the swinging pendulum must remain constant. But the amplitude decreases over time because of friction, and thus so does the period. Huygens discovered that by placing cycloid-shaped baffles on either side of the pendulum to constrain its swing, it follows a cycloid-shaped path, yielding a constant period. (This was a brilliant theory,

FIGURE 4. The fastest track from A to B is a cycloidal arc.

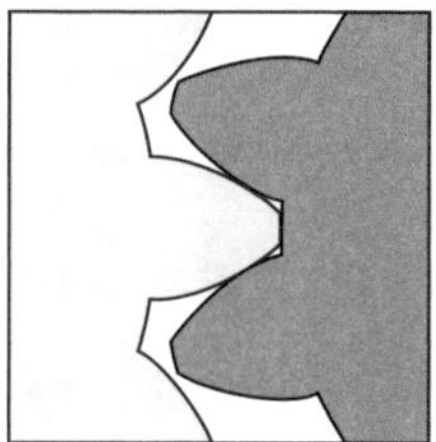

FIGURE 5. Gear teeth shaped using arcs of a cycloid.

but the clock was a disappointment: The friction from the baffles negated the benefits from the improved path.)

Trochoids were even useful in machine design. In the late nineteenth century, it was discovered that if the sides of gear teeth and the bottoms of the valleys between the teeth are a portion of a cycloidal curve, the gears roll around each other with less friction and thus more smoothly and with much less resistance and wear. See Figure 5 or woodgears.ca/gear/cycloid.html.

Schilling model three produces the family of trochoids known as hypotrochoids in which the roulette rolls around the inside of the circle. The roulette forms epitrochoids when it rolls around the outside of the circle. In Figure 6, using the same nomenclature as the cylcoids, the red curve is a prolate hypotrochoid, the blue curve is a hypotrochoid, and the green curve is a curate hypotrochoid.

But what does all this have to do with the Spirograph? The Spirograph consists of several plastic disks (also known as roulettes) with toothed edges and variously placed holes in the interior (Figure 7). The toy also comes with two plastic toothed rings of different radii, a toothed bar, and some other funky shapes just for fun. The toothed bar can be used to draw a curate cycloid, while the toothed rings can be used to draw trochoids. My new Spirograph set comes with adhesive putty to secure the pieces to the paper, which works better than the old method of using pins. Now I don't prick myself.

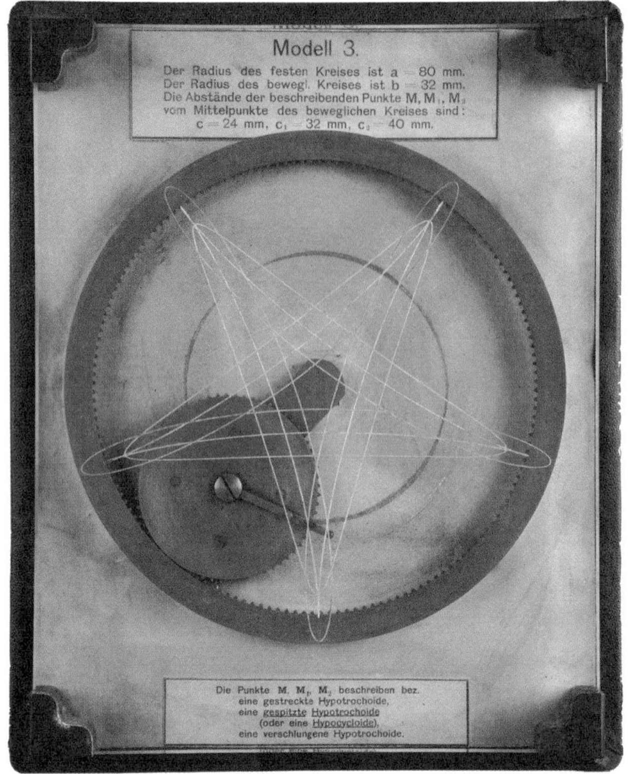

FIGURE 6. Schilling model three shows all three types of hypotrochoids. Smithsonian negative DOR2013-50214

Once all is ready, the artist places a pen through a hole in the roulette and rolls it around the ring or bar. Because the pen is inside the rolling disk, it creates only curate trochoids. The roulette rolling around the outside of the ring in Figure 7 forms a curate epitrochoid (shown in green), while the roulette rolling around the inside of the ring forms a curate hypotrochoid (shown in blue). The complexity of the drawing depends on the radius of the roulette and the position of the pen along the radius—the closer to the edge of the roulette, the more loopy the drawing.

Although most of these models have disappeared from the classroom and can be found only in a dusty closet in a college math department or in the archives of a science museum, some of them are seeing the virtual light of day in online museum collections such as the Smithsonian's. To

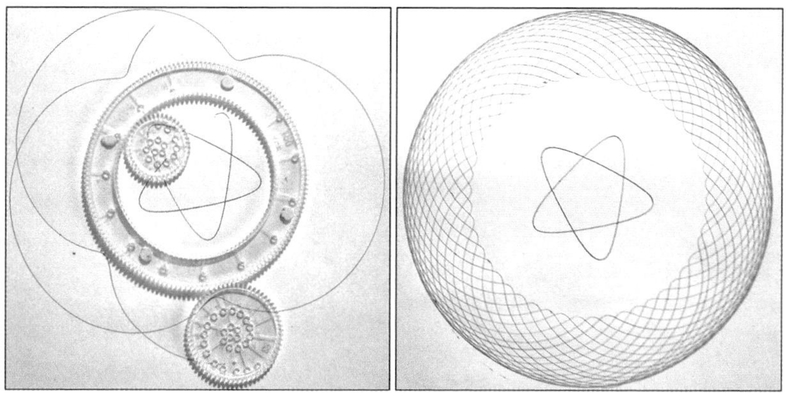

FIGURE 7. The author's Spirograph drawing in progress (left) and completed (right).

a math historian like me, working with the mathematical items housed at the Smithsonian's National Museum of American History has been like being a kid in a toy shop!

Further Reading

You can find these Schilling kinematic models and many more scientific and mathematical items on the Smithsonian Institution's website, americanhistory.si.edu/collections.

For information on all kinds of physical models used in teaching, see *Tools of American Mathematics Teaching, 1800–2000*, by P. A. Kidwell et al. (Baltimore: Johns Hopkins University Press, 2008).

For a history of the cycloid, see "The Helen of Geometry," by John Martin (*College Math Journal*, January 2010).

What Does "Depth" Mean in Mathematics?

John Stillwell

1. Introduction

Mathematicians often call theorems "deep," but the term has never been precisely defined. It is not easy even to explain the intuition behind it, though it probably includes the idea of being hidden, buried, and hard to uncover. A deep result, or a deep connection, is *un*covered only by digging through several layers of concepts and proofs. Viewed from the other direction, a deep result *emerges* or *comes to light* only after the intermediate layers of concepts and proofs have been elucidated; so deep theorems are generally the work of several generations of mathematicians.

The history of mathematics bears this out. Most results considered deep today were found by "standing on the shoulders of giants," as Newton said:

> What Descartes did was a good step. You have added in several ways . . . If I have seen a little further it is by standing on the shoulders of giants.
>
> —Isaac Newton, letter to Robert Hooke, February 5, 1676

In this article, I have selected examples of depth initially by the historical criterion—according to the number of giants involved in their discovery. Later, I try to find criteria from mathematics or logic that support the judgment of depth in other ways. The latter criteria are very tentative and, as will be seen, they "explain" depth only for a limited class of examples. Nevertheless, I think any possibility of formalizing the concept of depth is worth pointing out.

I give examples of "large," "medium," and "small" depth. One reason is that the relation "deeper than" very likely makes more sense than the property "deep" (just as the relation "larger than" for numbers makes more sense than the property "large"). Another reason is that the small and medium examples are easier to understand than the large examples because the large examples are still mysterious to most, if not all, mathematicians. Also, investigating theorems at the levels of medium and small depth leads us to distinguish "historical depth" from at least two other notions of depth.

In Sections 3 and 4, I contrast historical depth with "foundational depth," which I attribute to fruitful theorems often called "fundamental." Such theorems are deep in the sense that they *underlie* many other theorems in their areas. Finally, in Sections 5 and 6, I search for notions of "formal depth" in mathematical logic. Logic has demonstrated that there are hierarchies of provability, complexity, and unsolvability, but so far these notions do not seem to capture the concept of depth in mainstream mathematics very well. Nevertheless, there are formal concepts that seem to reflect the depth of some of our medium and small examples.

I begin with the examples of greatest depth, since they are the most generally accepted examples of depth and they best illustrate "standing on the shoulders of giants." Then I discuss examples of medium depth, accessible to undergraduates. These are less mysterious, but at the same time it is clearer what makes them deep—it is because the real numbers play an essential role. Finally, I give an example from geometry which is only slightly deep, but the reason for the difference in depth is very clear: the deeper case is the one that essentially involves infinity.

2. Some Theorems That Are Considered Deep

Here are four theorems that are widely regarded as deep. The first is much earlier than the others, and by today's standards not as deep. But it was notable as probably the deepest theorem of its time, and the harbinger of analytic methods in number theory. In fact, a proof of Dirichlet's theorem *not* using analysis (a so-called "elementary" proof) was obtained only in the 1940s, by Selberg.

The other three theorems, like Dirichlet's, are about discrete objects, but their proofs were first obtained only with the help of clues from the continuous world.

1. **Dirichlet's theorem on primes in arithmetic progressions**: When $gcd(a, b) = 1$, the sequence a, $a + b$, $a + 2b$, $a + 3b$, . . . contains infinitely many primes.
2. **The Poincaré conjecture**: Any compact, simply-connected 3-dimensional manifold is homeomorphic to the 3-sphere.
3. **Fermat's last theorem**: The equation $x^n + y^n = z^n$ has no solution in positive integers when $n > 2$ is an integer.
4. **The classification of finite simple groups**: If G is a finite simple group, then G is either: cyclic of prime order, A_n for some $n \geq 5$, of Lie type, or one of 26 "sporadic" groups.

2.1. Dirichlet's Theorem (1839)

The "depth" of Dirichlet's theorem is indicated by its reliance on analysis. The proof rests upon

- The Euler [1748] formula for $\zeta(s)$

$$\sum \frac{1}{n^s} = \prod \frac{1}{1 - p^{-s}},$$

  relating a product over the primes to a sum over all positive integers.
- Analysis: Newton and Leibniz, late seventeenth century.
- Proof that there are infinitely many primes: Euclid (c. 300 BCE).

Euclid originated the very idea that there are infinitely many prime numbers; so the *Elements* launched a thousand proofs of the infinitude of primes. Euler's formula gave another, and Dirichlet's was the most sophisticated of his time.

However, the step from Euler's theorem to Dirichlet's was not straightforward, and Dirichlet found the way only as a result of trying to do something else: find a formula for the *class number for quadratic forms*. His discovery of such a formula in 1837 depended upon

- Theory of quadratic forms [Gauss, 1801].
- Definition of class number [Lagrange, 1773].
- Theorems on primes of forms: $x^2 + y^2$, $x^2 + 2y^2$, $x^2 + 3y^2$, $x^2 + 5y^2$, which hint (very subtly) at the existence of class number (Fermat, c. 1640).

2.2. The Poincaré Conjecture (1904); Proved in 2002

A compact 3-manifold M can be defined as a space, divisible into finitely many polyhedral cells, in which each point has a neighborhood homeomorphic to $\mathbb{R}^3$.

M is *simply connected* if each closed path (of edges) bounds a topological disk, and $\mathbb{S}^3$ is a specific 3-manifold homeomorphic to the hypersurface

$$w^2 + x^2 + y^2 + z^2 = 1 \text{ in } \mathbb{R}^4.$$

Thus the conjecture of Poincaré [1904], that any compact simply connected 3-manifold is homeomorphic to $\mathbb{S}^3$, is in principle a combinatorial problem. The many attempts to prove it, from 1908 up to the 1970s, were all combinatorial, but they all failed. The conjecture was proved in 2002 by Grigory Perelman, who famously turned down a Fields Medal and the Clay Institute million dollar prize for his work.

Perelman's proof rests upon

- Ideas from differential geometry and partial differential equations—Ricci curvature and the Ricci flow—introduced by Richard Hamilton in the 1980s, in an attempt to prove *Thurston's geometrization conjecture.*
- Which was inspired by the *geometrization of compact 2-manifolds*: each compact surface can be given a geometry of constant curvature.
- Geometrization of 2-manifolds was used by Nielsen in the 1920s and 1930s to prove results about mappings of surfaces. (Thurston in the 1970s rediscovered some of Nielsen's results).
- Dehn (1910–1912) used geometrization to discover combinatorial results, such as *Dehn's algorithm.*
- Poincaré himself used geometrization in his 1904 paper to study curves on surfaces.

Thus Poincaré had the beginning of a successful idea, but it needed much more analysis to make it work.

2.3. Fermat's Last Theorem

Fermat conjectured his theorem, that $x^n + y^n \neq z^n$ for positive integers x, y, z, and $n > 2$ when considering whether it was possible to generalize Euclid's solution of $x^2 + y^2 = z^2$ ("Pythagorean triples"). Fermat's own

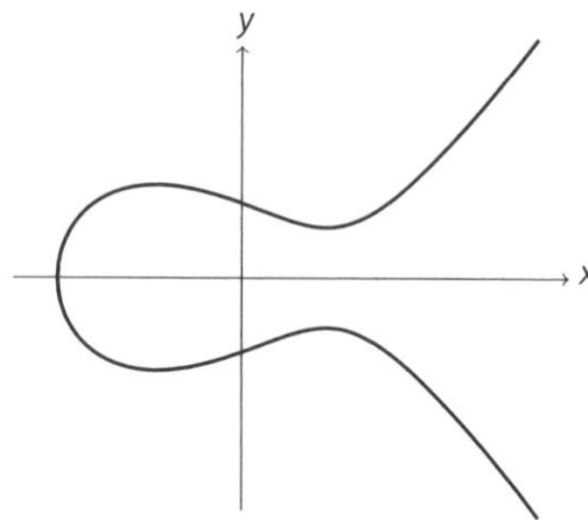

FIGURE 1. An elliptic curve.

contribution (around 1640), $x^4 + y^4 \neq z^4$, rested on Euclid's solution of $x^2 + y^2 = z^2$, and the first results on the problem were for isolated values of n.

Euler [1770] proved $x^3 + y^3 \neq z^3$, by an argument that brought in algebraic number theory to factorize $x^3 + y^3$. This idea was extended to many values of n by Kummer in the 1840s, with the help of a much more sophisticated theory of algebraic numbers (introducing *ideal theory* to get around the problem of nonunique prime factorization). Hilbert made Kummer's results the climax of his *Zahlbericht* of 1897, but the approach did not get much further.

The proof of Fermat's last theorem, which was eventually found by Wiles [1995], rested on an entirely different nineteenth-century development: *elliptic curves* (Figure 1).

- These concepts developed in the nineteenth-century theory of elliptic curves and modular functions (Jacobi, Clebsch, Hermite, Klein).
- It was made possible by the theory of elliptic functions (Gauss 1790s, Abel and Jacobi 1820s).
- Which was developed to clarify the theory of elliptic integrals (Fagnano 1718, Euler 1751).
- The original example of which is the integral for the arc length of the ellipse, found by Wallis in 1655.

2.4. FINITE SIMPLE GROUPS

Group theory is a relatively new area of mathematics, beginning its development implicitly in Lagrange's theory of algebraic equations in 1770, and explicitly in Galois's theory of equations around 1830. Galois introduced the concept of normal subgroup and solvable group to explain which equations are solvable by radicals (see also Section 4.2 here). He saw that the general quintic equation is *not* solvable by radicals because its group is not solvable. The group in question, A_5, is in fact as far from solvable as possible, because it is nonabelian and *simple*.

Simple groups are those that cannot be *simplified*, in the sense of admitting a homomorphism onto a smaller group, whereas solvable groups always can be simplified. Galois found a whole infinite family of finite, nonabelian simple groups: the alternating groups A_n for $n \geq 5$. Before then, only trivial examples were known: namely the cyclic groups of prime order.

This problem raised the question (hopefully, the first step toward classifying all finite groups): how can we classify the finite simple groups? This turned out to depend upon:

- The classification of certain *continuous* groups, the simple Lie groups. Lie groups were introduced by Lie around 1880, and the simple ones were classified by Killing around 1890. This was itself a very deep theorem, not foreseen by Lie.
- It was relevant to finite groups because many Lie groups are matrix groups with real or complex entries. They have finite analogs (matrix groups over finite fields), which were classified (mainly by Chevalley) in the 1950s, following the Lie group classification.
- After this, only finitely many *sporadic* groups remained to be found (1960s and 1970s), but finding them was really hard! It depended on theorems, e.g., the Feit–Thompson theorem that odd-order nonabelian groups are not simple, that are themselves very deep.

In 1911, Burnside conjectured that simple nonabelian groups of odd order do not exist. This was proved by Feit and Thompson [1963] in a paper of around 250 pages. In 2012, a computer-checkable proof of the theorem was obtained by Georges Gonthier after six years of work using the Coq proof assistant.

Number of lines ~ 170,000
Number of definitions ~ 15,000
Number of theorems ~ 4,200

These figures capture what one expects depth to be, I think. The proof is not merely long, but also a highly nested structure with dependency of concepts and proofs on many levels.

3. *Criteria for Depth*

The above theorems are "historically deep" in the sense that it took a long time to uncover them, and many other theorems had to be uncovered first. But how well do they match some of the other criteria that have been thought to indicate depth?

They are certainly *difficult, surprising* (in what it took to prove them), and their proofs are *explanatory* (of things that had to be understood in order to construct the proofs).

We could also say they are *important* (in the sense that good problems are important) because their proofs enriched whole fields of mathematics: analytic number theory, geometric topology, elliptic curves, group theory.

But it is debatable how far the theorems *themselves* are

fruitful
elegant
fundamental

The latter criteria are better satisfied by theorems that are less deep (though still *somewhat* deep) but better understood. The latter have what I call "foundational depth" because they turn out to lie at the foundations of some area of mathematics, supporting and explaining a large body of other facts. At the same time, they have a certain "conceptual depth," in the sense that they all involve the real numbers. Whether this is related to their foundational depth is not clear, but the role of the real numbers in deep theorems seems to be a recurring theme.

4. *Deep Theorems at the Undergraduate Level*

The following theorems are sometimes covered at the undergraduate level because they are fundamental, fruitful, or the answer to natural questions. But they are deep enough to cause problems for students and others; so they are often glossed over or mentioned without proof. Even for those who become mathematicians, I suspect that one or more of these theorems was perplexing at first encounter. The probable cause of most of these difficulties lies in the nature of the real numbers.

1. Independence of the parallel postulate.
2. Fundamental theorem of algebra.
   Any polynomial equation with real coefficients has a solution in the complex numbers.
3. $ab = ba$ for $a, b \in \mathbb{R}$ or $\mathbb{C}$.
4. Riemann integrability of continuous functions.
   For any function f, *continuous on an interval* [a, b], *the Riemann integral* $\int_a^b f(x)\,dx$ *exists.*
5. Uncountability of $\mathbb{R}$.
   (particularly Cantor's 1891 proof by the diagonal argument).

These theorems also have considerable historical depth, which has to do with the gradual uncovering of properties of $\mathbb{R}$. Mathematics has had two bruising encounters with $\mathbb{R}$—in ancient Greek geometry after the discovery of irrationals, and in the seventeenth century with the development of calculus and the attempt to base it on infinitesimals—before the resolution of many (but not all) of the difficulties with $\mathbb{R}$ in the nineteenth century.

4.1. Independence of the Parallel Postulate

A parallel axiom was postulated by Euclid, but it has a more complicated character than his other axioms, which was regarded as a flaw. Over the next two thousand years, there were determined attempts to *prove* it.

1733 Saccheri gave the ultimate attempt in his *Euclid freed from every flaw*, reducing the negation of the parallel axiom to the apparent absurdity of asymptotic lines with a common perpendicular at infinity. It seemed absurd, yes, but it was not contradictory.

1820s Bolyai and Lobachevsky took the opposite tack, assuming the negation of the parallel axiom and deriving from it many interesting and apparently noncontradictory consequences.

1830s Minding found that Bolyai's and Lobachevsky's "hyperbolic trigonometry" held locally on surfaces of constant negative curvature. So, there was a "local" model of non-Euclidean geometry.

However, no one found a complete simply connected surface of constant negative curvature in $\mathbb{R}^3$. This problem frustrated the search for a "global" model of non-Euclidean geometry. What was needed was an abstract concept of manifold, not dependent on an embedding in $\mathbb{R}^n$.

1854 Riemann proposed an abstract manifold concept, formalized today using charts.

1868 Beltrami, inspired by Riemann's work, gave abstract models of non-Euclidean geometry.

Thus the first proof of independence relied on differential geometry.

1871 Noticing some hints of non-Euclidean geometry in the work of Cayley, Klein reinterpreted one of Beltrami's models in projective geometry. This placed non-Euclidean geometry in a setting that was already familiar.

1882 Poincaré interpreted two Beltrami models in the plane of complex numbers, with isometries that are certain linear fractional transformations. Like Klein's result, this placed non-Euclidean geometry in a familiar setting, where it could in fact help to solve problems in complex analysis (by bringing geometric intuition to bear on them).

Both the Klein and Poincaré models depend on a little analysis, since distance is defined using the logarithm function.

Once the independence of the parallel axiom was known, it became clear that it is in fact the *characteristic* axiom of Euclidean geometry—the axiom that distinguishes Euclidean geometry from the other geometries of constant curvature. Thus the independence proof reveals the parallel axiom to be *fundamental* to geometry.

It is typical of these medium-deep theorems to be fundamental theorems of some area of mathematics. They have "foundational depth" because they underlie many other theorems in their areas. This is not necessarily the case for some of the extremely deep theorems of Section 2, but it probably holds for the more general theorems used to prove them. For example, the Poincaré conjecture may not underlie many theorems about 3-manifolds, but the Thurston's geometrization conjecture certainly does.

4.2. The Fundamental Theorem of Algebra

The search for solutions of polynomial equations began in Italy around 1505–1545, when del Ferro and Tartaglia solved $x^3 = px + q$ "by radicals":

$$x = \sqrt[3]{\frac{q}{2} + \sqrt{\left(\frac{q}{2}\right)^2 - \left(\frac{p}{3}\right)^3}} + \sqrt[3]{\frac{q}{2} - \sqrt{\left(\frac{q}{2}\right)^2 - \left(\frac{p}{3}\right)^3}}.$$

Tartaglia disclosed the solution to Cardano, whose student Ferrari solved the quartic by radicals. This gave hope that every polynomial equation might be solved by radicals, leading to a *fundamental theorem of algebra* that every polynomial equation has a solution in the complex numbers.

However, this approach stalled on the quintic, and the fundamental theorem of algebra was eventually proved quite differently.

1799 Gauss took a new approach: *prove existence of a solution from general properties of continuous functions*. He gave several attempted proofs; the most nearly correct one assumed the *intermediate-value theorem* for continuous functions.

1817 Bolzano saw that the problem reduced to proving the intermediate-value theorem, which he tried to do by assuming the *least-upper-bound property* of real numbers. But he lacked a *definition* of the real numbers.

1858 Dedekind proposed a definition of R, by *Dedekind cuts*, which allowed the least-upper-bound property to be proved. (It was influenced by the Eudoxean "theory of proportions" expounded in Book V of Euclid's *Elements*.)

This work established the basic properties of continuous functions, such as the intermediate-value and extreme-value theorems. The gaps in the proof of the fundamental theorem of algebra were now filled. The key conceptual advance was Dedekind's definition of $\mathbb{R}$, which implied the *completeness* of R necessary for the intermediate-value theorem.

Thus, the quest for the "fundamental theorem of algebra" was actually fundamental for the theory of real numbers and continuous functions.

Meanwhile, the original approach—solving equations by radicals—had been ruled out by the discovery of Galois that an equation is solvable by radicals if and only if its group is solvable. His discovery, however, initiated the study of finite simple groups, as we saw in Section 2.4.

4.3. $ab = ba$—Deeper Than It Looks

The meaning of this equation depends on the interpretation of a and b, and on the interpretation of multiplication. For Euclid, a and b were line segments, and ab was the *rectangle* with adjacent sides a and b, as shown in Figure 2. Then $ab = ba$ goes without saying (or perhaps by saying that "and" is commutative).

With Dedekind's 1858 definition of $\mathbb{R}$, published in [Dedekind, 1872],

$$ab = ba \text{ for } a, b \in \mathbb{R} \text{ follows from } ab = ba \text{ for } a, b \in \mathbb{Q}$$
$$\text{hence from } ab = ba \text{ for } a, b \in \mathbb{N}.$$

Before 1860, mathematicians seem to have assumed that $ab = ba$ is obvious for natural numbers. They continued to have the rectangle picture in mind. For example, it occurs at the beginning of Dirichlet's lectures on number theory, published by Dedekind in 1863. The first to look for a deeper explanation of $ab = ba$ was Grassmann, who discovered that *induction* provides it; not only for $ab = ba$ but for all the basic laws of arithmetic. In fact, Grassmann [1861] proved all the basic properties of $\mathbb{N}$ by induction (anticipating *Peano arithmetic*, whose fundamental axiom is the induction axiom). The explanation of the commutative law of multiplication is in fact one of the deeper ones, taking until Proposition 72 in the book! (See Figure 3.)

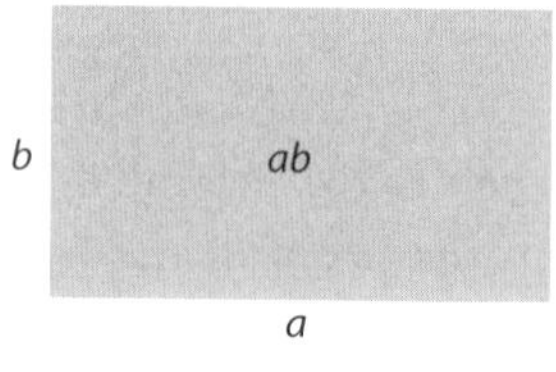

b ab

a

Figure 2. Rectangle interpretation of the product of lengths.

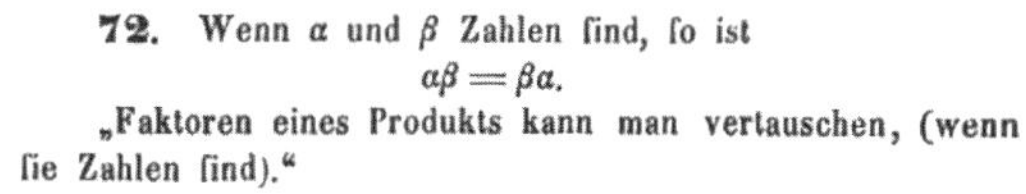

72. Wenn α und β Zahlen ſind, ſo iſt
$$\alpha\beta = \beta\alpha.$$
„Faktoren eines Produkts kann man vertauschen, (wenn ſie Zahlen ſind)."

FIGURE 3. Grassmann's Proposition 72.

ab = ba in Projective Geometry

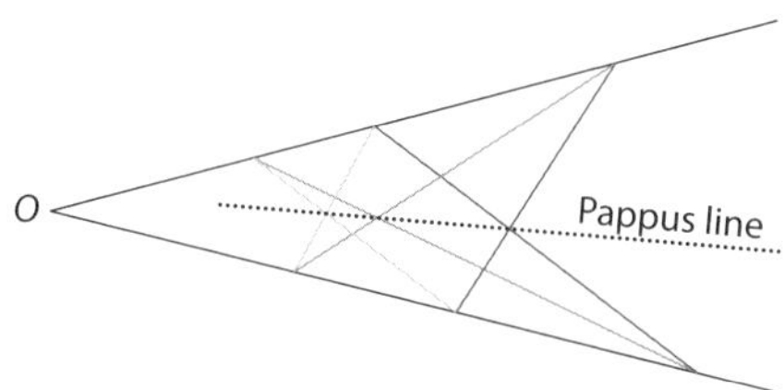

FIGURE 4. The setup for the Pappus theorem.

There is another geometric explanation of $ab = ba$, far less obvious than that of Euclid. It occurs in the geometric context of *projective geometry*, and in the algebraic context of *fields*. Using some ideas of von Staudt [1847], Hilbert [1899] showed that the field concept (including $ab = ba$) can be based on the *Pappus theorem*. This theorem says that if a hexagon has vertices alternately on two lines, then the intersections of its opposite sides lie on a line (the dotted *Pappus line* shown in Figure 4, where the "hexagon" is the sequence of six line segments zigzagging between the two lines with origin O).

The idea is to add Pappus's theorem, as an extra axiom, to the three "obvious" axioms for a projective plane:

1. There exist four points, no three of which are in a line.
2. Any two points lie on a line.
3. Any two lines have a common point.

When this is done, the resulting four axioms allow us to define the sum and product of points on a line and to prove that all the (nine) field axioms hold. Conversely, in any plane that admits a sum and product with the field properties, one can define coordinates and use them to prove the Pappus theorem.

Thus the Pappus theorem provides an entirely different, and deeply hidden, foundation for the laws of arithmetic (playing a role like that of induction in the Peano axioms). In fact, it is a foundation for all fields, not just the field of rational numbers whose properties arise from induction.

4.4. Integrability of a Continuous Function f on $[a, b]$

In seventeenth-century calculus, the real problem of integration was thought to be *evaluation* of the integral. Its existence was thought to be obvious, just as it was (wrongly) thought obvious that any continuous function is differentiable, e.g., by Newton [1671]. The problem of existence came into focus only gradually, as did the concept of continuity.

Around 1750, the vibrating string problem led d'Alembert, Daniel Bernoulli, and Euler to more general concepts of "function" and "continuity." It was debated whether every function could be expressed by a formula, and whether infinite formulas (such as Fourier series) were valid. Even when Cauchy gave essentially the modern definition of continuous function in the 1820s, it was not noticed that integration needs the more refined concept of *uniform* continuity.

Full clarity was not attained until the mid-nineteenth century, along with the definition of R and the concept of *compactness*. Compactness of closed intervals ensures the integrability of a continuous f on $[a, b]$ by implying its uniform continuity, via the Heine-Borel theorem. The latter theorem was first appreciated by Borel [1898]. He called it the "first fundamental theorem of measure theory" because it implies, among other things, that a covering of the closed unit interval by open intervals necessarily has total length ≥ 1.

Now we know that several aspects of compactness are equivalent (in the sense of *reverse mathematics*, see Section 5.1) and hence "equally deep":

- Heine-Borel theorem;
- Continuous function on $[a, b]$ is uniformly continuous;
- Continuous function on $[a, b]$ attains a maximum;
- Brouwer fixed-point theorem;
- Jordan curve theorem.

Reverse mathematics gives a formal sense in which all of the above theorems are deeper than the elementary calculus which calculates derivatives and integrals, because it shows that they are provable only with the help of a new axiom. This is one of the few places where the intuitive relation of "deeper than" has been found to have a formal counterpart.

4.5. Uncountability of $\mathbb{R}$

The proof by the diagonal argument gets a high score for elegance and simplicity. It qualifies as deep also for its *fruitfulness*: yielding consequences that were previously obtained only with difficulty, or not foreseen at all.

- Existence of transcendental numbers. Cantor derived this immediately from uncountability of $\mathbb{R}$, using Dedekind's result that the set of algebraic numbers is countable. (In doing so, Cantor showed that existence of transcendental numbers was *less deep* than previously thought—something that could happen, I suppose, to other theorems we now consider deep.)
- Unsolvability of the halting problem for Turing machines. Formally, Turing [1936] proved "unsolvability by Turing machine," but this became *absolute* unsolvability, which is arguably deep in a new way because it rests on a probably unprovable assumption: the *Church-Turing thesis* that computability is identical with Turing-machine computability.
- Gödel incompleteness. No consistent formal system T can settle all instances of the halting problem; otherwise, we could solve the halting problem by enumerating the theorems of T. This is not as strong as Gödel's own incompleteness theorem, which guarantees the incompleteness of Peano arithmetic.

Unsolvability and Incompleteness

The route from diagonalization to unsolvability to incompleteness was discovered by Emil Post in the 1920s, but he held it back because he saw the problem of proving the Church-Turing thesis. In the meantime, Gödel independently discovered incompleteness *without* using unsolvability in 1930, by the *arithmetization* of formal systems. His method shows that any sufficiently strong formal system has unprovable *arithmetic* sentences.

In discovering incompleteness, Gödel had found that there is no absolute concept of proof: any consistent axiom system for arithmetic can be extended by adding one of its unprovable theorems as an axiom, just as any list of real numbers can be extended by adding its diagonal number. Coming to incompleteness from this direction, Gödel was inclined to doubt that there was an absolute concept of computation, since it would appear that any proposed class of computable functions can be extended by diagonalization. He was not convinced that there is an absolute concept of computation, and hence of solvable problem, until he saw Turing's 1936 paper. (This is another indication that the Church-Turing thesis is a deep idea!)

Deeper Unsolvable Problems

By modeling Turing machines in various mathematical structures, various problems about discrete objects have been shown to be unsolvable.

1. The word problem for finitely presented groups, proved unsolvable by Novikov [1955]. This problem and the next were proposed as candidates for unsolvability by Church in the 1930s. The word problem seems deeper than the halting problem, because it requires the development of a new area of group theory.
2. The homeomorphism problem (for finitely presented 4-manifolds), proved unsolvable by Markov [1958]. This depends on the unsolvability of the word problem, and a little more.
3. Hilbert's tenth problem, proved unsolvable by Matijasevič [1970]. This problem—which asks for an algorithm to decide solvability of polynomial equations in integers—was conjectured to be unsolvable by Post in the 1940s, but it was proved so only after a long series of partial results by Davis, Putnam, and Julia Robinson.

In claiming that these results are deeper than the halting problem, we are reverting to the informal "shoulder of giants" concept of depth. They are considered deeper because it took considerable time and effort to uncover them. At present, there is no technical concept of logic that supports this claim about depth. Indeed, according to the technical concept of *degree of unsolvability*, the three problems are of exactly the same degree of unsolvability as the halting problem. We know of unsolvable problems of higher degree (for example, deciding the truth of arbitrary sentences of Peano arithmetic), so such problems might be considered technically of greater depth. But it is questionable whether such problems are mathematically as natural as the three problems above.

Do Unsolvable Problems Guarantee Deep Theorems?

It might be fruitful to study the depth of algorithmic problems rather than the depth of theorems, because we already have concepts of *reducibility* between problems that have been extensively studied and found to yield many levels of depth. However, there is still a dearth of results about natural mathematical problems and natural concepts of relative depth or difficulty, exemplified by the failure to prove P ≠ NP.

Therefore, let us continue to investigate the question of depth for theorems, rather than for problems. Even so, it is worth asking whether deep theorems can be extracted from unsolvable problems. Take Hilbert's tenth problem, for example. This problem seeks an algorithm to answer all questions of the form:

> Q_p: *Does the polynomial equation* $p(x_1, x_2, \dots, x_n) = 0$, *with integer coefficients, have an integer solution?*

A formal system for mathematics can generate theorems mechanically. Also, any competent system, such as Peano arithmetic (PA), can test whether

$$p(a_1, a_2, \dots, a_k) = 0 \text{ for given integers } a_1, a_2, \dots, a_k.$$

Hence PA proves all theorems of the form

$$p(x_1, x_2, \dots, x_n) = 0 \text{ has a solution.}$$

It follows that PA *cannot* prove all theorems of the form

$$p(x_1, x_2, \dots, x_n) = 0 \text{ has no solution.}$$

Otherwise, we could solve Hilbert's tenth problem by looking through the theorems of PA.

Thus, there are theorems of the form "$p(x_1, x_2, \dots, x_k) = 0$ has no integer solution" that PA cannot prove. In this sense, they are deeper than PA. So if we add such a theorem, as a new axiom, to PA we get a stronger system PA^+. Applying the same argument to PA^+, we get a stronger system PA^{++}, and so on. In this way, we obtain an infinite sequence of theorems of the form

$$p(x_1, x_2, \ldots, x_n) = 0 \text{ has no solution}$$

that are deeper and deeper in the provability sense. However, we cannot say that the polynomials p in these theorems are of any independent interest.

5. Is There a Natural Hierarchy of Axiom Systems?

The systems PA, PA^+, PA^{++}, . . . just described form a hierarchy of axiom systems that prove deeper and deeper theorems in the provability sense. We could measure the depth of a theorem by the height of the system required to prove it. However, this is a very rough and unsatisfactory way to measure depth. It is rough because even the lowest system PA has theorems of considerable depth. We know that Dirichlet's theorem on primes in arithmetic progressions can be proved in PA, and it is strongly suspected that Fermat's last theorem can be proved there too. The depth of these theorems, if it is not illusory, must be due to the structure of their proofs in PA, not to the axioms of PA.

In any case, the hierarchy PA, PA^+, PA^{++}, . . . is unsatisfactory because it is unnatural and unlikely to measure the height of arbitrary sentences of PA. There are probably sentences of PA not decided in any system of the hierarchy. This is why we raise the question in the title of this section. For PA, we do not yet know of any natural hierarchy of extensions of PA. Indeed, though Gödel's incompleteness theorem gives an infinite supply of true sentences of PA not provable in PA, all the known examples of such sentences have been devised by logicians. These include some fascinating theorems, such as *Goodstein's theorem* [1944], but only a logician could have discovered Goodstein's theorem.

Still, I do not know where to look *except* logic for clues about the depth of sentences in PA. Another quite plausible idea is that proofs of deep theorems in PA, such as Dirichlet's theorem, can be drastically *simplified* by the use of higher level concepts, such as real numbers or functions. As we mentioned in Section 2.1, Dirichlet's theorem was first proved with the help of ideas from analysis (real numbers and functions), and it was a long time before a proof in PA was discovered. Moreover, the analytic proof is still preferred, hence it is presumably simpler.

It so happens that there is a theorem of logic, *Gödel's speedup theorem*, that offers a possible explanation of such phenomena. The speedup theorem says that there are infinitely many theorems of PA whose proofs can be drastically shortened (in a precise sense) by extending PA with higher order concepts, such as real numbers. Alas, this theorem also has the "for logicians only" defect. The known examples of speedup do not include any theorems of independent interest.

Returning to the idea that depth of a theorem might be measured by the strength of the axiom system needed to prove it, there is some good news with regard to theorems about real numbers or, equivalently, about infinite sequences of natural numbers or other finite objects. *Reverse mathematics* has been quite successful here, sorting classical theorems of analysis into at least five different levels, which seem to correspond to levels of depth.

It also has something to say about some deep modern theorems of graph theory, concerning infinite sequences of graphs.

5.1. A Little Reverse Mathematics

Reverse mathematics, developed by Harvey Friedman and Steve Simpson, measures the strength of theories needed to prove theorems about *real* numbers. As its name suggests, the aim of reverse mathematics is to find the minimum axioms needed to prove a given theorem. It turns out, rather surprisingly, that a large number of theorems need one of just five particular axiom systems.

It would take us too far afield to describe these systems in detail. Instead, I refer the reader to the definitive book [Simpson, 2009] and simply mention the systems by name. Here are the five main systems, in order of increasing strength, with some of the theorems they (and no system of lower strength) can prove. Thus, the theorems are listed in order of increasing depth:

- RCA_0 proves the intermediate-value theorem.
- WKL_0 proves the extreme-value theorem, Heine-Borel theorem, Brouwer fixed-point theorem, and the Jordan curve theorem.
- ACA_0 proves the Bolzano-Weierstrass theorem.
- ATR_0 proves determinacy of open games.
- $\prod_1^1 - CA_0$ proves the Cantor-Bendixson theorem.

The last two theories, and the theorems they prove, seem more like set theory than analysis; however, there are some theorems of *graph theory* (not devised by logicians) that they cannot prove. These theorems of graph theory could therefore be considered deeper than any of the above theorems of analysis.

5.2. Kruskal's Theorem and the Graph Minor Theorem

Theorem of Kruskal [1960]: *Any infinite sequence of distinct finite trees contains an infinite subsequence which is increasing under the homeomorphic embedding relation.*

For example, here is an increasing sequence of trees:

"Homeomorphic embedding" is a relaxation of graph embedding in which vertices of degree 2 can be ignored (as it has been in the third graph). A further relaxation, which allows any edge to collapse to a single vertex, gives the *graph minor* relation. For example,

We then get the theorem of Robertson and Seymour [2004]: *Any infinite sequence of distinct finite graphs contains an infinite subsequence which is increasing under the graph minor relation.* (This paper was twentieth in a series that began in 1983.)

5.3. Historical Depth of the Graph Minor Theorem

The Robertson-Seymour theorem not only has a long proof, expressible only in strong axiom systems, but also a long history, mainly about map coloring on surfaces.

1852 Guthrie posed the four-color problem, solved in 1976.

1890 Heawood posed the problem of coloring maps on other surfaces. This led to the question of embedding graphs (or their minors) in surfaces.

1930 Kuratowski proved that a graph is planar if and only if K_5 or $K_{3,3}$ does not homeomorphically embed in it.

1937 Wagner proved a variant of Kuratowski's theorem: a graph is planar if and only if it has neither K_5 nor $K_{3,3}$ as a minor. They are *forbidden minors*.

1979 Glover, Huneke, and Wang found the forbidden homeomorphic subgraphs for the projective plane. It follows that it has thirty-five forbidden minors.

2004 The graph minor theorem implies that *the number of forbidden minors for any compact surface is finite*. (The number of forbidden minors for the torus is not known, but greater than sixteen thousand.)

6. A Minimal Example of "Provable Depth"

$$\textit{Volume of a tetrahedron} = \frac{1}{3}\,\textit{base area} \times \textit{height}.$$

Compare this with the theorem that

$$\textit{Area of a triangle} = \frac{1}{2}\,\textit{base length} \times \textit{height}.$$

The latter is not deep with the ancient Greek concept of area, because the triangle is *equidecomposable* with a rectangle with the same base length and half the height:

That is, with finitely many straight cuts, the triangle can be decomposed and reassembled to form the rectangle (Figure 5). On the other hand, Euclid needed infinitely many cuts to find the volume of a tetrahedron. This led to:

Hilbert's third problem: *Is the regular tetrahedron equidecomposable with a cube?*

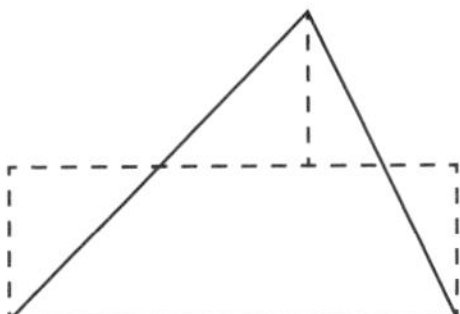

Figure 5. Area of a triangle

This question was raised by Gauss and asked by Hilbert in 1900. Soon after the problem was posed, Dehn [1900] showed that the answer was no: *The tetrahedron cannot be cut into finitely many polyhedra that reassemble to form a cube.*

Thus, the formula for volume of a tetrahedron is provably deeper than the formula for area of a triangle, in the sense that it cannot be proved by a finite construction. (Its historical depth is also quite impressive, considering the time interval between Euclid and Dehn.)

7. Conclusion

History offers ample evidence of the concept of depth in mathematics, in the form of theorems depending on large, multilevel structures of concepts and lemmas. The trouble is: logic has so far offered only slight evidence that this concept can be formalized.

It is true that there are several theorems of logic that offer *possibilities* of proving certain theorems of ordinary mathematics to be deeper than others:

- Gödel incompleteness;
- Gödel speedup;
- Unsolvable algorithmic problems;
- Reverse mathematics.

Of these, I find Gödel speedup—the simplification of proofs by the introduction of higher level concepts—the phenomenon that best corresponds to the intuitive concept of depth. But so far, only reverse mathematics has been able to prove anything about the depth of ordinary mathematical theorems, and then only for theorems about the real numbers or equivalent objects (such as infinite sequences of graphs).

The world of discrete objects, such as the natural numbers, has barely been touched by such developments, since the apparently deep theorems about natural numbers found by logicians do not greatly interest ordinary mathematicians. Logic can establish unsolvability of algorithmic problems in ordinary mathematics, such as Hilbert's tenth problem about polynomial equations, but it has not yet told us anything about the relative depth of any interesting equations.

This result is disappointing, but perhaps a sign that logic and mathematics are not yet well understood, and we may hope for developments that completely change the face of mathematics in the future. Certainly there remain simply stated mathematical questions that are completely baffling. We do not know whether the following questions are deep, or inaccessible, or both:

- Is $\sqrt{2}$ (or π, or e) a normal number?

  Borel [1909] proved that almost all real numbers are normal; that is, their digits (in any base) occur with the same frequency. Indeed, on probability grounds, the theorem seems obvious: why should one digit occur with greater frequency than another? Nevertheless, no natural example of a normal number is known.
- The $3n + 1$ problem (posed by Collatz in 1937):

  For any positive integer n, form $n/2$ if n is even, $3n + 1$ if n is odd, and repeat indefinitely. Does this process always reach the number 1?

  Paul Erdős said "mathematics is not yet ripe for such problems" and, on another occasion, "Hopeless, absolutely hopeless."

Acknowledgments

I would like to thank Pen Maddy for encouraging me to speak on this topic at UC Irvine and to write up my talk subsequently. This led me to glimpse a few vague ideas on the subject of depth, which were further clarified by the incisive comments of three anonymous reviewers. Naturally, I take responsibility for any confused ideas that remain, though I tend to think that the concept of depth itself is still struggling to escape from a state of confusion.

References

Borel, E. [1898]: *Leçons sur la théorie des fonctions*. Paris: Gauthier-Villars.

———. [1909]: "Les probabilités dénombrables et leurs applications arithmétiques." *Rendiconti del Circolo Matematico di Palermo* **27**, 247–51.

Dedekind, R. [1872]: *Stetigkeit und irrationale Zahlen*. Braunschweig, Germany: Vieweg und Sohn. English translation by Wooster Woodruff Beman in *Essays on the Theory of Numbers*. New York: Dover, 1963.

Dehn, M. [1900]: "Über raumgleiche Polyeder." *Gött. Nachr.*, 345–54.

Euler, L. [1748]: *Introductio in analysin infinitorum, I.* Lausanne, Switzerland: Marcum-Michaelem Bousquet. Vol. 8 of his *Opera Omnia*, series 1. English translation by John Blanton, *Introduction to the Analysis of the Infinite. Book I.* New York: Springer-Verlag, 1988.

———. [1770]: *Elements of Algebra.* Translated from the German by John Hewlett. Reprint of the 1840 edition, with an introduction by C. Truesdell. New York: Springer-Verlag, 1984.

Feit, W., and J. G. Thompson. [1963]: "Solvability of groups of odd order." *Pacific J. Math.* **13**, 775–1029.

Gauss, C. F. [1801]: *Disquisitiones Arithmeticae.* Translated and with a preface by Arthur A. Clarke. Revised by William C. Waterhouse, Cornelius Greither, and A. W. Grootendorst and with a preface by Waterhouse. New York: Springer-Verlag, 1986.

Goodstein, R. L. [1944]: "On the restricted ordinal theorem" *J. Symbolic Logic*, **9**, 33–41.

Grassmann, H. [1861]: *Lehrbuch der Arithmetic.* Berlin: Enslin.

Hilbert, D. [1899]: *Grundlagen der Geometrie.* Leipzig, Germany: Teubner. English translation by Leo Ungar, *Foundations of Geometry.* Chicago: Open Court, 1971.

Kruskal, J. B. [1960]: "Well-quasi-ordering, the tree theorem, and Vazsonyi's conjecture." *Trans. Amer. Math. Soc.* **95**, 210–25.

Lagrange, J. L. [1773]: "Recherches d'arithmétique." *Nouv. mém. de l'acad. sci. Berlin*, pp. 265 ff. Also in his *Œuvres*, J.-A. Serret, ed., Vol. 3, pp. 695–795. Paris: Gauthier-Villars.

Markov, A. [1958]: "The insolubility of the problem of homeomorphy." *Dokl. Akad. Nauk SSSR* **121**, 218–20.

Matijasevič, J. V. [1970]: "The Diophantineness of enumerable sets." *Dokl. Akad. Nauk SSSR* **191**, 279–82.

Newton, I. [1671]: "De methodis serierum et fluxionum." *Mathematical Papers*, Vol. 3, pp. 32–353. Cambridge, U.K.: Cambridge University Press.

Novikov, P. S. [1955]: *Ob algoritmičeskoĭ nerazrešimosti problemy toždestva slov v teorii grupp.* Trudy Mat. Inst. im. Steklov, No. 44. Moscow: Izdat. Akad. Nauk SSSR.

Poincaré, H. [1904]: "Cinquième complément à l'analysis situs." *Palermo Rend.* **18**, 45–110. Also in his *Œuvres*, Vol. 6, pp. 435–98. Paris: Gauthier-Villars.

Robertson, N., and P. D. Seymour. [2004]: "Graph minors. XX. Wagner's conjecture." *J. Combin. Theory Ser. B* **92**, 325–57.

Simpson, S. G. [2009]: *Subsystems of Second Order Arithmetic,* 2nd ed., Perspectives in Logic. Cambridge, U.K.: Cambridge University Press; Poughkeepsie, NY: Association for Symbolic Logic.

Turing, A. [1936]: "On computable numbers, with an application to the *Entscheidungsproblem*." *Proc. Lond. Math. Soc.* (2) **42**, 230–65.

von Staudt, K. G. C. [1847]: *Geometrie der Lage.* Nuremberg, Germany: Bauer und Raspe.

Wiles, A. [1995]: "Modular elliptic curves and Fermat's last theorem." *Ann. of Math.* (2) **141**, 443–551.

Finding Errors in Big Data

Marco Puts, Piet Daas, and Ton de Waal

No data source is perfect. Mistakes inevitably creep in. Spotting errors is hard enough when dealing with survey responses from several thousand people, but the difficulty is multiplied hugely when that mysterious beast Big Data comes into play.

Statistics Netherlands is about to publish its first figures based on Big Data—specifically road sensor data, which count the number of cars passing a particular point. Later, we plan to use cell phone data for statistics on the daytime population and tourism, and we are considering an indicator to capture the "mood of the nation" based on sentiment expressed through social media (1).

Statistics derived from unedited data sets of any size would be biased or inaccurate. But the challenge Statistics Netherlands faces in dealing with Big Data sets is to find data editing processes that scale up appropriately to allow quick and efficient cleaning of a huge number of records.

How huge? For the sentiment indicator, we plan to use three billion public messages predominantly gathered from Facebook and Twitter (2), and for the road sensor data there are 105 billion records. But size is not the only distinguishing characteristic of a Big Data set.

A clear, generally accepted definition of "Big Data" does not exist, though descriptions often refer to the three Vs: volume, velocity, and variety (3). So not only do we have a large amount of data to deal with (volume), but also the frequency of observations is very high (velocity). For the road sensor data, for example, we have data on a minute-by-minute basis. Big Data also tends to be "messy" in comparison to traditional data (variety). Again, for the road sensor data, we only know how many vehicles passed by. We do not know who drove the cars. In addition, background characteristics, which are important for data editing and estimation methods, are lacking, thus making such methods difficult to apply.

A Big Problem

Our experience with cleaning large data sets started a few years before we began to study the use of Big Data for statistical purposes. In those days, we were investigating how to edit and impute large amounts of administrative data. Administrative data can have high volume, but they differ from Big Data with respect to velocity and variety. We learned that finding errors in large administrative data sets is already a challenge. Automatic editing techniques and graphical macro-editing techniques (see box) work best for such data sets.

Cleaning Survey Data: A Small Data Perspective

In the (distant) past, *manual editing* was used with the intention of correcting all data in every detail. Data were checked and adjusted in separate steps. The editing process thus consisted of cycles where records often had to be examined and adjusted several times, which made for a time-consuming and costly process.

Interactive editing is also a manual activity, where, in principle, all records are examined, and if necessary, corrected. The difference with respect to manual editing is that the effects of adjusting the data can be seen immediately on a computer screen. This immediate feedback directs one to potential errors in the data and enables one to examine and correct each record only once. Interactive editing typically uses edit rules, that is, rules capturing the subject-matter knowledge of admissible (combinations of) values in each record—a male cannot be pregnant, for example—to guide the editing process.

Efficiency is further increased by *selective editing*: identify the records with potentially influential errors and restrict interactive editing to those records only. The most common form of selective editing is based on score functions. A record score is a combination of local scores for each of a number of important target parameters. Local scores are generally products of a risk component and an influence component. The risk component is measured by comparing a raw value with an "anticipated" value, often based on information from previous data. The influence component is measured as the (relative) contribution of the anticipated value to the estimated total. Only records with scores above a certain threshold are directed to interactive editing.

In *automatic editing*, data are edited by computers without any human intervention. We distinguish between correcting systematic errors and random errors, and different kinds of techniques are used to edit these errors. Once detected, systematic errors can often easily be corrected because the underlying error mechanism can usually be deduced. Random errors can be detected by outlier detection techniques, by deterministic checking rules that state which variables are considered erroneous when a record violates the edit rules in a certain way, or by solving an optimization problem, for example by minimizing the number of fields to change so that the adjusted data satisfy all edit rules (7). With the introduction of automatic editing, one was able to clean relatively large amounts of survey data in a reasonable time.

Macro-editing can be used when (most of) the data set has been collected. It checks whether the data set as a whole is plausible. We distinguish between two forms: the aggregation method and the distribution method. The *aggregation method* consists of verifying whether figures to be published seem plausible by comparing them to related quantities from other sources. This method is often used as a final check before publication. In the *distribution method*, the available data are used to characterize the distribution of variables. Then individual values are compared with this distribution.

In order to apply graphical macro-editing to large administrative data sets, we applied and (further) developed visualizations. An example of such a visualization is the *tableplot*. A tableplot can be applied in two ways: to detect implausible or incorrect values, or to monitor the effects of the editing process on the quality of the data. In a tableplot, a quantitative variable is used to order the data for all variables shown. The ordered records are divided into a certain number of equally sized bins. For each bin, the mean value is calculated for numerical variables, and category fractions are determined for categorical variables, where missing values are considered as a separate category. These results are subsequently plotted. A disruptive change in the distribution in a tableplot can indicate the presence of errors. Moreover, a nonuniform distribution over the columns can indicate selectivity. Finally, the distribution of correlated variables can be examined by looking at the value distribution in the unsorted columns.

Figures 1 and 2 show tableplots for the Dutch annual Structural Business Statistics (SBS), based on unedited and edited data, respectively. These relatively small data sets—in comparison to Big Data, that is—are used to illustrate the benefits of applying visualization methods for monitoring the editing process. The SBS survey covers the economic sectors of industry, trade, and services. Survey data are received from approximately fifty-two thousand respondents annually. Topics covered in the questionnaire include turnover, number of employed persons,

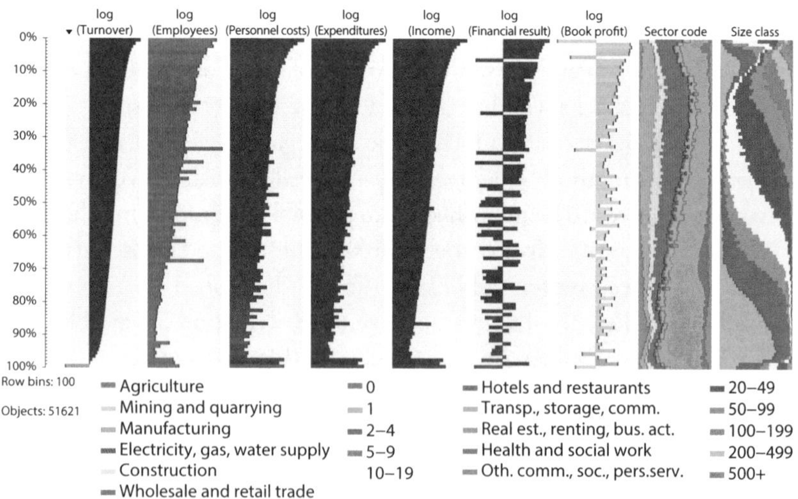

FIGURE 1. Tableplot of unedited SBS data. When sorted on turnover (leftmost column) a considerable number of the other numeric variables display a clear—and predictable—downward trend occasionally distorted with large values (4).

total purchases, and financial results. Figure 1 was created by sorting on the first column, "turnover," and dividing the 51,621 observed units into one hundred bins, so that each row bin contains approximately 516 records. A subset of approximately forty-nine thousand records was deemed suitable for publication purposes. The tableplot for the corresponding edited data is shown in Figure 2.

The distributions of the numerical variables in Figure 2 are much smoother than in Figure 1; they are less disturbed by row bins with large values. In particular, the difference between the distributions for "results" stands out. The same is true for the categorical variables "sector" and "size." Both display a much smoother distribution in Figure 2, and in "size" the remarkable disturbance displayed in the upper part of the column in Figure 1 is completely gone. This is very likely the result of corrections for so-called "thousand errors": businesses have to report their amounts in thousands of euros, but many neglect to do so. Also, note that "book profit" no longer suffers from missing data and the negative "turnover" values are gone. These are all indications that editing has improved the quality of the data (4).

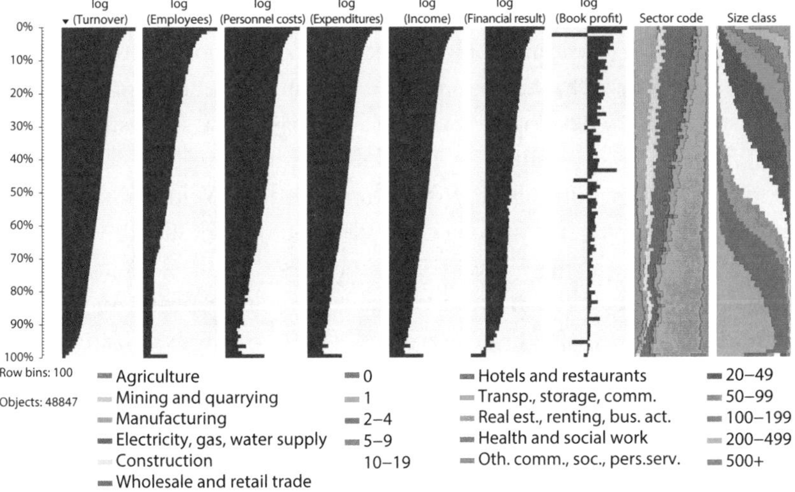

FIGURE 2. Tableplot of edited SBS data. After data editing and sorting on turnover, the majority of the quality issues have been solved as indicated by the smooth distribution of the variables shown (4).

A Bigger Problem

Having gained such experience editing large administrative data sets, we felt ready to process Big Data. However, we soon found out that we were unprepared for the task. Owing to the lack of structure (variety) and the large amounts of data (volume), we discovered that several editing techniques developed for survey data cannot be applied efficiently to Big Data, including interactive editing and selective editing (see box for definitions).

Even automatic editing methods are hard to apply to Big Data as they often require subject-matter knowledge in the form of a detailed set of edit rules. Obtaining and applying such knowledge is challenging for many Big Data sources. The most promising traditional kind of automatic editing methods are those based on statistical modeling as these do not require user-specified edit rules. However, even these are hampered by the selectivity of many Big Data sources since not all parts of the target population may be equally well represented. This negatively affects the estimation of model parameters.

The aggregation method of the macro-editing approach, where the plausibility of publication figures is checked by comparing these figures to related quantities from other sources, can be applied to Big Data. The aggregation method is, however, only suited as a last final check before publication of the figures and should almost always be supplemented by other editing techniques that can be applied earlier in the cleaning process.

Visualizations developed for "merely" large data sets, such as the tableplot, do hold promise for Big Data and its three Vs. Volume can be dealt with by binning or aggregating the data. Velocity can be addressed by making animations or by developing a dashboard. Variety can be handled through interactive interfaces that allow visualizations to be adapted quickly. Besides the tableplot, other promising visualizations are "treemaps" and "heatmaps" (4, 5). Such visualizations can often be used to monitor the effects of the editing process. However, to correct errors in Big Data sources, new approaches are needed.

Cleaning Big Data

The approach we describe here has been developed specifically for road sensor data. The sensors work as follows: whenever a vehicle passes by, information about traffic flows is generated, such as vehicle counts and mean speed of vehicles passing. In the Netherlands, for about sixty thousand sensors, the number of passing cars in various vehicle length categories is available on a minute-by-minute basis.

The most important issue we ran into while studying road sensor data was that the quality of the data fluctuates tremendously. For some sensors, data for many minutes are not available and, because of the stochastic, or random, nature of the arrival times of vehicles at a road sensor, it is hard to directly derive the number of vehicles missing during these minutes (6).

The high frequency at which the data are generated severely hampers the use of traditional data editing techniques. Even traditional automatic editing and graphical macro-editing failed in this case. The breakthrough was the realization that the high frequency of the data enables us to apply signal processing techniques for editing and imputation purposes. In particular, one can estimate a Markov chain model for each road sensor (Figure 3).

In such a Markov chain model, a road sensor can be in a certain state at time *t*, where a state is the number of vehicles that passed over the road sensor during the last minute. A Markov chain is a random process that undergoes a transition from one state at time *t* to another state at time $t + 1$ with a certain probability. The most characteristic aspect of a Markov chain is that it is memoryless: the probability of transitioning from the current state to the next depends only on the current state and not on the preceding states.

In Figure 3, Y_t ($t = 1, 2, \ldots$) denotes the observed signal at time *t*, that is, the observed (but possibly incorrect) number of vehicles that

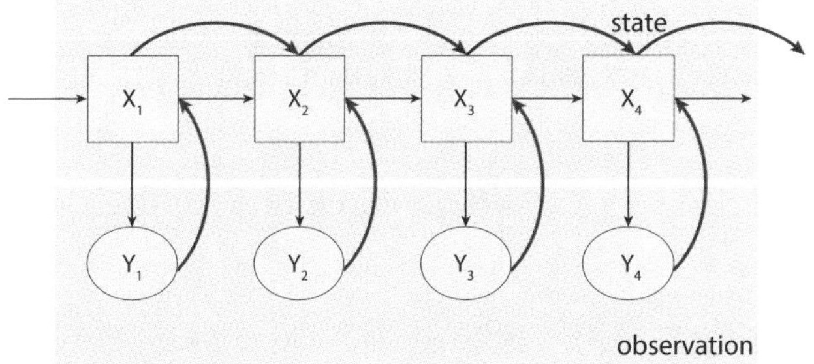

FIGURE 3. A Markov chain model for road sensor data.

passed the sensor during the last minute before time t, and X_t the true (unobserved) signal, that is, the true number of vehicles that actually passed the sensor. The observed data Y_t ($t = 1, 2, \ldots$) is used to estimate the transition probabilities to go from one state X_t to the next X_{t+1}.

The most common kind of error that occurs in road sensor data is that observations are missing due to the fact that the sensor is temporarily not working properly. The Markov chain model can be used to automatically correct for this kind of error. Namely, in cases where the observed signal Y_t is missing, the Markov chain draws a value for X_t using the previous true state X_{t-1} and the estimated transition probabilities. The Markov chain model makes it possible to automatically edit and correct exceedingly large amounts of data. We applied this successfully to 105 billion records.

Growing Up

The use of Big Data for statistical purposes is still in its infancy, particularly in the development of efficient editing techniques. One of the big challenges for Big Data is monitoring the quality of the data without the need to inspect the data in its most granular form. As a result, one needs technological and methodological aids to inspect quality at an aggregated level.

An even bigger challenge is to detect and correct errors in Big Data quickly and automatically. The most promising direction appears to be the development of tailor-made automatic editing techniques such as the Markov chain approach we applied to road sensor data.

It is an exciting period for statistics, and official statistics in particular. Big Data offers the possibility of producing statistics in new ways by thinking "outside the box," and it will inevitably stimulate the development of new editing approaches. It might be a new era, but the old requirements for robust, clean, and reliable data remain.

References

1. Lansdall-Welfare, T., Lampos, V., and Cristianini, N. (2012) Nowcasting the mood of the nation. *Significance*, **9**(4), 26–28.
2. Daas, P. J. H., and Puts, M. J. H. (2014) *Social Media Sentiment and Consumer Confidence*. Statistics Paper Series No. 5, European Central Bank, Frankfurt, Germany.

3. Daas, P., and Puts, M. (2014) Big Data as a source of statistical information. *The Survey Statistician*, **69**, 22–31.
4. Tennekes, M., de Jonge, E., and Daas, P. (2013) Visualizing and inspecting large datasets with tableplots. *Journal of Data Science*, **11**, 43–58.
5. Tennekes, M., de Jonge, E., and Daas, P. (2012) Innovative visual tools for data editing. Presented at the United Nations Economic Commission for Europe Work Session on Statistical Data Editing, Oslo. http://www.unece.org/fileadmin/DAM/stats/documents/ece/ces/ge.44/2012/30_Netherlands.pdf.
6. Daas P. J. H., Puts, M. J., Buelens, B., and van den Hurk, P. A. M. (2013) Big Data and Official Statistics. *Journal of Official Statistics* **31**(2), 1–15.
7. Fellegi, I. P., and Holt, D. (1976) A systematic approach to automatic edit and imputation. *Journal of the American Statistical Association*, **71**, 17–35.

Programs and Probability

Brian Hayes

Randomness and probability are deeply rooted in modern habits of thought. We meet probabilities in the daily weather forecast and measures of uncertainty in opinion polls; statistical inference is central to all the sciences. Then there's the ineluctable randomness of quantum physics. We live in the Age of Stochasticity, says David Mumford, a mathematician at Brown University.

Ours is also an age dominated by deterministic machines—namely, digital computers—whose logic and arithmetic leave nothing to chance. In digital circuitry, strict causality is the rule: Given the same initial state and the same inputs, the machine will always produce the same outputs. As Einstein might have said, computers don't play dice.

But in fact, they do! Probabilistic algorithms, which make random choices at various points in their execution, have long been essential tools in simulation, optimization, cryptography, number theory, and statistics. How is randomness smuggled into a deterministic device? Although computers cannot create randomness de novo, they can take a smidgen of disorder from an external source and amplify it to produce copious streams of *pseudorandom* numbers. As the name suggests, these numbers are not truly random, but they work well enough to fool most probabilistic algorithms. (In other words, computers not only play dice; they also cheat.)

A recent innovation weaves randomness even more deeply into the fabric of computer programming. The idea is to create a probabilistic programming language (often abbreviated PPL and sometimes pronounced "people"). In a language of this kind, random variables and probability distributions are first-class citizens, with the same rights and privileges as other data types. Furthermore, statistical inference—the essential step in teasing meaning out of data—is a basic, built-in operation.

Most of the probabilistic languages are still experimental, and it's unclear whether they will be widely adopted and prove effective in handling large-scale problems. But they have already provided an intriguing new medium for expressing probabilistic ideas and algorithms.

Counting vs. Sampling

Suppose you throw three standard dice, colored red, green, and blue for ease of identification. What are the chances of scoring a 9? No less a luminary than Galileo Galilei took up this question four hundred years ago—and got the right answer! His method was enumeration, or counting all the cases. Each die can land in six ways, so the number of combinations is 6^3, or 216 (Figure 1). If the dice are fair, all these outcomes are equally likely. Go through the table of 216 combinations and check off those with a sum of 9. You'll find that there are twenty-five of them, giving a probability of 25/216, or 11.6 percent.

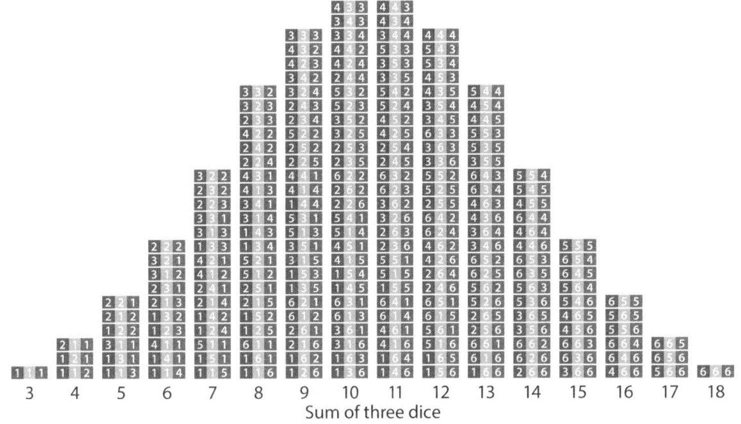


```
(define (roll-die)
  (uniform-draw'(1 2 3 4 5 6)))

(enumeration-query
 (define r (roll-die))
 (define g (roll-die))
 (define b (roll-die))
 (+ r g b))
```

FIGURE 1. The 216 possible outcomes of rolling three dice stack up to form an approximation to the normal bell curve. The snippet of computer code at left, written in a programming language called Church, gives the probability of each three-die sum from 3 to 18. Church is one of a new generation of languages designed to model probabilistic reasoning.

Another way to answer the same question is simply to run the experiment: Roll the dice a few million times and see what happens. This sampling procedure is easier with computer-simulated dice than with real ones (an option unavailable to Galileo). In 216 million simulated rolls, I got 24,998,922 scores of 9, for a probability within 0.01 percent of the true value.

Enumeration is often taken as the gold standard for probability calculations. It has the admirable virtue of exactness; it is also deterministic and repeatable, whereas sampling gives only approximate answers that may be different every time. But sampling also has its defenders. They argue that randomized sampling offers a closer connection to the way nature actually works; after all, we never see the idealized, exact probabilities in any finite experiment.

In the end, these philosophical quibbles are usually swept aside by practical constraints: Enumerating all possible outcomes is not always feasible. You can do it with a handful of dice, but you can't list all possible sequences of English words or all possible arrays of pixels in an image. When it comes to exploring these huge search spaces, sampling is the only choice.

The size of the solution space is not the only challenge in solving probability problems; another factor is the complexity of the questions being asked. You may need to estimate joint probabilities (how often do *x* and *y* occur together?) or conditional probabilities (how likely is *x*, given that *y* is observed?). The most interesting queries are often matters of inference, where the aim is to reason "backward" from observed effects to unknown causes. In medical diagnosis, for example, the physician records a set of symptoms and must identify the underlying disease. The conclusion depends not only on the probability that a disease will give rise to the observed symptoms but also on the probability of the disease itself in the population. (The formal statement of this principle is known as Bayes' rule. It's also implicit in an old saying among diagnosticians: Uncommon presentations of common diseases are more common than common presentations of uncommon diseases.)

A Random Walk in Monte Carlo

The idea of solving probability problems by running computer experiments had its genesis at the Los Alamos laboratory soon after World

War II. The mathematician Stanislaw Ulam was playing solitaire while recuperating from an illness and tried to work out the probability that a random deal of the cards would yield a winning position.

> After spending a lot of time trying to estimate [the probability] by pure combinatorial calculations, I wondered whether a more practical method than "abstract thinking" might not be to lay it out say one hundred times and simply observe and count the number of successful plays. This was already possible to envisage with the beginning of the new era of fast computers . . .

Ulam's technique was named the Monte Carlo method, after the famous randomizing devices in that Mediterranean capital. In 1948, the scheme was put to work on a graver task than solving solitaire. A program running on the ENIAC, the early vacuum tube computer, calculated the probability that a neutron moving through a cylinder of uranium or plutonium would be absorbed by a fissionable nucleus before wandering away. (I don't know if Ulam ever got back to the solitaire problem.)

The most basic algorithm for sampling is straightforward: Just make repeated trials and tally the results. Where the process gets dicey (so to speak) is with conditional probabilities. Suppose you want to know the probability of rolling a 14 with three dice, on the condition that at least two of the dice display the same number. The simplest approach to such a problem is called *rejection sampling*. The program simulates many rolls of the dice and discards cases in which the three dice are all different. This procedure is clearly correct; it follows directly from the definition of conditional probability. For the dice problem, it works fine, but it becomes woefully inefficient for studying rare phenomena. The program spends most of its time generating events that it immediately throws away. Think of trying to stumble on meaningful English sentences by assembling random sequences of words.

Ideally, we would like to generate only "successes"—states that satisfy the imposed condition—without wasting time on the failures (Figure 2). That goal is not always attainable, but a family of techniques known as Markov chain Monte Carlo (MCMC) can often get much closer than rejection sampling. The basic idea is to construct a random walk that visits each state in proportion to its probability in the observed distribution.

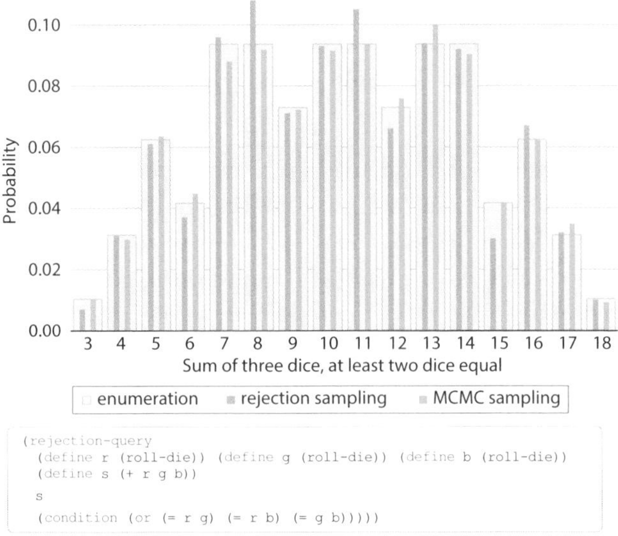


```
(rejection-query
  (define r (roll-die)) (define g (roll-die)) (define b (roll-die))
  (define s (+ r g b))
  s
  (condition (or (= r g) (= r b) (= g b)))))
```

FIGURE 2.Imposing a condition on the dice game—requiring at least two dice to show the same value—yields a probability distribution quite different from the normal curve. (Be careful if you're betting on this game.) The Church program here computes the distribution by rejection sampling, in which cases that fail to satisfy the condition are discarded. Variants of the same program use enumeration or Markov chain Monte Carlo (MCMC) sampling. The graph shows all three results.

Let's take a walk through some states of the two-or-more-equal dice game. Assume a roll of the dice comes up *red* = 2, *green* = 2, *blue* = 3, which for brevity we can denote 223. This combination satisfies the two-or-more-equal condition, so we write it down as the initial state of the system. Now, instead of picking up all three dice and throwing them again, choose one die at random and roll that one alone, leaving the others undisturbed. If the result is a state that satisfies the condition, note it down as the next state. If not, return to the previous state and write that down as the second state as well. Continue by choosing another die at random and repeat the procedure. A sequence of states might run 223, 323, 323, 343, 443, 444 . . .

This "chain" of states eventually reaches all combinations that meet the two-or-more-equal criterion. In the long run, moreover, the states in the random walk have the same probability distribution as the one produced by rejection sampling. In subtler ways, however, the MCMC sequence is not quite random. Nearby elements are closely correlated: 223 can make an immediate transition to 233 but not to 556. The remedy is to collect only every *n*th state, with a value of *n* large enough to let the correlations fade away. This policy imposes an *n*-fold penalty in efficiency, but in many cases MCMC still beats rejection sampling.

A Language Called Church

Monte Carlo simulations and other probabilistic models can be written in any programming language that offers access to a pseudorandom number generator. What a PPL offers is an environment where probabilistic concepts can be expressed naturally and concisely, and where procedures for computing with probabilities are built into the infrastructure of the language. Variables can represent not just ordinary numbers but entire probability distributions. There are tools for building such distributions, for combining them, for imposing constraints on them, and for making inquiries about their content.

A PPL called Church illustrates these ideas. Church was created by Noah D. Goodman, Vikash K. Mansinghka, Daniel M. Roy, Keith Bonawitz, and Joshua B. Tenenbaum when they were all at MIT. Goodman and Tenenbaum's online book, *Probabilistic Models of Cognition* (https://probmods.org), provides an engrossing introduction to Church and to probabilistic programming in general. Readers can modify and run programs directly in the book's web interface.

Church is named for Alonzo Church, an early twentieth-century logician who developed a model of computation called the lambda calculus. The Church language is built atop Scheme, a dialect of the Lisp programming language that has deep roots in the lambda calculus. Like other variants of Lisp, Church surrounds all expressions with parentheses and adopts a "prefix" notation, writing `(+ 1 2)` instead of `1 + 2`.

Here is the Church procedure for rolling a single die:

```
(define (roll-die)
        (uniform-draw '(1 2 3 4 5 6)))
```

When this function is executed, it returns a single number chosen uniformly at random from the list shown. In the larger context of a probabilistic query, however, the result of the `roll-die` procedure can represent a distribution over all the numbers in the list.

The query expression below has three parts: first a list of `define` statements, creating a model of a probabilistic process; second the expression that will become the subject of the query (in this case simply the variable *g*); and finally a condition imposed on the outcome:

```
(enumeration-query
        (define r (roll-die))
        (define g (roll-die))
        (define b (roll-die))
        (define score (+ r g b))
        g
        (condition (> score 13)))
```

Again we are rolling three simulated dice, designated *r, g,* and *b*. The query computes the probability distribution of *g* subject to the condition that the sum of *r, g,* and *b* is greater than 13. The result of running the query is a list giving the possible values of *g* (2, 3, 4, 5, 6) and their probabilities (ranging from 0.029 to 0.429).

The keyword `enumeration-query` in this program signals that it counts all cases and returns exact results. Substituting the keyword `rejection-query` produces what you would expect: a sampling from the same distribution, performed by discarding cases that fail to satisfy the greater-than-13 condition. A third option is the keyword `mh-query`, which invokes a Markov chain Monte Carlo algorithm. (The initials `mh` stand for Metropolis-Hastings; Nicholas Metropolis was a member of the early Monte Carlo community at Los Alamos, and W. Keith Hastings is a Canadian statistician.)

Behind the Curtain

The query expressions of Church specify *what* is to be computed but not *how*; all the algorithmic magic happens behind the curtains. That's part of the point of a PPL—to allow the programmer to work at a higher level of abstraction, without being distracted by details of counting cases or collecting samples. Those details still have to be attended to, but the necessary code is written by the implementer of the language rather than the user of it.

Embedding such technology inside a programming language requires algorithms general enough that they work in a variety of problem domains, from playing dice to following neutrons to recognizing faces in images. MCMC simulations often exploit specific features of the problem definition, but a PPL must somehow construct a random walk through the solution space without knowing anything about the nature of the individual states. The strategy adopted in Church is to regard all possible execution paths of the program itself as the elements of the solution space. Each random choice in the program establishes a new branch point in the network of execution paths. The Church MCMC algorithm works by altering one of those choices and checking to see if the new output still satisfies any conditions imposed on the model.

The generic MCMC algorithm is not quite foolproof. Consider another variation on the dice game with a slightly different condition: not `(> score 13)` but `(= score 13)`. That is, we want to find the distribution of values of the green die when the sum of the three dice is exactly 13. The Church program for this task works fine with an enumeration query or a rejection query, but the MCMC program fails. The reason is that the algorithm always proposes to alter the value of a single die, an action that cannot leave the sum of the three dice unchanged; thus none of the proposals are ever accepted. There are work-arounds for this problem, such as defining a "noisy equal" operator that occasionally judges two numbers to be equal even when they're not. But the need to understand the cause of the failure and take steps to correct it suggests that we have not yet reached the stage where we can blithely hand off probabilistic models to an automated solver.

Sprouting Languages

PPLs are not the only approach to computing with probabilities. The main alternatives are *graphical models*, in which networks of nodes are connected by lines called edges. Each node represents a random variable along with its probability distribution; an edge runs between two nodes if one variable depends on the other. For example, a network for a model of three dice might look like Figure 3. Information flows from each of the dice to the sum, but not between the dice because they are independent.

Advocates of PPLs suggest at least two reasons for favoring a linguistic over a graphical representation. First, certain concepts are easier to

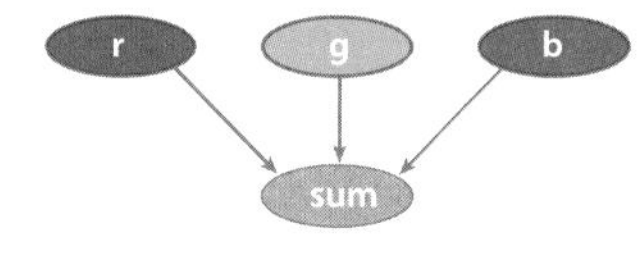

FIGURE 3. In a graphical model of dice rolling, arrows called directed edges indicate that the sum of the roll depends on each of the individual dice (r, g, b). But the dice are independent of one another; no arrows connect them.

encode in a language than in a diagram; an important example is *recursion*, where a program invokes itself. Second, over the entire history of digital computers, programming languages have proved to be the most versatile and expressive means of describing computations.

Almost twenty years ago, Daphne Koller of Stanford University, with David McAllester and Avi Pfeffer, examined the prospects for a probabilistic programming language and showed that programs could be well-behaved, running with acceptable efficiency and terminating reliably with correct answers. A few years later, Pfeffer designed and implemented a working probabilistic language called IBAL, then later created another named Figaro.

The movement toward PPLs has gathered momentum in the past five years or so. Interest was fortified in 2013 when the U.S. Defense Advanced Research Projects Agency announced a four-year project funding work on probabilistic programming for advancing machine learning. At least a dozen new projects have sprouted up. Here are thumbnail sketches of three of them:

WebPPL was created by Noah Goodman (a member of the Church collaboration) and Andreas Stuhlmüller, both now of Stanford University. Like the web version of Church, it is a language anyone can play with in a browser window, but its roots in web technology are even deeper: The base language is JavaScript, which every modern browser has built in. An online tutorial, *The Design and Implementation of Probabilistic Programming Languages* (http://dippl.org), introduces the language.

Stan—which is named for Stanislaw Ulam—comes from the world of statistics. The developers are Bob Carpenter, Andrew Gelman, and a large group of collaborators at Columbia University. Unlike many other PPLs, Stan is not just a research study or a pedagogical exercise but an attempt to build a useful and practical software system. The project's website (http://mc-stan.org) lists two dozen papers reporting on uses of the language in biology, medicine, linguistics, and other fields.

Meanwhile, some members of the Church group at MIT have moved on to a newer language called Venture, which aims to be more robust and provide a broader selection of inference algorithms, as well as facilities allowing the programmer to specify new inference strategies (http://probcomp.csail.mit.edu/venture/).

Into the Mainstream?

The computing community has a long tradition of augmenting programming languages with higher level tools, absorbing into the language tasks that would otherwise be the responsibility of the programmer. For example, doing arithmetic with matrices involves writing fiddly routines to comb through the rows and columns. In programming languages that come pre-equipped with those algorithms, multiplying matrices is just a matter of typing `A * B`. Perhaps multiplying probability distributions will someday be just as routine.

However, PPLs add much more than a new data type. Algorithms for probabilistic inference are more complex than most built-in language facilities, and it's not clear that they can be made to run efficiently and correctly over a broad range of inputs without manual intervention. One precedent for building such intricate machinery into a language is Prolog, a "logic language" that caused much excitement circa 1980. The Prolog programmer does not specify algorithms but states facts and rules of inference; the language system then applies a built-in mechanism called *resolution* to deduce consequences of the input. Prolog has not disappeared, but neither has it moved into the mainstream of computing. It's too soon to tell whether probabilistic programming will become a utility everyone takes for granted, like matrix multiplication, or will remain a niche interest, like logic programming.

Some of the algorithms embedded in PPLs may also be embedded in people. Probabilistic reasoning is part of how we make sense of the world: We predict what will probably happen next, and we assign probable causes to what we observe. Most of this mental activity lies somewhere below the level of consciousness, and we don't necessarily know how we do it. Getting a clearer picture of these cognitive mechanisms was one of the original motivations for studying PPLs. Ironically, though, when we write PPL programs to do probabilistic inference, most of us won't know how the programs do it either.

Bibliography

Eckhardt, R. 1987. "Stan Ulam, John von Neumann, and the Monte Carlo method." *Los Alamos Science* 15:131–37.

Goodman, N. D. 2013. "The principles and practice of probabilistic programming." *Proceedings of the 40th ACM SIGPLAN-SIGACT Symposium on Principles of Programming Languages (POPL 13)*, New York: Association for Computing Machinery, 399–402.

Goodman, N. D., V. K. Mansinghka, D. M. Roy, K. Bonawitz, and J. B. Tenenbaum. 2008. "Church: A language for generative models." *Proceedings of the 24th Conference on Uncertainty in Artificial Intelligence*, Corvallis, OR: AUAI Press, 220–29.

Goodman, N. D., and A. Stuhlmüller. 2015. *The Design and Implementation of Probabilistic Programming Languages*. Online document: http://dippl.org.

Goodman, N. D., and J. B. Tenenbaum. 2015. *Probabilistic Models of Cognition*. Online document: https://probmods.org.

Koller, D., D. McAllester, and A. Pfeffer. 1997. "Effective Bayesian inference for stochastic programs." *Proceedings of the Fourteenth National Conference on Artificial Intelligence (AAAI-97)*, Palo Alto, CA: Association for the Advancement of Artificial Intelligence, 740–47 (At the time of publication, it was the American Association for Artificial Intelligence).

Mansinghka, V., D. Selsam, and Y. Perov. 2014 preprint. "Venture: A higher-order probabilistic programming platform with programmable inference." arXiv:1404.0099v1.

Mumford, D. 1999. "The dawning of the age of stochasticity." *Mathematics: Frontiers and Perspectives*. Providence, RI: American Mathematical Society, 197–217.

Pfeffer, A. 2001. "IBAL: A probabilistic rational programming language." *Proceedings of the International Joint Conference on Artificial Intelligence 2001*, Palo Alto, CA: AAAI Press/ International Joint Conferences on Artificial Intelligence, 733–40.

Stan Development Team. 2014. *Stan Modeling Language Users Guide and Reference Manual*, Version 2.5.0. http://mc-stan.org/.

Lottery Perception

Jorge Almeida

"Perception of randomness" is a subject of research in experimental psychology (1). One of the beliefs shared by many of those working on this theme is that, when assessing outcomes of repeated tosses of a fair coin, people tend to underestimate probabilities of runs of consecutive heads (H) or tails (T) and, conversely, overestimate probabilities of alternations (e.g., HTTHTHHHTHT has one run of three, one run of two, six runs of one, and seven alternations). For example, given 21 tosses of a fair coin, the most frequent class of outcomes will be that with 10 alternations. However, people tend to judge as more likely the class of sequences with 12 to 14 alternations. Of course, work in this field has to bring together two disciplines as it requires confronting people's beliefs (psychology) and calculating probabilities of sequences or classes of sequences (mathematics). One such calculation was presented by McPherson and Hodson (2), who addressed the problem of the occurrence of runs of consecutive numbers in lotteries in which people bet on 6 out of 49 numbers.

McPherson and Hodson (2) estimated the probabilities of drawing combinations of all unattached numbers or combinations consisting of possible types of runs of the six numbers and then compared their results to data on past winning combinations. No significant deviation from expectation, on the assumption of randomness, was found, attesting to the fairness of the lottery draw. For example, out of 1,028 lottery draws, sequences with a run of three, a run of two, and a single have occurred five times, very close to the predicted value of 5.84, and sequences with "all unattached" numbers (six singles), which are expected to occur 519 times, have occurred 555 times. Of course, given that the probability of occurrence of a run of six is only 44/13,983,816 and that the number of draws in the dataset (1,028) is very small compared to the number of possible lottery results (13,983,816), we expect runs of six never to have occurred and, indeed, so it is. (Note that the probability that a run of six does not occur in 1,028 draws is given by $(1 - \frac{44}{13983816})^{1028} = 0.99677$. In other words, out of one thousand sets each of 1,028 draws, we expect runs of six to occur in fewer than four of the sets.)

A parallel can be readily established between alternations in repeated coin tosses and alternations in the lottery by considering that, in the string of 49 numbers, the six chosen numbers in a combination are Hs and the remaining numbers are Ts (or ones and zeros). For example, a combination such as {2, 7, 11, 12, 13, 18} can be expressed as

{THTTTTHTTTHHHTTTTHT, followed by all Ts up to position 49}.

Given that an alternation is a transition from H to T or T to H, this combination would have eight alternations. Using the equation of McPherson and Hodson for relating numbers of combinations to run structure, it can be determined that the average number of alternations in the 6/49 lottery is almost 10.53 (for a range of 1 to 12 alternations). Combinations with a run of six have only one alternation if at the ends of the 1–49 range or two alternations otherwise, that is, far below the 10.53 average. Therefore, according to psychologists' views, most people would be inclined to believe that it is obviously a bad idea to bet on runs of six. That this belief is illusory, if founded on probabilities of occurrence of lottery combinations, was acknowledged by the French mathematician Pierre-Simon de Laplace (1749–1827), a founder of probability theory. Quoting from Hawking's commentary on Laplace's work in the compilation *God Created the Integers* (3):

> . . . when playing a lottery in which six numbers get picked out of fifty . . . many people would avoid playing the set {01, 02, 03, 04, 05, 06} supposing that that set of six numbers shows much more regularity than the recent winning set of numbers {06, 13, 15, 15, 32, 36} [one of the 15s is certainly a lapse, my comment]. But, Laplace would argue, a thorough analysis of the lottery process would reveal the independence of each number's selection. Thus, the set {01, 02, 03, 04, 05, 06} shows neither more nor less regularity than the set {06, 13, 15, 15, 32, 36}. . . .

To account for the illusion brought to light by Laplace, it is perhaps useful to view people's perception of probability as a two-step process. First, people inevitably partition the set of all possible outcomes according to given equivalence relations such as alternations/run-lengths. Second, following the partition, the illusion arises through some mechanism leading to confounding probabilities of subsets resulting from the partition, and probabilities of particular elements of subsets. For example, McPherson and Hodson (2) calculated that the probability of the class of runs of six is only 44/13,983,816, whereas the probability of the class "all unattached" is 7,059,052/13,983,816, that is, more than 50%. Thus, one would rather bet on "all unattached" than on a run of six. But this is an illusion since if a lottery draw yields an "all unattached," highly probable result, the probability that a particular "all unattached" ticket is the winner is only 1/7,059,052, whereas if a draw gives an improbable run of six, the probability that a particular ticket with a six-run wins is a relatively high 1/44. Hence, the probability of a particular element of the "all unattached" class is

$$\frac{7{,}059{,}052}{13{,}983{,}816} \times \frac{1}{7{,}059{,}052} = \frac{1}{13{,}983{,}816},$$

which is exactly the same as the probability of a particular element of the six-run class ($\frac{44}{13{,}983{,}816} \times \frac{1}{44} = \frac{1}{13{,}983{,}816}$). Of course, all this is saying is that any two outcomes are equally likely. The somewhat convoluted way of arriving at such equivalence serves, however, the purpose of establishing a parallel between calculation of probabilities and the proposed two-step process of probability perception. Should people see in a single step that any two particular results out of the 14 million are equally likely, there would be no problem of probability perception.

It therefore seems that it is not such a bad idea to bet on a run of six. Neither, at first sight, would it be a good idea. Consider, however, the problem of betting in the hope of becoming not just a lottery winner but the sole lottery winner, that is, one who, in the very unlikely event of winning, does not have to share the prize with other contenders. Under these conditions, suppose that all but one of the players follow the advice "have your numbers all unattached." Would the odd player be acting smartly? Clearly, if there is only one player betting on the subset "runs of six," this player is guaranteed to be a sole winner in the very unlikely event of winning. Laplace himself acknowledged this in following through the idea mentioned above. Quoting again from Hawking's comment

> . . . Given the fact that a lottery payout depends on the number of winners, the very fact that many people would avoid playing {01, 02, 03, 04, 05, 06} is, in fact, a very good reason to play it. There would be less likelihood of anyone else having the same winning set of numbers. . . .

One problem with Laplace's advice is that it is self-defeating. A combination perceived as unpopular, i.e., a good combination, may attract a crowd of Laplace's followers, thus turning bad. For example, lottery organizers do not systematically divulge numbers on players' preferences, but rumor has it that thousands of people bet on the combination {1, 2, 3, 4, 5, 6} (4, 5). Notwithstanding the veracity of this claim, analysis of what little information lottery organizers release (e.g., numbers of winners for jackpots and other categories) indicates that unlike lottery draws, which seem to be random, people do indeed make biased choices (4). This can be explained, as in the two-step process of probability perception proposed above, if people tend to confound probabilities of subsets resulting from partitions and probabilities of particular elements of subsets (e.g., perceiving a particular element of the six-run class as less likely than a particular element of the all unattached class). In this view, betting randomly amounts to not making such confusion. One sure way of not making such confusion is in turn to partition the set of all possible combinations (say, a set of size C) into C subsets of a single element each, as recognized, for example, by Haigh (4). It is hard to see how most people would make their choices according to a partition into nearly 14 million subsets, if not with widespread and systematic use of devices for generating random numbers, hence, the inevitability for people to base their decisions on partitions into "manageable" numbers of subsets.

Run length is but one of diverse criteria that people use for partitioning the set of nearly 14 million combinations in 6/49 lotteries. For example, people may avoid numbers at the upper end of the range 1–49 (especially numbers greater than 31, which cannot be birthdates) or may even respond to the particular layout of the numbers on the ticket (5). We may hope that research combining probability theory and experimental psychology will reveal with increasing accuracy how people make decisions when confronted with uncertainty. But rather than expecting this to produce a guide for action at betting counters, it is perhaps more interesting to expect deeper insight into the workings of the human mind.

References

1. R. Falk and C. Konold, "Making sense of randomness: Implicit encoding as a basis for judgment," *Psychological Rev.* **104** (1997), 301–318.
2. I. McPherson and D. Hodson, "Lottery combinatorics," *Math. Spectrum* **41** (2009), 110–115.
3. S. W. Hawking, ed., *God Created the Integers* (Penguin, London, 2006).
4. J. Haigh, "The statistics of the National Lottery," *J. R. Statist. Soc. A* **160** (1997), 187–206.
5. S. J. Cox, G. J. Daniell, and D. A. Nicole, "Using maximum entropy to double one's expected winnings in the UK National Lottery," *The Statistician* **47** (1998), 629–641.

Why Acknowledging Uncertainty Can Make You a Better Scientist

Andrew Gelman

Top journals in psychology routinely publish ridiculous, scientifically implausible claims, justified based on "$p < 0.05$." Recent examples of such silliness include claimed evidence of extrasensory perception (published in the *Journal of Personality and Social Psychology*), claims that women at certain stages of their menstrual cycle were three times more likely to wear red or pink clothing (published in *Psychological Science*), and a claim that people react differently to hurricanes with male and female names (published in the *Proceedings of the National Academy of Sciences*).

All these studies had serious flaws, to the extent that I (and others) found the claims to be completely unconvincing from a statistical standpoint, matching their general implausibility on substantive grounds.

It is easy to dismiss these particular studies, one at a time. But to the extent that they are being conducted using standard statistical methods, this calls into question all sorts of more plausible, but not necessarily true, claims—claims that are supported by this same sort of evidence. To put it another way: we can all laugh at studies of ESP, or ovulation and voting, but what about MRI studies of political attitudes, or embodied cognition, or stereotype threat, or, for that matter, the latest potential cancer cure? If we cannot trust p-values, does experimental science involving human variation just have to start over?

Figure 1 demonstrates what can happen with classical hypothesis testing. A study is performed in which the underlying parameter of interest (typically a causal effect or some other sort of comparison in the general population) is relatively small, and measurements are noisy and biased (not uncommon in a psychology setting in which the underlying constructs are often not clearly defined).

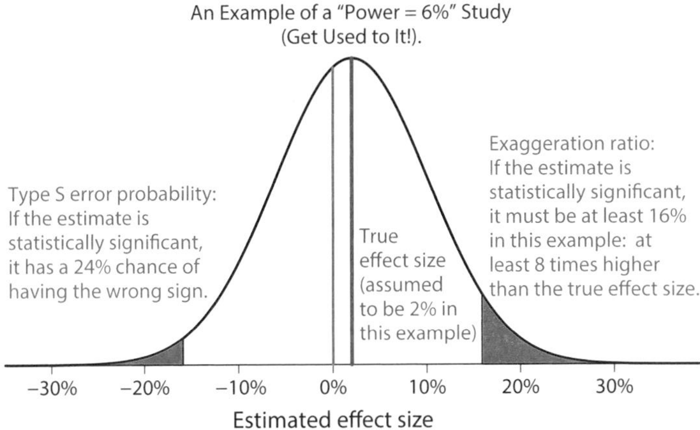
An Example of a "Power = 6%" Study
(Get Used to It!).
Type S error probability:
If the estimate is
statistically significant,
it has a 24% chance of
having the wrong sign.
True
effect size
(assumed
to be 2% in
this example)
Exaggeration ratio:
If the estimate is
statistically significant,
it must be at least 16%
in this example: at
least 8 times higher
than the true effect size.
−30%
−20%
−10%
0%
10%
20%
30%
Estimated effect size

Figure 1. A study has low power when the population difference or effect size is small, while variation and measurement are also small. In low-power studies, the "Type S (sign) error rate"—the probability that the observed difference is in the opposite direction to the true effect or population difference—can be high, even if the estimate is statistically significant. And the "exaggeration ratio"—the factor by which the observed estimate exceeds the true parameter value being estimated—can be huge. The particular numbers in this graph come from a study of a difference in political attitude, comparing women at different times in their menstrual cycles, for which we know, based on substantive grounds, that the true population effect size could be at most two percentage points. The bell-shaped curve represents the distribution of estimates that could occur in a study with this precision. The shaded red areas indicate the probability of obtaining a statistically significant effect (the "power," which in this case is six percent). Given the precision of this particular study, for an estimate to be statistically significant it would have to be at least sixteen percentage points (that's two standard errors away from zero), hence at least eight times larger than any true effect. And the probability that an estimate in this example is the wrong sign, if it is statistically significant, is twenty-four percent—the proportion of the red shaded areas on the negative side of the graph.

The particular example we were considering when constructing this graph is a published study claiming that, in the 2012 U.S. presidential election, "Ovulation led single women to become more liberal, less religious, and more likely to vote for Barack Obama. In contrast, ovulation led married women to become more conservative, more religious, and more likely to vote for Mitt Romney."

This dramatic set of claims was supported by a statistically significant comparison: an interaction effect estimated at about twenty percentage points that was more than two standard errors away from zero (a standard error being 8.1 percentage points in this example). Based on pre-election survey data, however, we believe that very few people changed their vote intentions during this campaign. A more plausible size of this menstrual-cycle effect would be two percentage points or less.

Hence, in Figure 1, the blue line indicating true effect size is at two percentage points, which is at the high end of any plausible effect here, and the bell-shaped curve shows the distribution of possible differences in the data that could be observed given this assumed effect size. Due to the high level of variation between people, the distribution is broad, indicating a wide range of possible data that could arise in such a study. The areas shaded red under the curve indicate the probability that the observed difference is "statistically significant"—that is, more than two standard errors away from zero. As the diagram indicates, a statistically significant finding here actually has a high probability of being in the wrong direction (a "Type S (sign)" error) and in any case will be at least sixteen percentage points—that is, at least eight times higher than the assumed true effect of two. In this sort of problem, classical hypothesis testing is a recipe for exaggeration.

When applied to the scientific process more generally, the result of all these hypothesis tests is a flow of noisy claims which bear only a weak relation to reality, but which attain statistical significance, which is, conventionally, a necessary and sufficient condition for publication, if said result is paired with any story that is qualitatively justified by a substantive theory.

Various researchers in psychology and medicine have made the following linked points: statistical significance cannot generally be taken at face value (1); a scientific publication system based on null hypothesis significance tests leads to large-scale errors in reporting; and these problems are particularly severe in the context of low signal and high noise (2).

Psychology is particularly subject to such problems, for several reasons:

- The objects of study (mental states, personality traits, and cognitive and social abilities) are inherently latent and typically cannot be precisely defined.

- Theories are correspondingly vague (in comparison with physics or chemistry, say, or even medicine), in that the magnitude and even the direction of effects cannot always be predicted based on theoretical grounds.
- Variation between people is typically large, as is variation across repeated measurements within people; indeed, analysis of this variation is often a central research goal.
- The stakes are low, so that it is easy to quickly do a small study and write up the conclusions. Unlike in medical research, there is no hurdle to performing a publishable study. This is not to say that psychology research is trivial; our point here is just that, compared to much medical research, typical studies in psychology have low, if any, risks to the participants, so the barriers to performing and publishing a study are minimal.

The resulting proliferation of studies with small effect sizes and high noise, along with a willingness of high-profile, prestigious journals such as *Psychological Science* and the *Proceedings of the National Academy of Sciences* to publish surprising, newsworthy findings based on statistically significant comparisons, has led us to a crisis in scientific replication.

Based on the considerations discussed above, I would say that the most important way that statistics can help solve the replication crisis is to recognize the fundamental nature of the problem: if effects are small and measurements are biased and noisy, there is no way out, other than to put effort into taking measurements that are more valid and reliable, most notably in psychology studies by using more carefully designed instruments and performing within-person comparison where possible to reduce variance.

Once better data have been collected, how can statistical inference help? Given the problems with classical significance testing, there should be something better. Some have suggested replacing hypothesis tests with confidence intervals, but this by itself will not solve any problems: checking whether a ninety-five percent interval excludes zero is mathematically equivalent to checking whether $p < 0.05$. And just as statistically significant results can be huge overestimates, confidence intervals can similarly contain wildly implausible effect sizes, estimates that happen to be consistent with the data at hand but make no sense in the context of subject-matter understanding.

One direction for statistical analysis that appeals to me is Bayesian inference, an approach in which data are combined with prior information (in this case, the prior expectation that newly studied effects tend to be small, which leads us to downwardly adjust large estimated effects in light of the high probability that they could be coming largely from noise). I do see a potential Bayesian solution using informative priors and models of varying treatment effects (3), but these steps will not be easy because they move away from the usual statistical paradigm in which each scientific study stands alone.

To resolve the replication crisis in science, we may need to consider each individual study in the context of an implicit meta-analysis. And we need to move away from a simplistic, deterministic model of science with its paradigm of testing and sharp decisions: accept/reject the null hypothesis and do/don't publish the paper. To say that a claim should be replicated is not to criticize the original study; rather, replication is central to science, and statistical methods should recognize this. We should not get stuck in the mode in which a "data set" is analyzed in isolation, without consideration of other studies or relevant scientific knowledge. We must embrace variation and accept uncertainty.

References

1. Simmons, J., Nelson, L. and Simonsohn, U. (2011). False-positive psychology: Undisclosed flexibility in data collection and analysis allows presenting anything as significant. *Psychological Science*, **22**, 1359–66.
2. Button, K. S., Ioannidis, J. P. A., Mokrysz, C., Nosek, B., Flint, J., Robinson, E. S. J., and Munafo, M. R. (2013). Power failure: Why small sample size undermines the reliability of neuroscience. *Nature Reviews Neuroscience*, **14**, 1–12.
3. Gelman, A. (2014). The connection between varying treatment effects and the crisis of unreplicable research: A Bayesian perspective. *Journal of Management*, **41**, 632–43.

For Want of a Nail: Why Unnecessarily Long Tests May Be Impeding the Progress of Western Civilization

HOWARD WAINER AND RICHARD FEINBERG

Standardized tests—whether to evaluate student performance, to choose among college applicants, or to license candidates for various professions—are often marathons. Tests designed to evaluate knowledge of coursework typically use the canonical hour; admissions tests are usually two to three hours, and licensing exams can take days.

Why are they as long as they are? The first answer that jumps immediately to mind is the inexorable relationship between a test's length and its reliability, which is merely a standardized measure of the stability and consistency of test results, ranging between a low of 0 (the score is essentially a random number) and a high of 1 (the score does not fluctuate at all).

Although a test score always becomes more reliable as the test generating it becomes longer, the law of diminishing returns sets in very quickly. In Figure 1, we show the reliability of a typical professionally prepared test as a function of its length. It shows that the marginal gain of moving from a thirty-item test to a sixty- or even ninety-item one is not worth the trouble unless such small additional increments in reliability are required.

But perhaps there are other uses for the information gathered by the test that require this additional length and accuracy. Here, the U.S. census provides an example.

The last decennial population count cost $13 billion, or approximately $42 per person, to estimate the number of people in the country at 308,745,538, give or take 31,000. If all the census gave us was that single number, it would be a colossal waste of taxpayer money.

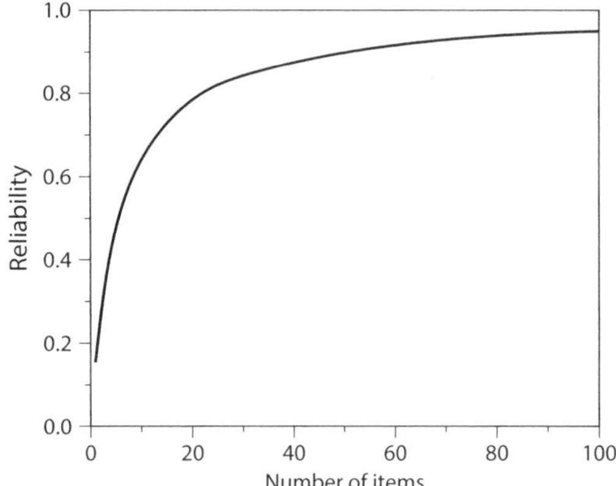

FIGURE 1. Spearman–Brown function showing the reliability of a test as a function of its length, if a one-item test has a reliability of 0.15.

However, the constitutionally mandated purpose of the census is far broader than that. It must also provide small-area estimates and answers to questions such as "How many households with two parents and three or more children live in the Bushwick section of Brooklyn, New York?" Such small-area estimates are crucial for the allocation of social services and for other purposes.

In tests, the equivalent of the small-area estimate is usually called a subscore. On a high school mathematics test, there might be subscores on algebra, arithmetic, geometry, and trigonometry. For a licensing exam in veterinary medicine, there might be subscores on the pulmonary system, the skeletal system, the renal system, and so on.

Thus the production of meaningful subscores would be a justification for tests that contain more items than would be required merely for an accurate enough estimate of total score. But what is a meaningful subscore? It is one that is reliable enough for its prospective use and one that has information that is not adequately contained in the total test score.

There are at least two prospective uses of such subscores: to aid examinees in assessing their strengths and weaknesses, often with an eye toward remediating the latter, and to aid individuals and institutions (e.g., teachers and schools) in assessing the effectiveness of their instruction, again with an eye toward remediating weaknesses.

In the first case, helping examinees, the subscores need to be reliable enough so that attempts to address shortcomings do not become just the futile pursuit of noise. And obviously, the subscore must contain information that is not available from the total score. Let us designate these two characteristics of a worthwhile subscore as "reliability" and "orthogonality." Subscore reliability is governed by the same inexorable rules of reliability as overall score—as test length decreases, so too does reliability. Thus, if we need reliable subscores, we must have enough items for that purpose. This requirement would mean that the overall test length would have to be greater than would be necessary for merely a single score.

For the second use, helping institutions, the test length would not have to increase, for the reliability would be calculated over the number of individuals from that institution who were administered the items of interest. If that number was large enough, the estimate could achieve high reliability.

So it would seem that one key justification for what appears at first to be the excessive length of most common tests is to provide feedback to examinees in subscores calculated from subsets of the tests.

But how successful are test developers in providing such subscores? Not particularly, for such scores are typically based on few items and hence are not very reliable (see box entitled "Substandard Subscores" on p. 330).

Justification

Where does this leave us? Unless we can find a viable purpose for which unreliable and nonorthogonal subscores have marginal value over the total test score, we are being wasteful (and perhaps unethical) to continue to administer tests that take more examinee time than is justified by the information yielded.

One such purpose might be the use of the test as a prod to motivate students to study all aspects of the curriculum and for the teachers to teach it. If tests were much shorter, fewer aspects of the curriculum would be well represented. But this problem is easily circumvented in a number of ways. If the curriculum is sampled cleverly, neither the teachers nor the examinees will know exactly what will be on the test and so have to include all of it in their study.

Another approach is to follow that taken by the U.S. National Assessment of Educational Progress and use some sort of a balanced incomplete block design in which all sectors of the curriculum are well covered but not each examinee takes all parts. That solution allows estimates of subarea mastery to be estimated in the aggregate and, through the statistical magic of score equating, still allows all examinee scores to rest on a common metric.

We should also keep in mind the result that has been shown repeatedly with adaptive tests in which we can give a test of half its usual length with no loss of motivation. Certainly there are fewer items of each sort, but examinees must still study all aspects because on a shorter test each item "counts" more toward the final score.

So, unless evidence can be gathered that shows a radical change in teaching and studying behavior with shorter tests, we believe that we can reject motivation as a reason for setting overly long tests.

Every Little Bit Helps?

Another justification for the apparent excessive length of tests is that the small increase in reliability is of practical importance. Of course, the legitimacy of such a claim would need to be examined on a case-by-case basis, but perhaps we can gain some insight through a careful study of one artificial situation that has similarities to a number of serious tests.

Let us consider the characteristics of a prototypical licensing examination that has, say, 300 items, which takes eight hours to administer and has a reliability of 0.95. Such characteristics show a marked similarity to a number of professional licensing exams (it might be for attorneys, veterinarians, physicians, nurses, or certified public accountants). The purpose of such an exam is to make a pass–fail decision, so let us assume that the pass score is 63%.

To make this demonstration dramatic, let us see what happens to the accuracy of our decisions if we make draconian reductions in test length. To begin with, let us eliminate 75% of the test items and reduce the test to just 75 items. Because of the gradual slope of the reliability curve shown in Figure 1, this kind of reduction would only shrink the reliability to 0.83. Is this still high enough to safeguard the welfare of future clients? The metric of reliability is not one that is close to our intuitions, so let us shift to something easier to understand: how many wrong pass–fail decisions would be made?

With the original test, 3.13% of the decisions would be incorrect, and these decisions would be divided between false positives (passing when they should have failed) of 1.18% and false negatives (failing when they should have passed) of 1.95%. How well does our shrunken test do? First, the overall accuracy rate declines to 6.06%, almost double what the longer test yielded. This result breaks down to a false positive rate of 2.26% and a false negative rate of 3.80%.

Is the inevitable diminution of accuracy sufficiently large to justify the fourfold increase in test length? Of course, that is a value judgment, but before making it we must realize that the cost in accuracy can be eased. The false positive rate for this test is the most important one, for it measures the proportion of incompetent practicing professionals who are incorrectly licensed. Happily, we can control the false positive rate by simply raising the pass score. If we increase the pass score to 65% instead of 63%, the false positive rate drops back to the same 1.18% we had with the full test. Of course, by doing this, the false negative rate grows to 6.6%, but this is a venial sin that can be ameliorated easily by adding additional items to those candidates who only barely failed.

Note that the same law of diminishing returns that worked to our advantage in total score (shown in Figure 1) also holds in determining the marginal value of adding more items to decrease the false negative rate. The parallel to the Spearman–Brown curve is shown in Figure 2.

The function in Figure 2 shows us that by adding only forty items for those examinees just below the cutoff score (those whose scores range from just below the minimal pass score of 65% to about 62%), we can attain false negative rates that are acceptably close to those obtained with the three hundred-item test. This extension can be accomplished seamlessly if a computer administers the test. Thus, for most of the examinees, their test would take only a quarter of the time it would previously have required, and even for the small number of examinees who had to answer extra items, these would be few enough so that even they still came out far ahead in time. The connections among test length, reliability, and error rates are summarized in Figure 3.

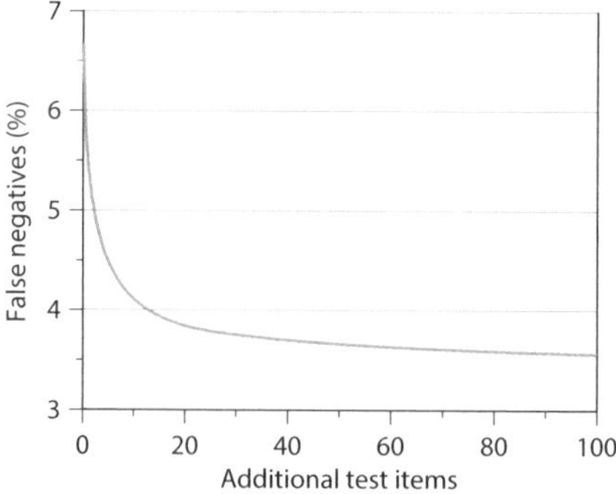

Figure 2. The improvement in the false negative rate yielded through the lengthening of the test for those who only marginally failed.

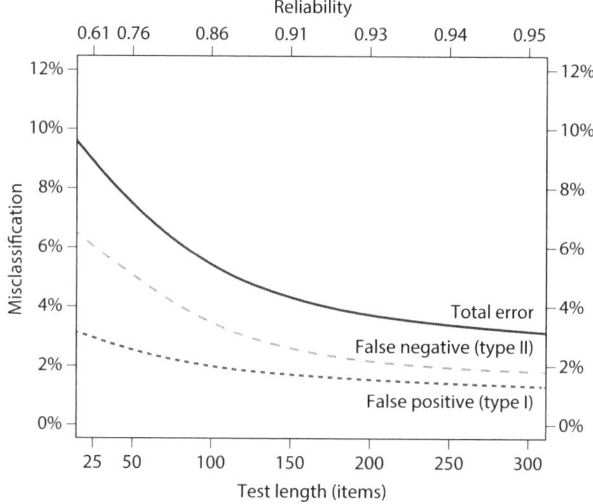

Figure 3. Illustration of how increasing test length (number of items) increases score reliability and decreases error rates, though with diminishing returns.

In Table 1, we show a summary of these results, as well as parallel results for an even more dramatically reduced test form with only forty items.

Thus, at least for simple pass–fail decisions, it seems that we can dismiss accuracy as a reason for the excessive test lengths used; for within plausible limits, we can obtain comparable error rates with much shorter tests, albeit with an adaptive stopping rule.

TABLE 1. Summary of Passing Statistics

Test Length	Pass Score (%)	Reliability	Total Error Rate (%)	False Positive Rate (%)	False Negative Rate (%)
300	63	0.95	3.13	1.18	1.95
75	63	0.83	6.06	2.26	3.80
75	65	0.83	7.78	1.18	6.60
40	63	0.72	8.23	2.59	5.64
40	66	0.72	11.62	1.14	10.48

Excess Costs

The costs associated with lengthy tests can be measured in various ways and, of course, they accrue differentially to different portions of the interested populations.

The cost to users of test scores is nil, since neither their time nor money is used to gather the scores.

The cost to the testing organization is likely to be substantial, since the cost of a single operational test question (item) is typically greater than $2,500. Add to this the costs of "seat time" paid to whoever administers the exam, grading costs, etc., and it adds up to a considerable sum. Fixed costs being what they are, giving a new test a quarter of the length of an older one does not mean a quarter of the cost, but it does portend worthwhile savings. We are also well aware of concerns that many tests, at their current lengths, do not allow enough time for some examinees. This worry could be ameliorated easily if the time allowed for the test was shrunk, but not quite as far as the number of items would suggest.

This point brings us to the examinees. Their costs are of two kinds: the actual costs paid to the testing organization, which could be reduced if the costs to that organization were dramatically reduced, and the opportunity costs of time.

If the current test takes eight hours, then a shortened form only a quarter as long might take only two hours, a saving of six hours per examinee. Multiplied by perhaps 100,000 examinees who annually seek licensure, this change would yield a time saving of 600,000 hours. Keeping in mind that the examinees taking a licensing exam are (or shortly will be) professionals for hire, it raises the question of what can be accomplished with 600,000 extra hours of their time?

If those being licensed were attorneys and this was a bar exam, consider how much good 600,000 annual hours of pro bono legal aid could do, or a like amount of effort from professional accountants at tax time, or engineers, or veterinarians. It is hardly an exaggeration to suggest that this amount of spare time from pre- or just-licensed professionals could accelerate the progress of our civilization.

Multiple Choice

The facts presented here leave us with two possibilities: to shorten our tests to the minimal length required to yield acceptable accuracy for the total score, and thence choose more profitable activities for the time freed up, or to re-engineer our tests so that the subscores that we calculate have the properties that we require.

Though the time savings offered by the first option would be of benefit (if only to the examinees and their work–life balance), users of

A Caveat

It may be that re-engineering tests will not solve the subscore problem if the examinee population is too homogeneous.

Consider a health inventory. Surely everyone would agree that recording both height and weight is important. Though the two variables are correlated, each contains important information not found in the other; a person who weighs 220 pounds is treated differently if he is 6 ft 5 in. than if he is 5 ft 6 in. Additionally, both can be recorded with great accuracy. So, by the rules of valuable subscores, both should be kept in the inventory. But suppose the examinees were all men from a regiment of marines. Their heights and weights would be correlated close to 1—if you know one, you do not need the other.

test scores almost universally desire more information than a simple pass–fail result. Thus, the second option—a redesign—is the one we find most attractive.

Sinharay, Haberman, and Wainer have shown that the shortcomings found in the subscores calculated on so many of our large-scale tests are due to flaws in the tests' designs.[7] A redesign is needed because we cannot retrieve information from our tests if the capacity to gather that information was not built in to begin with.

Happily, a blueprint for how to do this was carefully laid out more than a decade ago when Mislevy, Steinberg, and Almond provided the principles and procedures for what they dubbed "evidence-centred design."[8] It seems worth a shot to try it. In the meantime, we ought to stop wasting resources giving tests that are longer than the information they yield is worth.

References

1. Wainer, H., Sheehan, K., and Wang, X. (2000) Some paths toward making Praxis scores more useful. *Journal of Educational Measurement*, **37**, 113–140.
2. Thissen, D., and Wainer, H. (2001) *Test Scoring*. Hillsdale, NJ: Lawrence Erlbaum Associates.
3. Haberman, S. (2008) When can subscores have value? *Journal of Educational and Behavioral Statistics*, **33**(2), 204–229.
4. Sinharay, S. (2010) How often do subscores have added value? Results from operational and simulated data. *Journal of Educational Measurement*, **47**(2), 150–174.
5. Feinberg, R. A. (2012) A simulation study of the situations in which reporting subscores can add value to licensure examinations. Ph.D. dissertation, University of Delaware. Retrieved October 31, 2012, from ProQuest Digital Dissertations database (Publication No. 3526412).
6. Sinharay, S. (2014) Analysis of added value of subscores with respect to classification. *Journal of Educational Measurement*, **51**(2), 212–222.
7. Sinharay, S., Haberman, S. J., and Wainer, H. (2011) Do adjusted subscores lack validity? Don't blame the messenger. *Educational and Psychological Measurement*, **7**(5), 789–797.
8. Mislevy, R. J., Steinberg, L. S., and Almond, R. G. (2003) On the structure of educational assessments. *Measurement: Interdisciplinary Research and Perspectives*, **1**(1), 3–67.
9. Haberman, S. J., Sinharay, S., and Puhan, G. (2009) Reporting subscores for institutions. *British Journal of Mathematical and Statistical Psychology*, **62**, 79–95.

Substandard Subscores

The problem of reliability in subscores led to the development of empirical Bayes methods that would allow weak test items to borrow strength from other items that empirically yielded an increase in reliability. This methodology was proposed by Wainer et al.[1] in 2000 and was later elaborated the following year by Thissen and Wainer.[2]

This methodology often increased the reliability of subscores substantially, but at the same time the influence of items from the rest of the test reduced the orthogonality of those subscores to the rest of the test. Empirical Bayes gaveth, but it also tooketh away. What was needed was a way to measure value of an augmented subscore that weighed the delicate balance between increased reliability and decreased orthogonality.

Until such a measure became available, the instigating question, "How successful are test developers in providing useful subscores?" would remain unanswered.

Happily, the ability to answer this important question was improved markedly in 2008 with the publication of Shelby Haberman's powerful new statistic that combined both reliability and orthogonality.[3] Using this tool, Sandip Sinharay searched high and low for subscores that had added value over total score, but came up empty.[4,9]

Sinharay's empirical results were validated in simulations that matched the structure commonly encountered in different kinds of testing situations. These results were enriched and expanded by Richard Feinberg,[5] and again the paucity of subscores was confirmed.

This same finding, of subscores adding no marginal value over total score, was reconfirmed by Sinharay for tests whose goal was classification.[6] Although it is too early to say that there are no subscores that are ever reported that are worth having, it seems sensible that unless tests are massively redesigned, such subscores are likely rare.

How to Write a General Interest Mathematics Book

IAN STEWART

I've always wanted to write a book.
Then why don't you?

—*Common party conversation*

Popular science is a well-developed genre in its own right, and popular mathematics is an established subgenre. Several hundred popular mathematics books now appear every year, ranging from elementary introductions through school-level topics to substantial volumes about research breakthroughs. Writing about mathematics for the general public can be a rewarding experience for anyone who enjoys and values communication. Established authors include journalists, teachers, and research mathematicians; subjects are limited only by the imagination of authors and publishers' assessments of what booksellers are willing to stock. Even that is changing as the growth of e-books opens the way for less orthodox offerings. The style may be serious or lighthearted, preferably avoiding extremes of solemnity or frivolity. On the whole, most academic institutions no longer look down on "outreach" activities of this kind, and many place great value on them, both as publicity exercises and for their educational aspects. So do government funding bodies.

1. What Is Popular Mathematics?

For many people, the phrase is an oxymoron. To them, mathematics is *not* popular. Never mind: popularization is the art of making things popular when they were not originally. It is also the art of presenting advanced material to people who are genuinely interested but do not have the technical background required to read professional journals. Generally speaking, most popular mathematics books address this second audience. It would be wonderful to write a book that would open up the beauty, power, and utility of mathematics to people who swore off the subject when they were five, hate it, and never want to see it or hear about it again—but, by definition, very few of them would read such a book, so you would be wasting your time.

Already we see a creative tension between the wishes of the author and the practicalities of publishing. As e-books start to take off, the whole publication model is changing. One beneficial aspect is that new kinds of books start to become publishable. If an e-book fails commercially, the main thing wasted is the author's time and energy. That may or may not be an issue—an author with a track record can use his or her time to better effect by avoiding things that are likely to fail—but it will not bankrupt the publisher.

Popular mathematics is a genre, a specific class of books with common features, attractive to fans and often repellent to everybody else. In this respect it is on a par with science fiction, detective novels, romantic fantasy, and bodice rippers. Genres have their own rules, and although these rules may not be explicit, fans notice if you break them. If you want to write a popular mathematics book, it is good preparation to read a few of them first. Many writers started that way; they began as fans and ended up as authors, motivated by the books they enjoyed reading.

Most popular mathematics books fall into a relatively small number of types. Many fall into several simultaneously. The rest are as diverse as human imagination can make them. The main classifiable types are:

(1) Children's books.
  (a) Basic school topics.
  (b) More exciting things.
(2) History.
(3) Biography.
(4) Fun and games.
(5) Big problems.
(6) Major areas.
  (a) Classical.
  (b) Modern.

(7) Applications.
(8) Cultural links.
(9) Philosophy.

Children's books are a special case. They involve many considerations that are irrelevant to books for adults, such as deciding which words are simple enough to include—I will say no more about them, lacking experience. Histories and biographies have a natural advantage over more technical types of books: there is generally plenty of human interest. (If not, you chose the wrong topic or the wrong person.) Books about fun and games are lighthearted, even if they have a more serious side. Martin Gardner was the great exponent of this form of writing. Authors have written in depth about individual games, while others have compiled miscellanies of "fun" material.

Big problems, and major areas of mathematical research, are the core of popular mathematics. Examples are Fermat's last theorem, the Poincaré conjecture, chaos, and fractals. To write about such a topic, you need to understand it in more depth than you will reveal to your readers. That will give you confidence, help you find illuminating analogies, and generally grease the expository wheels. You should choose a topic that is timely, has not been exhausted by others, and stands some chance of being explained to a nonexpert. Within both of these subgenres you can present significant topics from the past—Fourier analysis, say—or you can go for the latest hot research area—wavelets, maybe. It is possible to combine both if there is a strong historical thread from past to present.

It is always easier to explain a mathematical idea if it has concrete applications. People can relate to the applications when the mathematics alone starts to become impenetrable. The same goes for cultural links, such as perspective in Renaissance art and the construction of musical scales.

Finally, there are deep conceptual issues, "philosophical" aspects of mathematics: infinity, many dimensions, chance, proof, undecidability, computability. Even simple ideas like zero or the empty set could form the basis of a really fascinating book, and have done so. The main thing is to have something to say that is worth saying. That is true for all books, but it is especially vital for philosophical ones, which can otherwise seem woolly and vague.

2. *Why Write a Popular Mathematics Book?*

Authors write for many reasons. When Frederik Pohl, a leading science fiction writer, was being inducted into the U.S. military, he was asked his profession and replied "writer." This was received with a degree of concern; writers are often impractical idealists who criticize everything and cause trouble. So Pohl was asked *why* he wrote. "To make money," he replied. This was received with relief as an entirely sensible and comprehensible reason. Another leading American writer, Isaac Asimov, produced more than three hundred books of science fiction, popular science, and other genres. When asked why he wrote so many books, he replied that he found it impossible to stop. He also went out of his way not to recommend his prolific approach to anyone else. Some authors write one book and are satisfied, even if it becomes a best seller; others keep writing whether or not they receive commercial success. Some write one book and vow never to repeat the experience. Some want to put a message across that strikes them as being of vital importance—a new area of mathematics, a social revolution, a political innovation. Some just like writing. There are no clear archetypes, no hard and fast rules; everything is diverse and fluid.

The best reason for writing a popular mathematics book, in my opinion, is that you desperately want to tell the world about something you find inspiring and interesting. Books work better when the author is excited and enthusiastic about the topic. The excitement and enthusiasm will shine through of their own accord, and it is best not to be too explicit about them. Far too many television presenters seem to imagine that if they keep telling the viewers how excited they are, viewers will also become excited. This is a mistake. Do not *tell* them you are excited; *show* them you are. It is the same with a book. Tell the story, bring out its inherent interest, and you are well on the way. Popularization is not about *making* mathematics fun (interesting, useful, beautiful, . . .); it is about showing people that it already *is* fun (interesting, useful, beautiful, . . .).

In the past, I have called mathematics the Cinderella science. It does all the hard work but never gets to go to the ball. Our subject definitely suffers, compared with many, because of a negative image among a large sector of the public. One reason for writing mathematics books for general readers is to combat this image. It is mostly undeserved, but as a profession, mathematicians do not always help their own cause. Even nowadays, when some area of mathematics attracts attention from the media, some mathematicians immediately go into demolition mode and complain loudly about "hype," exaggeration, and lack of precision.

When James Gleick's *Chaos* became a best seller, and the U.S. government noticed and began considering increasing the funding for nonlinear dynamics, a few distinguished mathematicians went out of their way to inform the government that it was all nonsense and the subject should be ignored. This might have been a good idea if they had been right—not about the popular image of "chaos theory," which was at best a vague approximation to the reality, but about the reality itself—but they were wrong. Nonlinear dynamics, of which chaos is one key component, is one of the great success stories of the late twentieth century, and it is powering ahead into the twenty-first. I remember one letter to a leading mathematics journal claiming that chaos and fractals had no applications whatsoever at a time when you could not open the pages of *Science* or *Nature* without finding papers that made excellent scientific use of these topics. Conclusion: many mathematicians have no idea what is going on in the rest of the scientific world, perhaps because they do not read *Science* or *Nature*.

To ensure that our subject is valued and supported, we mathematicians need to explain to ordinary people that mathematics is vital to their society, to their economic and social welfare, to their health, and to their children's future. No one else is going to do it for us. But we will not succeed if no one is allowed to mention a manifold without explaining that it has to be Hausdorff and paracompact as well as locally Euclidean. We have to grab our audience's attention with things they can understand; that necessarily implies using imprecise language, making broad-brush claims, and selecting areas that can be explained simply rather than others of equal or greater academic merit that cannot.

I am not suggesting that we should mislead the public about the importance of mathematics. But whenever a scientific topic attracts public attention, its media image is seldom a true reflection of the technical reality. Provided that the technical reality is useful and important, a bit of overexcitement does no serious harm. By all means, try to calm it down but not at the expense of ruining the entire enterprise. Grabbing public attention and then leveraging that (as the bankers would say) into a more informed understanding is fine. Grabbing public attention and then self-destructing because a few fine points have not quite been understood is silly.

3. Choosing a Topic

An article on popular writing is especially appropriate in a companion to applied mathematics because most people understand mathematics better if they can see what it is good for. This is one reason why chaos and fractals have grabbed public attention but algebraic *K*-theory has not. This is not a value judgment; algebraic *K*-theory is core mathematics, hugely important—it may even be more important than nonlinear dynamics. But there is little point in arguing their relative merits because each enriches mathematics. We are not obliged to choose one and reject the other. As research mathematicians, we probably do not want to work in both, but it is not terribly sensible to insist that your own area of mathematics is the only one that matters. Think how much competition there would be if everyone moved into your area. This happens quite a lot in physics, and at times it turns the subject into a fashion parade.

The key to most popular mathematics books is simple: *tell a story*.

In refined literary circles, the role of narrative is often downplayed. Whatever you think of *Finnegans Wake*, few would consider it a rip-roaring yarn. But popular science, like all genre writing, does not move in refined literary circles. What readers want, what authors must supply, is a story. Well, usually: all rules in this area have exceptions. In genre writing, humans are not *Homo sapiens*, wise men. They are *Pan narrans*, storytelling apes. Look at the runaway success of *The Da Vinci Code*—all story and little real sense.

A story has structure. It has a beginning, a middle, and an end. It often involves a conflict and its eventual resolution. If there are people in it, that is a plus; it is what made Simon Singh's *Fermat's Last Theorem* a best seller. But the protagonist of your book could well be the monster simple group or the four-color theorem. Human interest helps, in some subgenres, but it is not essential.

4. How to Write for the General Public

I wish I knew.

The standard advice to would-be authors is to ask, Who am I writing for? In principle, that is sound advice, but in practice there is a snag:

it is often impossible to know. You may be convinced you are writing for conscientious parents who want to help their teenage children pass mathematics exams. The buying public may decide, by voting with their wallets, that the correct audience for your book is retired lawyers and bank managers who always regretted not knowing a bit more mathematics and have spotted an opportunity to bone up on the subject now that they have got the time.

Some books are aimed at specific age groups—young children, teenagers, adults—and of course you need to bear the age of your readers in mind if you are writing that kind of book. But for the majority of popular mathematics books, the audience turns out to be very broad, not concentrated in any very obvious demographic, and difficult to characterize. "The sort of people who buy popular math books" is about as close as you can get, so I do not think you should worry too much about your audience.

The things that matter most are audience independent. Write at a consistent level. If the first chapter of your book assumes that the reader does not know what a fraction is and chapter two is about p-adic cohomology, you may be in trouble. The traditional advice to colloquium lecturers—start with something easy so that everyone can follow; then plow into the technicalities for the experts—was always bad advice even for colloquia because it lost most of the audience after five minutes of trivia. It is a complete disaster for a popular mathematics book.

It is common for the level of difficulty to ramp up *gradually* as the book progresses. After all, your reader is gaining insight into the topic as your limpid prose passes through their eyeballs to their brains. Chapter ten *ought* to be a bit more challenging than chapter one; if it is not, you are not doing your job properly.

One useful technique—when writing anything, be it for the public or for the editors and readers of the *Annals of Mathematics*—is self-editing. You need to develop an editor's instincts and apply them to your own work. You can do it as you go, rejecting poor sentences before your fingers touch the keyboard, but I find that slows me down and can easily lead to "writer's block." I hardly ever suffer from that affliction because I leave the Maoist self-criticism sessions for later. The great mathematical expositor Paul Halmos always said that the key to writing a book was to write it—however scruffily, however badly organized. When you have got most of it down on paper, or in the computer, you can go through the text systematically and decide what is good, what is bad, what is in the wrong place, what is missing, or what is superfluous. It is much easier to sort out these structural issues if you have something concrete to look at.

Word processors have made this process much easier. I generally write ten to fifteen percent more words than the book needs and then throw the excess away. I find it is quicker if I do that than if I agonize over each sentence as I type it. As George Bernard Shaw wrote, "I'm sorry this letter is so long, I didn't have time to make it shorter." When editing your work, here are a few things to watch out for:

- If you are using a term that you have not explained already, and it seems likely to puzzle readers, find a place to set it up. It might be a few chapters back; it might be just before you use it. Whatever you do, do not put it in the middle of the thing you are using it for; that can be distracting: "Poincaré conjectured that if a three-dimensional manifold (that is, a space . . . [several sentences] . . . three coordinates, that is, numbers that . . . [several sentences] . . . a kind of generalized surface) such that every closed loop can be shrunk to a point . . ." is unreadable.
- If some side issue starts to expand too much, as you add layer upon layer of explanation, ask whether you really need it. David Tall and I spent ten years struggling to explain the basics of homology in a complex analysis book for undergraduates without getting into algebraic topology as such. Eventually we realized that we could omit that chapter altogether, whereupon we finished the book in two weeks.
- If you are really proud of the classy writing in some section, worry that it might be overwritten and distracting. Cut it out (saving it in case you decide to put it back later) and reread the result. Did you need it? "Fine writing" can be the enemy of effective communication.
- Above all, try to keep everything simple. Put yourself in the shoes of your reader. What would *they* want you to explain? Do you need the level of detail you have supplied? Would something less specific do the same job? Do not start telling them about the domain and range of a function if all they need to know is that a function is a rule for turning one number into another. They are not going to take an exam.

- Do not forget that many ideas that are bread and butter to you (increment cliché count) are things most people have never heard of. *You* know what a factorial is; they may not. *You* know what an equivalence relation is; most of your readers do not. They can grasp how to compose two loops, or even homotopy classes, if you tell them to run along one and then the other but not if you give them the formula.
- It is said that the publishers of Stephen Hawking's *A Brief History of Time* told him to avoid equations, every one of which would allegedly "halve his sales." To some extent, this advice was based on the publisher's hang-ups rather than on what readers could handle. Look at Roger Penrose's *The Road to Reality*, a huge commercial success with equations all over most pages. However, Hawking's publishers had a point: do not use an equation if you can say the same thing in words or pictures.

Your main aims, to which any budding author of popular science should aspire, are to keep your readers interested, entertain them, inform them, and—the ultimate—make them feel like geniuses. Do that, and they will think that *you* are a genius.

5. Technique

Every author develops his or her own characteristic style. The style may be different for different kinds of books, but there is not some kind of universal house style that is perfect for a book of a given kind. Some people write in formal prose, some are more conversational in style (my own preference), some go for excitement, some like to keep the story smooth and calm. On the whole, it is best to write in a natural style—one that you feel comfortable with—otherwise you spend most of the time forcing words into what for you are unnatural patterns when you should be concentrating on telling the story.

There are, however, some useful guidelines. You do not need to follow them slavishly; the main point here is to illustrate the kinds of issues that an author should be aware of.

Use correct grammar. If in doubt, consult a standard reference such as *Fowler's Modern English Usage*. Bear in mind that some aspects of English usage have changed since his day. Be aware that informal language is often grammatically impure, but even when writing informally, be a little conservative in that respect. For example, it is impossible nowadays to escape phrases like “the team are playing well.” Technically, “team” is singular, and the correct phrase is “the team is playing well.” Sometimes the technically correct usage sounds so pedantic and awkward that it might be better not to use it, but on the whole it is better to be correct. At all costs avoid being inconsistent: “the team is playing well and they have won nine of its last ten games.”

I have some pet hates. “Hopefully” is one. It can be used correctly, meaning “with hope,” but much more often it is used to mean “I hope that,” which is wrong. “The fact that” is another: it is almost always a sign of sloppy sentence construction, and most of the time it is verbose and unnecessary. Replace “in view of the fact that” and similar phrases by the simple English word “because.” Try deleting “the fact” and leaving just “that.” If that fails to work, you can usually see an easy way to fix things up. Another unnecessarily convoluted phrase is “the way in which.” Plain “how” usually does the same job, better.

Avoid Latin abbreviations: e.g., i.e., etc. They are obsolete even in technical mathematical writing—Latin is no longer the language of science—and they certainly have no place in popular writing. Replace with plain English. The writing will be easier to understand and less clumsy. Avoid clichés (Wikipedia has links to lists), but bear in mind that it is impossible to avoid them altogether. Attentive readers will find some in this article: “bear in mind” for example! Stay away from crass ones like “run it up the flagpole,” and keep your cliché quotient small.

Metaphors and analogies are great . . . provided they work. “DNA is a double helix, like two spiral staircases winding around each other” conveys a vivid image—although it is perilously close to being a cliché. Do not mix metaphors: I once wrote “the Galois group is a vital weapon in the mathematician’s toolkit” and a helpful editor explained that weapons belong in an armory and “vital” they are not. The human mind is a metaphor machine; it grasps analogies intuitively, and its demand for understanding can often be satisfied by finding an analogy that goes to the heart of the matter. A well-chosen analogy can make an entire book. A poor one can break it.

Those of us with an academic background need to work very hard to avoid standard academic reflexes. “First tell them what you are going to tell them, then tell them, then tell them what you have told them" is fine for teaching—and a gesture in that direction can help readers make sense of a chapter, or even an entire book—but the reflex can easily become formulaic. Worse, it can destroy the suspense. Imagine if *Romeo and Juliet* opened with "Behold! Here comes Romeo! He will take a sleeping potion and fair Juliet will think him dead and kill herself."

Some sort of road map often helps readers understand what you are doing. My recent *Mathematics of Life* opens with this:

> Biology used to be about plants, animals, and insects, but five great revolutions have changed the way scientists think about life.
>
> A sixth is on its way.
>
> The first five revolutions were the invention of the microscope, the systematic classification of the planet's living creatures, evolution, the discovery of the gene, and the structure of DNA. Let's take them in turn, before moving on to my sixth, more contentious, revolution.

The reader, thus informed, understands why the next few chapters, in a book ostensibly about mathematics, are about historical high points of biology. Notice that I did not tell them what revolution six *is*. It is mathematics, of course, and if they have read the blurb on the back of the book they will know that, but you do not need to rub it in.

Writing a popular science book is not like writing a textbook; it is closer to fiction. You are telling a story, not teaching a course. You need to think about pacing the story and about what to reveal up front and what to keep up your sleeve. Your reader may need to know that a manifold is a multidimensional analog of a surface, and you may have to simplify that to "curved many-dimensional space." Do not start talking about charts and atlases and C^∞ overlap maps; avoid all mention of "paracompact" and "Hausdorff." To some extent, be willing to tell lies: lies of omission, white lies that slide past technical considerations that would get in the way if they were mentioned. Jack Cohen and I call this technique "lies to children." It is an educational necessity: what experts need to know is different from what the public needs to know.

Above all, remember that you are not trying to teach a class. You are trying to give an intelligent but uninformed person some idea of what is going on.

6. How Do I Get My Book Published?

Experienced writers know how to do this. If you are a new writer, it is probably better at present to work with a recognized publisher. However, this advice could well become obsolete as e-books grow in popularity. A publisher will bear the cost of printing the book, organize publicity and distribution, and deal with the publication process. In return, it will keep most of the income, passing on about ten percent to the author in the form of royalties. An alternative is to publish the book yourself through websites that print small quantities of books at competitive prices. The stigma that used to be attached to "vanity publishing" is fast disappearing as more and more authors cut out the intermediary. You will have to handle the marketing, probably using a website, but distribution is no longer a great problem thanks to the Internet. An even simpler method is to publish your work as an e-book. Amazon offers a simple publishing service; the author need do little more than register, upload a Word file, proofread the result, and click the "publish" button. At the moment, the author gets seventy percent of all revenue. Many authors' societies are starting to recommend this method of publication.

Assuming that you follow the traditional route, you will need to make contact with one or more publishers. An agent will know which publishers to approach and will generally make this process quicker and more effective. The agency fee (between ten and twenty percent) is usually outweighed by improved royalty rates or advances, where the publisher pays an agreed sum that is set against subsequent royalties. The advance is not refundable provided the book appears in print, but you will not receive further payment until the book "earns out." Advice about finding an agent (and much else) can be obtained from authors' societies. In the absence of an agent, find out which publishers have recently produced similar types of books. Send an outline and perhaps a sample chapter, with a short cover letter. If the publisher is interested, an editor will reply, typically asking for further information. This may take a while; if so, be patient, but not if it is taking months.

If a publisher accepts your book, it will send a contract. It always looks official, but you should not hesitate to tear it to pieces and scribble all over it. By all means, negotiate with the publisher about these changes, but if you do not like it, do not sign it. Read the contract carefully, even if your agent is supposed to do this for you. Delete any clauses that tie your hands on future books, especially "this will be the author's next book." A few publishers routinely demand an option on your next work: authors should equally routinely cross out that clause. Try to keep as many subsidiary rights (translations, electronic, serializations, and so on) as you can, but be aware that you may have to grant them all to the publisher before it agrees to accept your book. Some compromises are necessary.

7. Organization

Different authors write in different ways. Some set themselves a specific target of, say, two thousand words per day. Some write when they are in the mood and keep going until the feeling disappears. Some start at the front and work their way through the book page by page. Some jump in at random, writing whichever section appeals to them at the time, filling in the gaps later.

A plan, even if it is a page of headings, is almost indispensable. Planning a book in outline helps you decide what it is about and what should go in it. Popular science books usually tell a story, so it is a good idea to sort out what the main points of the story are going to be. However, most books evolve as they are being written, so you should think of your plan as a loose guide, not as a rigid constraint. Be willing to redesign the plan as you proceed. Your book will talk to you; listen to what it says.

To avoid writer's block—a common malady in which the same page is rewritten over and over again, getting worse every time, until the author grinds to a halt, depressed—avoid rewriting material until you have completed a rough draft of the entire book. You will have a far better idea of how to rewrite chapter one when you have finished chapter twenty. When you are about halfway through, the rest is basically downhill and the writing often gets easier. Do not put off that feeling by being needlessly finicky early on. Once you know you have a book, tidying it up and improving it becomes a pleasure. Terry Pratchett called this "scattering fairydust."

8. What Else You Will Have to Do

Your work is not finished when you submit the final manuscript (typescript, Word file, LaTeX file, whatever). It will be read by an editor, who may suggest broad changes: "move chapter four after chapter six," "cut chapter fourteen in half," "add two pages about the origin of topology," and so on. There will also be a copy editor, whose main job is to prepare the manuscript for the printer and to correct typographical, factual, and grammatical errors. Some of them may suggest minor changes: "Why not put in a paragraph telling readers what a Möbius band is?" Many will change your punctuation, paragraphing, and choice of words. If they seem to be overdoing this—this is happening when you feel they are imposing their own style, instead of preparing yours for the printer—ask them to stop and change everything back the way it was. It is your book.

After several months, proofs will arrive, to be read and corrected. Often the time allowed will be short. The publisher will issue guidelines about this process, such as which symbols to use and which color of ink. (Many still work with paper copy. Others will send a PDF file, or a Word file with "track changes" enabled, and explain how they want you to respond.) Avoid making further changes to the text, unless absolutely necessary, even if you have just thought of a far better way to explain what cohomology is. You may be charged for excessive alterations that are not typesetting errors—see your contract, about twenty-five sections in.

With the proofs corrected, you might think that you have done your bit and can relax: you would be mistaken. Nowadays, books are released alongside a barrage of publicity material: podcasts, webcasts, blogs, tweets, and articles for printed media like *New Scientist* or newspapers. Your publisher's publicity department will try to get you on radio or television, and it will expect you to turn up for interviews if anyone bites. You may be asked to give public lectures, especially at literary festivals and science festivals. Your contract may oblige you to take part in such activities unless you have good reason not to or the demands become excessive. It is part of the job, so do not be surprised. If it seems to be getting out of hand, talk to your publisher. It won't be in the publisher's best interest to wear you out or take away too much time from writing another book it can publish.

Always wanted to write a book? Then do so. Just be aware of what you are letting yourself in for.

Contributors

Derek Abbott was born in London in 1960. He received the B.Sc. (Hons.) degree in physics from Loughborough University, Loughborough, Leicestershire, U.K., in 1982, and the Ph.D. degree in electrical and electronic engineering from the University of Adelaide, Adelaide, SA, Australia, in 1995, under K. Eshraghian and B. R. Davis. He was a research engineer with the GEC Hirst Research Centre, London, from 1978 to 1986. From 1986 to 1987, he was a VLSI design engineer with Austek Microsystems, Australia. Since 1987, he has been with the University of Adelaide, where he is currently a full professor with the School of Electrical and Electronic Engineering. He co-edited *Quantum Aspects of Life* (London: Imperial College Press, 2008) and coauthored the book entitled *Stochastic Resonance* (Cambridge: Cambridge University Press, 2012). His interests are in the area of multidisciplinary physics and electronic engineering applied to complex systems. His research programs span a number of areas of stochastics, game theory, photonics, biomedical engineering, and computational neuroscience. Abbott is a fellow of the Institute of Physics (IoP) and a fellow of the Institute of Electrical & Electronic Engineers (IEEE). He has won a number of awards, including the South Australian Tall Poppy Award for Science (2004), the Premier's SA Great Award in Science and Technology for outstanding contributions to South Australia (2004), an Australian Research Council Future Fellowship (2012), and the David Dewhurst Medal (2015). He is currently on the editorial boards of *Scientific Reports* (Nature), *Royal Society Online Science* (RSOS), and *Frontiers in Physics.*

David Acheson is an emeritus fellow at Jesus College, Oxford. He is the author of *Elementary Fluid Dynamics* (1990), *From Calculus to Chaos* (1997) and the best-selling *1089 and All That* (2002), which has now been translated into eleven languages. He was Oxford University's first winner of a National Teaching Fellowship, and in 2013 was awarded an Honorary D.Sc. by the University of East Anglia for his work in the public understanding of mathematics.

Jorge Almeida is associate professor of Genetics and Molecular Biology at the Institute of Agronomy of the University of Lisbon (Ph.D. from John Innes Institute/University of East Anglia, UK). His main research interest has been about how flower symmetry patterns arise during plant development and evolution (e.g., http://dev.biologists.org/content/develop/124/7/1387.full.pdf, http://genesdev.cshlp.org/content/16/7/880.full.pdf). From his research and, especially, from teaching biology students, he became interested in popular mathematics; his latest publication is about how infinite sums relate to demographics and gender preferences (https://plus.maths.org/content/population-growth).

Hyman Bass, Samuel Eilenberg Distinguished University Professor of Mathematics and Mathematics Education at the University of Michigan, received his Ph.D. from the University of Chicago. He has worked mainly in commutative homological algebra, algebraic *K*-theory, and geometric group theory. In the 1990s he was enlisted by Deborah Ball to study the mathematics in elementary classrooms, in part to develop a practice-based understanding of what kinds of mathematical understandings schoolteachers need in their work. Currently, he is also interested in mathematical practices, and specifically, in whether the problem-solving/theory-building duality in the way mathematicians work has, or could have, any meaningful expression in school mathematics.

Katharine Beals specializes in the educational needs of children with autism. She is an adjunct professor at the Drexel University School of Education, having designed two of the courses for Drexel's Graduate Certificate in Autism Spectrum Disorders. Also a lecturer at the University of Pennsylvania Graduate School of Education, she teaches courses on the education of special needs children. She is the author of *Raising a Left-Brain Child in a Right-Brain World: Strategies for Helping Bright, Quirky, Socially Awkward Children to Thrive at Home and at School.*

Viktor Blåsjö is a historian of mathematics who recently received his Ph.D. from Utrecht University, Netherlands, with a dissertation on the Leibnizian calculus. Having thus a "Ph.D. in calculus," he has put the educational import of his research to use by writing a historically informed calculus textbook, available at intellectualmathematics.com. In an age where the history of mathematics is increasingly taken over by humanists, he is also a rarity in the field in that he did his Ph.D. in a department of mathematics. In keeping with this trend, he has in a number of publications taken it upon himself to defend old-fashioned, mathematically oriented historiographical approaches.

Joshua Bowman is an assistant professor of mathematics in the Natural Science Division at Pepperdine University. He received his Ph.D. from Cornell in 2009. He is the translator of *The Scientific Legacy of Poincaré*, which was published by the American Mathematical Society and the London Mathematical Society in 2010. His research interests are in geometry, topology, and dynamics. He recently moved to Los Angeles, following several years of living, teaching, and researching on the East Coast in New York and Massachusetts. He blogs occasionally at http://thalestriangles.blogspot.com/.

Davide Castelvecchi is a senior reporter at *Nature*, where he covers physics, astronomy, computer science, and mathematics. He received an undergraduate degree at the Sapienza University of Rome in 1994 and a Ph.D. in mathematics at Stanford University in 2000, with a thesis on Morse theory and noncommutative invariants of foliations. After three years in academia, he went back to school to study science writing at the University of California, Santa Cruz. Before joining *Nature*, he was a reporter at *Science News* and a senior editor at *Scientific American*. He lives in London.

Hannah Christenson is an undergraduate student at Pomona College in Claremont, CA. She is studying computer science and minoring in mathematics, and she expects to graduate in 2017. She is originally from Gibsonia, PA.

Piet J. H. Daas obtained his Ph.D. degree in 1996 at the University of Nijmegen (Netherlands) cum laude. He started working at Statistics Netherlands in 2000, where he currently is senior methodologist and project leader for big data research. His research focuses on the statistical analysis and methodology of big data with specific attention to selectivity, feature extraction, and quality. He was one of the team members responsible for producing the first big data based official statistics in the world, i.e., traffic intensity statistics, and is involved in several European big data projects. More information on his work is available at pietdaas.nl/pubs.

Joseph W. Dauben is Distinguished Professor of History and History of Science at the City University of New York. He is the author of biographies of Georg Cantor and Abraham Robinson, and most recently, a three-volume Chinese-English dual-language edition of the ancient Chinese classic, *The Nine Chapters on the Art of Mathematics* (2013), written in collaboration with Guo Shuchun and Xu Yibao. In January 2012, he received the American Mathematical Society's Albert Leon Whiteman Memorial Prize for the history of mathematics.

Ton de Waal studied mathematics at Leiden University and obtained his Ph.D. degree in 2003 at Erasmus University in Rotterdam. Since 1993, he has worked at Statistics Netherlands, where his current position is senior methodologist. Since 2014, he has also been a professor of Methodology of Official Statistics at Tilburg University. His main fields of expertise are statistical disclosure control, statistical data editing, imputation, and data integration. He is coauthor of two books on statistical disclosure control (*Elements of Statistical Disclosure Control* and *Statistical Disclosure Control in Practice*) and a book on statistical data editing and imputation (*Handbook of Statistical Data Editing and Imputation*).

Richard Feinberg is a senior psychometrician at the National Board of Medical Examiners and an assistant professor at the Philadelphia College of Osteopathic Medicine. He received his Ph.D. from the University of Delaware in research methodology and evaluation. His research interests include psychometric applications in the fields of educational and psychological testing.

Marc Frantz received a B.F.A. in painting from the Herron School of Art in 1975, followed by a thirteen-year career with art galleries. He earned an M.S. in mathematics from IUPUI in 1990 and is currently a research associate in mathematics at Indiana University. He is coauthor with Annalisa Crannell of the textbook *Viewpoints: Mathematical Perspective and Fractal Geometry in Art* (Princeton University Press, 2011).

Stephan Ramon Garcia grew up in San Jose, CA, before attending UC Berkeley for his B.A. and Ph.D. After graduating, he worked at UC Santa Barbara before moving to Pomona College in 2006. He is the author of two books (both with Cambridge University Press) and more than sixty research articles in operator theory, complex analysis, matrix analysis, number theory, discrete geometry, and other fields. He has coauthored more than two dozen articles with students; some of these have appeared in journals such as the *Notices of the AMS*, the *Proceedings of the AMS*, the *Journal of Functional Analysis*, and the *American Mathematical Monthly*. He is on the editorial boards of the *Proceedings of the AMS* (2016–), *Involve* (2011–), and the *American Mathematical Monthly* (2017–). Garcia received three NSF research grants as principal investigator and five teaching awards from three different institutions. He was twice nominated by Pomona College for the CASE U.S. Professors of the Year Award.

Barry Garelick has written articles on math education for *The Atlantic*, *AMS Notices*, *Education Next*, and *Education News*. He majored in mathematics at the University of Michigan and worked in the field of environmental protection. After retiring five years ago, he sought a second career and obtained a teaching credential in secondary mathematics. He is teaching math at the middle school levels in California. He is also the author of three books; the most recent is *Math Education in the U.S.: Still Crazy after All These Years*.

Andrew Gelman is a professor of statistics and political science at Columbia University. His books include *Bayesian Data Analysis*, *Teaching Statistics: A Bag of Tricks*, *A Quantitative Tour of the Social Sciences*, and *Red State, Blue State, Rich State, Poor State: Why Americans Vote the Way They Do.*

Brian Greene is a theoretical physicist and string theorist. He has been a professor at Columbia University since 1996 and chairman of the World Science Festival since he cofounded it in 2008. Greene has become known to a wide audience through his books for the general public, *The Elegant Universe*, the board book *Icarus at the Edge of Time*, *The Fabric of the Cosmos*, *The Hidden Reality*, and related television specials.

Kevin Hartnett writes "Brainiac," a column about new ideas, for the *Boston Globe*. He's also a regular contributor to *Quanta Magazine*, where he covers research in math and physics. A native of Maine, Kevin lives in South Carolina with his wife and three sons and writes *Growing Sideways*, a blog about fatherhood. Follow him on Twitter @kshartnett.

Brian Hayes is a science writer and a former editor of *American Scientist* and *Scientific American*. He is also the author of *Infrastructure: A Guide to the Industrial Landscape* (W. W. Norton, 2005) and *Group Theory in the Bedroom and Other Mathematical Diversions* (Hill and Wang, 2008). His web site http://bit-player.org focuses on computational and mathematical topics. He has been journalist-in-residence at the Mathematical Sciences Research Institute in Berkeley and a visiting scientist at the Abdus Salam International Centre for Theoretical Physics in Trieste, Italy. He is the winner of a National Magazine Award.

Tanya Khovanova is a lecturer at MIT and a freelance mathematician. She received her Ph.D. in mathematics from the Moscow State University in 1988. Her current research interests lie in recreational mathematics, including puzzles, magic tricks, combinatorics, number theory, geometry, and probability theory. Her website is located at tanyakhovanova.com, her highly popular math blog at blog.tanyakhovanova.com, and her *Number Gossip* website at numbergossip.com. Khovanova works with gifted children in a variety of settings. The paper in this volume was a project done at MIT-PRIMES, a highly competitive program that helps high school students to conduct research.

Erica Klarreich is a mathematics journalist based in Berkeley, CA, who has more than fifteen years of experience writing about mathematics for the general public. Her work has appeared in *Quanta Magazine*, *Nature*, *New Scientist*, *Science News*, *Wired.com*, and many other publications.

Rachel Levy is an associate professor of mathematics and associate dean for faculty development at Harvey Mudd College. She serves as vice president for education for the Society for Industrial and Applied Mathematics (SIAM). She is a coauthor of the textbook *Partial Differential Equations: An Introduction to Theory and Applications* (Princeton, 2015). She has served as a lead writer for national reports on applied mathematics programs and industrial internships in mathematics as well as the *GAIMME report* (Guidelines for Assessment and Instruction in Mathematical Modeling Education). She is a recipient of the MAA Alder Award for teaching.

Eric Nie is a graduating senior at Westborough High School in Massachusetts. He will be studying computer science and mathematics at Carnegie Mellon University this fall. He enjoys math and programming competitions, playing piano in his school jazz band, and playing tennis on his school team.

Burkard Polster is the author of a number of books, including *The Mathematics of Juggling*, *Q.E.D: Beauty in Mathematical Proof* and *Math Goes to the Movies* (with M. Ross). Living in Australia, Burkard serves as Monash University's resident mathematical juggler, origami expert, bubble master, and mathemagician. On YouTube, Burkard preaches math as the *Mathologer*.

Alok Puranik is a freshman at MIT studying mathematics and computer science. He has participated for three years in MIT's PRIMES, a program for high school mathematics research, during which he studied cellular automata as well as the Lasserre hierarchy of semidefinite programming. This year, he participated in United States of America Mathematical Olympiad and won first place in the Massachusetts Math Olympiad. He has also competed in the highest division of USA Computing Olympiad. Recently, he presented his joint work on the Ulam-Warburton automaton and its connection to the Sierpinski sieve at the Museum of Math's 2015 MOVES Conference.

Marco J. H. Puts is a statistical researcher at Statistics Netherlands, with a focus on big data processing and methodology. He has a background in computer science and cognitive science. His special interests lie in the field of artificial intelligence and data filtering. He was one of the team members responsible for producing the first big data based official statistics in the world, i.e., traffic intensity statistics of the Netherlands, and was involved in the UNECE Big Data Sandbox.

Jennifer J. Quinn is a professor of mathematics and accidental administrator at the University of Washington Tacoma. She has served as executive director of the Association for Women in Mathematics and second vice president of the Mathematical Association of America. Together with Arthur Benjamin, she co-edited *Math Horizons*, and they received a Beckenbach Book Prize from MAA for *Proofs That Really Count: The Art of Combinatorial Proof.* Jenny thinks that beautiful proofs are as much art as science. Simplicity, elegance, transparency, and *fun* should be the driving principles.

David Richeson is a professor of mathematics at Dickinson College and the editor of *Math Horizons*, the undergraduate magazine of the Mathematical Association of America (MAA). His research interests include topology, dynamical systems, the history of mathematics, and recreational mathematics. He graduated from Hamilton College in 1993 and received his Ph.D. in mathematics from Northwestern University in 1998. He is the author of *Euler's Gem: The Polyhedron Formula and the Birth of Topology* (Princeton University Press, 2008), which won the MAA's 2010 Euler Book Prize.

Alan Schoenfeld is the Elizabeth and Edward Conner Professor of Education and Affiliated Professor of Mathematics at the University of California at Berkeley. He has won numerous awards for his research on mathematical thinking, teaching, and learning, including the International Commission on Mathematics Instruction's Klein Medal, the highest international distinction in mathematics education.

Marjorie Senechal is the Louise Wolff Kahn Professor Emerita in Mathematics and History of Science and Technology at Smith College, Northampton, MA, and editor-in-chief of the international quarterly journal, *The Mathematical Intelligencer*. Her research concerns the geometry of quasicrystals, described in *Quasicrystals and Geometry* (Cambridge University Press, 1996) and updated in articles. Her most recent books are *I Died for Beauty: Dorothy Wrinch and the Cultures of Science* (Oxford University Press), and *Shaping Space: Exploring Polyhedra in Nature, Art, and the Geometrical Imagination*) (Springer), both 2013. Marjorie is a fellow of the American Mathematical Society.

Isabel M. Serrano is an undergraduate student at California State University, Fullerton, where she is pursuing degrees in history and applied mathematics. She has interdisciplinary research interests, which is demonstrated in her work pertaining to the history of mathematics. In addition to intertwining history and mathematics, she is intrigued by research in the field of computational biology. Through the research-based MARC Program (Minority Access to Research Careers), she will conduct research in applied mathematics in connection to biology.

Amy Shell-Gellasch is an associate professor of mathematics at Montgomery College in Rockville, MD. Her area of research is the history of mathematics and its uses in teaching, and she is a volunteer researcher at the Smithsonian National Museum of American History. Her publications include a book-length biography *In Service to Mathematics: The Life and Work of Mina Rees* and a history of mathematics textbook *Algebra in Context: Introductory Algebra from Origins to Applications*. She has edited several resource volumes about incorporating history into the mathematics classroom and has served on several editorial boards, including the Mathematical Association of America's Spectrum book series.

Daniel S. Silver is a professor of mathematics at the University of South Alabama. Much of his published research explores the relationship between knots and dynamical systems. Other active interests include the history of science and the psychology of invention. He has contributed articles of general interest to *American Scientist* and *Notices of the American Mathematical Society*.

Ian Stewart is an emeritus professor of mathematics at the University of Warwick in Coventry, UK. He is an active research mathematician with more than a hundred eighty published papers on pattern formation, chaos, network dynamics, and biomathematics. He has published more than a hundred books, including *Professor Stewart's Cabinet of Mathematical Curiosities*, *Nature's Numbers, Does God Play Dice?*, and the *Science of Discworld* series with Terry Pratchett and Jack Cohen. His awards include the Royal Society's Faraday Medal, the Gold Medal of the IMA, the Public Understanding of Science Award of the AAAS, the LMS/IMA Zeeman Medal, and the Lewis Thomas Prize. He is a fellow of the Royal Society.

John Stillwell is a professor of mathematics at the University of San Francisco. He grew up in Australia and received an M.Sc. from the University of Melbourne and a Ph.D. from MIT in 1970. He has written on many areas of mathematics, from geometry to logic, and received the Chauvenet Prize in 2005 for his article "The Story of the 120-Cell." His books include *Mathematics and Its History*, *Yearning for the Impossible* (winner of the AJCU Book Prize in 2009), *Roads to Infinity*, and most recently, *Elements of Mathematics: From Euclid to Gödel*.

Gilbert Strang is a professor of mathematics at MIT. He is the author of twelve books, and linear algebra is his most important subject; he aims to open up this beautiful subject to many more students who need it. His books and articles and video lectures on ocw.mit.edu have helped to make clear the central ideas to millions of readers and viewers. He is a past president of SIAM and a member of the U.S. National Academy of Sciences.

Steven Strogatz is the Jacob Schurman Professor of Applied Mathematics at Cornell University. Among his honors are membership in the American Academy of Arts and Sciences (2012) and the Lewis Thomas Prize for Writing about Science (2015). A frequent guest on WNYC's "Radiolab," he is the author, most recently, of *The Joy of x*. Follow him on Twitter @stevenstrogatz.

Bogdan D. Suceavă is a professor of mathematics at California State University at Fullerton, where he has taught since 2002, after he completed his Ph.D. program in mathematics at Michigan State University. His B.Sc. and M.Sc. are from the University of Bucharest. His research interests lie mainly in differential geometry. Since 2011, Suceavă has contributed to the development of the Fullerton Mathematical Circle, a program for the gifted students in southern California. He is also an award-winning novelist, writing in Romanian; see, for example, *Coming from an Off-Key Time* (Northwestern Univ. Press, 2011) and *Miruna, a Tale* (Twisted Spoon Press, 2014).

Peter Turner is currently the dean of arts and sciences at Clarkson University, having previously been chair of mathematics and computer science. He was educated in the UK, where there is a long history of applied math and modeling, earning his Ph.D. from the University of Sheffield and then serving for fifteen years on the faculty at the University of Lancaster. In 1987, he came to the U.S. Naval Academy in Annapolis, MD, and then to Clarkson in 2002. Turner was SIAM's vice president for education for six years. While in that role, he started the Modeling across the Curriculum initiative. Turner is the inaugural chair of SIAM's Activity Group in Applied Mathematics Education, and of the (new) NCTM-SIAM Committee on Modeling.

Howard Wainer is Distinguished Research Scientist at the National Board of Medical Examiners. He received his Ph.D. from Princeton in 1968 and, in addition to more than four hundred fifty articles and chapters, he has authored, coauthored, or edited twenty-one books. His most recent ones include *Truth or Truthiness: Distinguishing Fact from Fiction by Learning to Think like a Data Scientist*; *Medical Illuminations: Using Evidence, Visualization and Statistical Thinking to Improve Healthcare*; *A Statistical Guide for the Ethically Perplexed*; *Uneducated Guesses: Using Evidence to Uncover Misguided Education Policies*; and *Picturing the Uncertain World: How to Understand, Communicate, and Control Uncertainty through Graphical Display*.

Notable Writings

At some time or another over the past year I considered for inclusion in this volume many of the pieces mentioned below but they fell on the wayside as we selected the contents of the book by consulting outside reviewers and considering constraints of length and copyright. I encourage readers with an interest in the many aspects posed by mathematics in the contemporary world to explore on their own the broad range of leads I offer here.

I give the full title of some publications but for select titles I use the following abbreviations: *AMM* = *The American Mathematical Monthly*; *ASc* = *American Scientist*; *ASt* = *The American Statistician*; *AHES* = *Archive for the History of the Exact Sciences*; *BSHMB* = *BSHM Bulletin: Journal of the British Society for the History of Mathematics*; *BAMS* = *Bulletin of the American Mathematical Society*; *CMJ* = *The College Mathematics Journal*; *CACM* = *Communications of the Association for Computing Machinery*; *ESM* = *Educational Studies in Mathematics*; *EJMSTE* = *Eurasia Journal of Mathematics, Science and Technology Education*; *FLM* = *For the Learning of Mathematics*; *HM* = *Historia Mathematica*; *HPL* = *History and Philosophy of Logic*; *IJMTL* = *International Journal for Mathematics Teaching and Learning*; *IJMEST* = *International Journal of Mathematical Education in Science and Technology*; *IJSME* = *International Journal of Science and Mathematics Education*; *JRME* = *Journal for Research in Mathematics Education*; *JHM* = *Journal of Humanistic Mathematics*; *JMB* = *The Journal of Mathematical Behavior*; *LMSN* = *London Mathematical Society Newsletter*; *MH* = *Math Horizons*; *MI* = *The Mathematical Intelligencer*; *MERJ* = *Mathematics Education Research Journal*; *ME* = *The Mathematics Enthusiast*; *MT* = *Mathematics Teacher*; *NAMS* = *Notices of the American Mathematical Society*; *PRIMUS* = *Problems, Resources, and Issues in Mathematics Undergraduate Studies*; *SS* = *Statistical Science*; *SHPS* = *Studies in History and Philosophy of Science, Part A*; *TMIA* = *Teaching Mathematics and Its Applications*; *ZDM* = *Zentralblatt für Didaktik der Mathematik*.

Abrahamson, Dor. "Reinventing Learning: A Design-Research Odyssey." *ZDM* 47(2015): 1013–26.

Acker, Amelia. "Toward a Hermeneutics of Data." *IEEE Annals of the History of Computing* 37.3(2015): 70–75.

Adu-Gyamfi, Kwaku, Michael J. Bossé, and Kayla Chandler. "Situating Student Errors: Linguistic-to-Algebra Translation Errors." *IJMTL* (2015).

Alexanderson, Gerald L., and Leonard F. Klosinski. "Gaspar Schott's 'Cursus Mathematicus.'" *BAMS* 52.3(2015): 497–501.

Alibali, Martha W., and Pooja G. Sydney. "Variability in the Natural Number Bias: Who, When, How, and Why." *Learning and Instruction* 37(2015): 56–61.

Alladi, Krishnaswami, and Gabriela Asli Rino Nesin. "The Nesin Mathematics Village in Turkey." *NAMS* 62.6(2015): 652–58.

Alvarez, Josefina. "Want Less Traffic? Build Fewer Roads!" *Plus Magazine* September 3, 2015.

Andrá, Chiara, et al. "Reading Mathematics Representations." *IJSME* 13.S2(2015): S237–59.

Andrews-Larson, Christine. "Roots of Linear Algebra: An Historical Exploration of Linear Systems." *PRIMUS* 25.6(2015): 507–28.

Apostol, Tom M., and Mamikon A. Mnatsakanian. "Volume/Surface Area Relations for *n*-Dimensional Spheres, Pseudospheres, and Catenoids." *AMM* 122.8(2015): 745–56.

Ardourel, Vincent. "A Discrete Solution for the Paradox of Achilles and the Tortoise." *Synthese* 192(2015): 2843–61.

Aydin, Nuh, and Lakhdar Hammoudi. "Root Extraction by Al-Kashi and Stevin." *AHES* 69(2015): 291–310.

Bajri, Sanaa, John Hannah, and Clemency Montelle. "Revisiting Al-Samaw'al's Table of Binomial Coefficients: Greek Inspiration, Diagrammatic Reasoning and Mathematical Induction." *AHES* 69(2015): 537–76.

Baldi, Brigitte, and Jessica Utts. "What Your Future Doctor Should Know about Statistics." *ASt* 69.3(2015): 231–40.

Ball, Derek. "'Thick-Rinded Fruit of the Tree of Knowledge': Mathematics Education in George Eliot's Novels." *BSHMB* 30.3(2015): 217–26.

Barany, Michael J. "The Myth and the Medal." *NAMS* 62.1(2015): 15–20.

Barrow-Green, June. "'Anti-Aircraft Guns All Day Long': Karl Pearson and Computing for the Ministry of Munitions." *Revue d'Histoire des Mathématiques* 21.1(2015): 111–50.

Barton, Bill, Greg Oates, Judy Paterson, and Mike Thomas. "A Marriage of Continuance: Professional Development for Mathematics Lecturers." *MERJ* 27(2015): 147–64.

Batchelor, Sophie, Matthew Inglis, and Camilla Gilmore. "Spontaneous Focusing on Numerosity and the Arithmetic Advantage." *Learning and Instruction* 40(2015): 70–88.

Bedenikovic, Tony. "An Analog of Bing's House in a Higher Dimension." *MI* 37.2(2015): 34–42.

Bellhouse, David R. "An Analysis of Errors in Mathematical Tables." *HM* 42(2015): 280–95.

Bellhouse, David R. "Mathematicians and the Early English Life Insurance Industry." *BSHMB* 30.2(2015): 131–42.

Bellhouse, David R., and Nicholas Fillion. "Le Her and Other Problems in Probability Discussed by Bernoulli, Montmort and Waldegrave." *SS* 30.1(2015): 26–39.

Bender, Andrea, Dirk Schlimm, and Sieghard Beller. "The Cognitive Advantages of Counting Specifically." *Topics in Cognitive Sciences* 7(2015): 552–69.

Benedictus, Fedde, and Dennis Dieks. "Reichenbach's Transcendental Probability." *Erkenntnis* 80(2015): 15–38.

Benétreau-Dupin, Yann. "Buridan's Solution to the Liar's Paradox." *HPL* 36.1(2015): 18–28.

Bennett, Deborah. "Drawing Logical Conclusions." *MH* 22.3(2015): 12–15.

Berkowitz, Talia, et al. "Math at Home Adds up to Achievement in School." *Science* 350(2015): 196–98.

Besson, Corine. "A Note on Logical Truth." *Logique et Analyse* 227(2015): 309–31.

Boellstorff, Tom. "Making Big Data, in Theory." *Data, Now Bigger and Better!* edited by Tom Boellstorff and Genevieve Bell. Chicago: Prickly Paradigm Press, 2015, 87–108.

Borwein, Jonathan M., and Scott T. Chapman. "I Prefer Pi: A Brief History and Anthology of Articles in the *AMM*." *AMM* 122.3(2015): 195–216.

Branchini, Erika, Roberto Burro, Ivana Bianchi, and Ugo Savardi. "Contraries as an Effective Strategy in Geometrical Problem Solving." *Thinking and Reasoning* 21.4(2015): 397–430.

Brantley, Kristina Leifeste. "A Forgotten Contrivance: A Study of the Diagonal Scale and Its Appearance in Mathematics Texts from 1714 to the Present." *BSHMB* 30.1(2015): 50–66.

Bressoud, David. "Insights from the MAA National Study of College Calculus." *MT* 109.3(2015): 179–85.

Bressoud, David, and Chris Rasmussen. "Seven Characteristics of Successful Calculus Programs." *NAMS* 62.2(2015): 144–46.

Brössel, Peter, and Franz Huber. "Bayesian Confirmation: A Means with No End." *British Journal for the Philosophy of Science* 66(2015): 737–49.

Bruza, Peter D., Zheng Wang, and Jerome R. Busemeyer. "Quantum Cognition: A New Theoretical Approach to Psychology." *Trends in Cognitive Sciences* 19.7(2015): 383–93.

Bullynch, Maarten, Edgar D. Daylight, and Liesbet De Mol. "Why Did Computer Science Make a Hero Out of Turing?" *CACM* 58.3(2015): 37–39.

Cable, John. "Mathematics Is Always Invisible, Professor Dowling." *MERJ* 27(2015): 359–84.

Camacho-Machin, M., and C. Guerro-Ortiz. "Identifying and Exploring Relationships between Contextual Situations and Ordinary Differential Equations." *IJMEST* 46.8(2015): 1077–95.

Capristo, Annalisa. "Volterra, Fascism, and France." *Science in Context* 28.4(2015): 637–74.

Chambers, Andrew. "Elliptic Cryptography." *Plus Magazine* July 20, 2015.

Chapman, Olive. "Reflective Awareness in Mathematics Teachers' Learning and Teaching." *EJMSTE* 11.2(2015): 313–24.

Chappell, James M., et al. "Geometric Algebra for Electrical and Electronic Engineers." *Proceedings of the IEEE* 102.9(2014): 1340–63.

Chemla, Karin, and Biao Ma. "How Do the Earliest Known Mathematical Writings Highlight the State's Management of Grains in Early Imperial China." *AHES* 69(2015): 1–53.

Chen, Yan, Rui Peng, and Zhong You. "Origami of Thick Panels." *Science* 349.6462(2015): 396–400.

Chen, Jian-Ping Jeff. "Trigonometric Tables: Explicating their Construction Principles in China." *AHES* 69(2015): 491–536.

Claessens, Guy A. J. "The Drawing Board of Imagination: Federico Commandino and John Philoponus." *Journal of the History of Ideas* 76.4(2015): 499–501.

Cook, Gareth. "The Singular Mind of Terry Tao." *The New York Times Online* July 24, 2015.

Cornock, Claire. "Teaching Group Theory Using Rubik's Cube." *IJMEST* 46.7(2015): 957–67.

Cromwell, Peter M. "Cognitive Bias and Claims of Quasiperiodicity in Traditional Islamic Patterns." *MI* 37.4(2015): 30–44.

Cupillari, Antonella. "Math in a Can." *MT* 108.6(2015):434–37.

Cvencek, Dario, Manu Kapur, and Andrew N. Meltzoff. "Math Achievement, Stereotypes, and Math Self-Concepts among Elementary-School Students in Singapore." *Learning and Instruction* 39(2015): 1–10.

Darrigol, Olivier. "The Mystery of Riemann's Curvature." *HM* 42(2015): 47–83.

Dawkins, Paul Christian. "Explication as a Lens for the Formalization of Mathematical Theory through Guided Reinvention." *JMB* 37(2015): 63–82.

De Freitas, Elizabeth, and Betina Zolkower. "Tense and Aspect in Word Problems about Motion: Diagram, Gesture, and the Felt Experience of Time." *MERJ* 27(2015): 311–30.

De Sterk, Hans, and Chris Johnson. "Data Science: What Is It and How Is It Taught?" *SIAM News* 48.6(2015).

Debnath, Lokenath. "A Brief History of the Most Remarkable Numbers e, i, and γ in Mathematical Sciences with Applications." *IJMEST* 46.6(2015): 853–78.

Debnath, Lokenath, and Kanadpriya Basu. "A Short History of Probability Theory and Its Applications."*IJMEST* 46.1(2015): 13–39.

Dejić, Mirko. "How the Old Slavs (Serbs) Wrote Numbers." *BSHMB* 29.1(2015): 2–17.

Desai, Adhaar Noor. "Number-Lines: Diagramming Irrationality in 'The Phoenix and Turtle.'" *Configurations* 23.3(2015): 301–30.

Deschamps, Isabelle, Galit Agmon, Yonathan Loewenstein, and Yosef Grodzinsky. "The Processing of Polar Quantifiers, and Numerosity Perception." *Cognition* 143(2015): 115–28.

Dewolf, Tinne, Wim Van Dooren, Frouke Hermens, and Lieven Verschaffel. "Do Students Attend to Representational Illustrations of Non-Standard Mathematical Word Problems, and, If So, How Helpful Are They?" *Instructional Science* 43(2015): 147–71.

Dietiker, Leslie. "Mathematical Story: A Metaphor for Mathematics Curriculum." *ESM* 90(2015): 285–302.

Drageset, Ove Gunnar. "Different Types of Student Comments in the Mathematics Classroom." *JMB* 38(2015): 29–40.

Dröscher, Ariane. "Gregor Mendel, Franz Unger, Carl Nägeli and the Magic of Numbers." *History of Science* 53.4(2015): 492–508.

Dunning, David E. "What Are Models For? Alexander Crum Brown's Knitted Mathematical Surfaces." *MI* 37.2(2015): 62–70.

Earnest, Darrell. "From Number Lines to Graphs in the Coordinate Plane: Investigating Problem Solving across Mathematical Representations." *Cognition and Instruction* 33.1(2015): 46–87.

Eder, Günter. "Frege's 'On the Foundations of Geometry' and Axiomatic Metatheory." *Mind* (2015): 1–36 [advanced online publication].

Edwards, Harold M. "Euler's Conception of the Derivative." *MI* 37.4(2015): 52–53.

Egenhoff, Jay. "Math as a Tool of Anti-Semitism." *ME* 11.3(2014): 649–64.

Ehrhardt, Caroline. "Tactics: In Search of a Long-Term Mathematical Project (1844–1896)." *HM* 42(2015): 436–67.

Ellenberg, Jordan. "The Mathematics of March Madness." *The New York Times* March 20, 2015.

Etingof, Pavel, Slava Gerovitch, and Tanya Khovanova. "Mathematical Research in High School." *NAMS* 62.8(2015): 910–18.

Ferrara, Katrina, and Barbara Landau. "Geometric and Featural Systems, Separable and Combined." *Cognition* 144(2015): 123–33.

Foster, Colin. "Exploiting Unexpected Situations in the Mathematics Classroom." *IJSME* 13(2015): 1065–88.

Franchella, Miriam. "Brouwer and Nietzsche: Views about Life, Views about Logic." *HPL* 36.4(2015): 367–91.

Fredua-Kwarteng, Eric. "How Prospective Teachers Conceptualized Mathematics: Implications for Teaching." *Mathematics Education* 10.2(2015): 77–95.

Freiberger, Marianne. "Introducing the Klein Bottle." *Plus Magazine* January 6, 2015.

Freiberger, Marianne. "Ramanujan Surprises Again." *Plus Magazine* November 3, 2015.

Freiberger, Marianne. "Why We Want Proof." *Plus Magazine* April 10, 2015.

Gelman, Andrew. "Disagreements about the Strength of Evidence." *Chance* 28.1(2015): 55–59.

Gerdes, Paulus. "African Dance Rattles and Plaiting Polyhedra." *MI* 37.2(2015): 52–61.

Gershman, Samuel J., Eric J. Horvitz, and Joshua B. Tenenbaum. "Computational Rationality: A Converging Paradigm for Intelligence in Brains, Minds, and Machines." *Science* 349.6245(2015): 273–78.

Goldstein, Harvey. "Jumping to the Wrong Conclusions." *Significance* 12(2015): 18–21.

Gough, John. "New Games That Stimulate Math Thinking." *Australian Mathematics Teacher* 71.1(2015): 24–27.

Grajales, Carlos A. Gómez. "Not Part of the Crowd." *Significance* 12.5(2015): 40–42.

Granville, Andrew. "A New Mathematical Celebrity [Yitang Zhang]." *BAMS* 52.2(2015): 335–37.

Grattan-Guinness, Ivor. "Where Is Logical Knowledge Located?" *Logique & Analyse* 229(2015): 3–24.

Grawe, Paul H. "Mathematics and Humor: John Allen Paulus and the Numeracy Crusade." *Numeracy* 8.2(2015): 1–16.

Gray, Mary W. "The Odds of Justice: Where Is the Reverend Bayes When Needed?" *Chance* 28.2(2015): 51–53.

Greenwald, Sarah J., Anne M. Leggett, and Jill E. Thomley. "The Association for Women in Mathematics." *MI* 37.3(2015): 11–21.

Greiffenhagen, Christian. "The Materiality of Mathematics: Presenting Mathematics at the Blackboard." *British Journal of Sociology* 65.3(2014): 502–28.

Grünberg, David, and Andreas Matt. "Using Surfer to Investigate Algebraic Surfaces." *MT* 109.3(2015); 221–26.

Guberman, Raisa, and Dvora Gorev. "Knowledge Concerning the Mathematical Horizon." *MERJ* 27(2015): 165–82.

Gunderson, Elizabeth A., et al. "Gesture as a Window onto Children's Number Knowledge." *Cognition* 144(2015): 14–28.

Haas, Robert. "On Mathematicians' Eccentricity." *JHM* 5.2(2015): 146–50.

Haciomeroglu, Erhan Selcuk. "The Role of Cognitive Ability and Preferred Mode of Processing in Students' Calculus Performance." *EJMSTE* 11.5(2015): 1165–79.

Haigh, Thomas. "Innovators Assemble: Ada Lovelace, Walter Isaacson, and the Superheroines of Computing." *CACM* 58.9(2015): 20–27.

Haigh, Thomas. "The Tears of Donald Knuth." *CACM* 58.1(2015): 40–44.

Harris, Diane, et al. "Mathematics and Its Value for Engineering Students: What Are the Implications for Teaching?" *IJMEST* 46.3(2015): 321–36.

Hawkins, Jane. "Dynamics of Mathematical Groups." *I, Mathematician*, edited by Peter Casazza, Steven G. Krantz, and Randi D. Ruden. Providence, RI: American Mathematical Society, 2015, 192–202.

Hawthorne, Casey, and Chris Rasmussen. "A Framework for Characterizing Students' Thinking about Logical Statements and Truth Tables." *IJMEST* 46.3(2015).
Hayes, Brian. "Computer Vision and Computer Hallucinations." *ASc* 103.6(2015): 380–83.
Hayes, Brian. "Crawling toward a Wiser Web." *ASc* 103.3(2015): 184–87.
Hayes, Brian. "Cultures of Code." *ASc* 103.1(2015): 10–13.
Hayes, Brian. "Playing in Traffic." *ASc* 103.4(2015): 260–63.
Hayes, Brian. "The 100-Billion-Body Problem." *ASc* 103.3(2015): 90–93.
Heyd-Metzuyamin, Einat. "Vicious Cycles of Identifying and Mathematizing: A Case Study of the Development of Mathematical Failure." *Journal of the Learning Sciences* 24.4(2015): 504–49.
Hoadley, Susan, et al. "Threshold Concepts in Finance: Conceptualizing the Curriculum." *IJMEST* 46.6(2015): 824–40.
Hoadley, Susan, et al. "Threshold Concepts in Finance: Student Perspectives." *IJMEST* 46.7(2015): 1004–20.
Hollings, Christopher. "The Acceptance of Abstract Algebra in the USSR, as Viewed through Periodic Surveys of the Progress of Soviet Mathematical Science." *HM* 42(2015): 193–222.
Holm, Tara. "The Real Reason Why the U.S. Is Falling Behind in Math." *Boston Globe* February 12, 2015.
Horton, Nicholas J. "Challenges and Opportunities for Statistics and Statistical Education: Looking Back, Looking Forward." *ASt* 69.2(2015): 138–45.
Howson, Colin. "What Probability Probably Isn't." *Analysis* 75.1(2015): 53–59.
Hudson, Brian, Sheila Henderson, and Alison Hudson. "Developing Mathematical Thinking in the Primary Classroom." *Journal of Curriculum Studies* 47.3(2015): 374–98.
Hüntelmann, Axel C. "Statistics, Nationhood, and the State." *Twentieth Century Population Thinking: A Critical Reader of Primary Sources*, edited by the Population Knowledge Network, Abingdon, UK : Routledge, 2016, 11–36.
Ivanova, Milena. "Conventionalism about What? Where Duhem and Poincaré Part Ways." *SHPS* 54(2015): 80–89.
Jakob, Wenzel, and Steve Marschner. "Geometric Tools for Exploring Manifolds of Light Transport Paths." *CACM* 58.11(2015): 103–11.
Jankvist, Uffe Thomas. "Changing Students' Images of 'Mathematics as a Discipline.'" *JMB* 38(2015): 41–56.
Jankvist, Uffe Thomas, and Mogens Niss. "A Framework for Designing a Research-Based 'Maths Counsellor' Teacher Programme." *ESM* 90(2015): 259–84.
Jankvist, Uffe Thomas, et al. "Analyzing the Use of History of Mathematics through MKT [Mathematical Knowledge for Teaching]." *IJMEST* 46.4(2015): 495–507.
Johnson, Timothy C. "Finance and Mathematics: Where Is the Ethical Malaise?" *MI* 37.4(2015): 8–11.
Johnson-Laird, P. N., Sangeet S. Khemlani, and Geoffrey P. Goodwin. "Logic, Probability, and Human Reasoning." *Trends in Cognitive Sciences* 19.4(2015): 201–14.
Jones, Michael A., Lon Mitchell, and Brittany Shelton. "Fractals and Mysterious Triangles." *MH* 23.1(2015): 22–25.
Jones, Steven R. "The Prevalence of Area-under-a-Curve and Anti-Derivative Conceptions over Riemann Sum-Based Conceptions in Students' Explanations of Definite Integrals." *IJMEST* 46.5(2015): 721–36.

Jones, Steven R., and Allison Dorko. "Students' Understandings of Multivariate Integrals and How They May Be Generalized from Single Integral Conceptions." *JMB* 40(2015): 154–70.

Jordan, M. I., and T. M. Mitchell. "Machine Learning: Trends, Perspectives, and Prospects." *Science* 349.6245(2015): 255–60.

Jungić, Ozren, and Veselin Jungić. "F. P. Ramsey: The Theory, the Myth, and the Mirror." *MI* 37.3(2015): 3–7.

Juslin, Peter, Marcus Lindskog, and Bastian Mayerhofer. "Is There Something Special with Probabilities? Insight vs. Computational Ability in Multiple Risk Combination." *Cognition* 136(2015): 282–303.

Kanovei, Vladimir, Karin U. Katz, Mikhail G. Katz, and Mary Schaps. "Proofs and Retributions, Or: Why Sarah Can't *Take* Limits?" *Foundations of Science* 20(2015): 1–25.

Kanovei, Vladimir, Karin U. Katz, Mikhail G. Katz, and David Sherry. "Euler's Lute and Edwards's Oud." *MI* 37.4(2015): 48–51.

Kaplan, Abdullah, and Mesut Öztürk. "Effect of Concept Cartoons to Academic Achievement in Instruction on the Topics of Divisibility." *Mathematics Education*, 10.2(2015): 67–76.

Kapur, Manu. "Preparatory Effects of Problem Solving versus Problem Posing on Learning from Instruction." *Learning and Instruction* 39(2015): 23–31.

Karaali, Gizem. "On Genius, Prizes, and the Mathematical Celebrity Culture." *MI* 37.3(2015): 61–65.

Kazak, Sibel, Rupert Wegerif, and Taro Fujita. "The Importance of Dialogic Processes to Conceptual Development in Mathematics." *ESM* 90(2015): 105–20.

Keane, Terese. "Partial Differential Equations versus Cellular Automata for Modeling Combat." *Journal of Defense Modeling and Simulation* 8.4(2010): 191–204.

Keffington, James. "Warren Buffett: Oracle or Orang-utan?" *Significance* 12(2015): 8–11.

Kërënxhi, Svjetllana, and Pranvera Gjoci. "Dual Treatments as Starting Point for Integrative Perceptions in Teaching Mathematics." *IJSME* 13(2015): 793–809.

Kettenring, Jon R., Kenneth J. Koehler, and John D. McKenzie Jr. "Challenges and Opportunities for Statistics in the Next 25 Years." *ASt* 69.2(2015): 86–90.

Khemlani, Sangeet S., Max Lotstein, and Philip N. Johnson-Laird. "Naive Probability: Model-Based Estimates of Unique Events." *Cognitive Science* 39(2015): 1216–58.

Khovanova, Tanya, and Joshua Xiong. "Cookie Monster Plays Games." *The College Mathematics Journal* 46.4(2015): 283–93.

Kichenassamy, Satyanad. "Continued Proportions and Tartaglia's Solution of Cubic Equations." *HM* 42(2015): 407–35.

Kidwell, Peggy Aldrich. "Useful Instruction for Practical People: Early Printed Discussions of the Slide Rule in the United States." *IEEE Annals of the History of Computing* 37.1(2015): 36–43.

Klarreich, Erica. "In Search of Bayesian Inference." *CACM* 58.1(2015): 21–24.

Klymchuk, Sergiy. "Provocative Mathematics Questions: Drawing Attention to a Lack of Attention." *TMIA* 34(2015): 63–70.

Kobiela, Marta, and Richard Lehrer. "The Codevelopment of Mathematical Concepts and the Practice of Defining." *JRME* 46.4(2015): 423–54.

Koichu, Boris, and Uri Leron. "Proving as Problem Solving: The Role of Cognitive Decoupling." *JMB* 40(2015): 233–44.

Kollár, János. "Is There a Course of the Fields Medal?" *NAMS* 62.1(2015): 21–25.
Koparan, Timur. "Difficulties in Learning and Teaching Statistics." *IJMEST* 46.1(2015): 94–104.
Kosmann-Schwarzbach, Yvette. "Women Mathematicians in France in the Mid-Twentieth Century." *BSHMB* 30.3(2015): 227–42.
Koss, Lorelei. "Differential Equations in Literature, Poetry, and Film." *Journal of Mathematics and the Arts* 9.1–2(2015): 1–16.
Kraemer, Daniel M. "Natural Probabilistic Information." *Synthese* 192(2015): 2901–19.
Krall, Jenne R., and Roger D. Peng. "The Volkswagen Scandal." *Significance* 12(2015): 12–15.
Kuehn, Christian. "The Curse of Instability." *Complexity* 20.6(2015): 9–14.
Landau, Susan. "NSA and Dual EC_DRBG: Déjà Vu All Over Again?" *MI* 37.4(2015): 72–83.
Lande, Daniel R. "Development of the Binary Number System and the Foundations of Computer Science." *ME* 11.3(2015): 513–40.
Lange, Marc. "Depth and Explanation in Mathematics." *Philosophia Mathematica* 23.2 (2015): 196–215.
Lange, Marc. "Explanation, Existence and Natural Philosophy in Mathematics—A Case Study: Desargues' Theorem." *Dialectica* 69.4(2015): 435–72.
Lazar, Nicole. "Now Trending on Twitter." *Chance* 28.3(2015): 34–36.
Lê, François. "'Geometrical Equations': Forgotten Premises of Felix Klein's *Erlanger Programm*." *HM* 42(2015): 315–42.
Leatham, Keith R. "Observations on Citation Practices in Mathematics Education Research." *JRME* 46.3(2015): 253–69.
Lewis, Gareth. "Motivational Classroom Climate." *FLM* 35.3(2015): 29–34.
Litvak, Nelly, and Frank van der Meulen. "Networks and Big Data." *Nieuw Archief voor Wiskunde* 16.2(2015): 138–39.
Lombardi, Olimpia, Sebastian Fortin, and Leonardo Vanni. "A Pluralist View about Information." *Philosophy of Science* 82.5(2015): 1248–59.
Lorenat, Jemma. "Figures Real, Imagined, and Missing in Poncelet, Plücker, and Gergonne." *HM* 42(2015): 155–92.
Lorenat, Jemma. "Polemics in Public: Poncelet, Gergonne, Plücker, and the Duality Controversy." *Science in Context* 28.4(2015): 545–85.
Lučić, Zoran. "Irrationality of the Square Root of 2." *MI* 37.2(2015): 26–32.
Lykins, Alicia N. "Using Repetition to Make Ideas Stick." *MT* 108.8(2015): 622–25.
Lynch, Matthew. "Matthew O'Brien, an Inventor of Vector Analysis." *Irish Mathematical Society Bulletin* 74(2014): 81–88.
Madison, Bernard L. "Quantitative Literacy and the Common Core State Standards in Mathematics." *Numeracy* 8.1(2015): 1–13.
Mahon, Basil. "How Maxwell's Equations Came to Light." *Nature Photonics* 9(2015): 2–4.
Mancosu, Paolo. "Grundlagen, Section 64: Frege's Discussion of Definitions by Abstraction in Historical Context." *HPL* 36.1(2015): 62–89.
Marcus, Solomon. "Starting from the Scenario Euclid-Bolyai-Einstein." *Synthese* 192 (2015): 2139–49.
Martin, Christie Lynn. "Writing as a Tool to Demonstrate Mathematical Understanding." *School Science and Mathematics* 115.6(2015): 302–13.

Mason, John. "Developing and Using an Applet to Enrich Students' Concept Image of Rational Polynomials." *TMIA* (2015) [advanced online publication].

Max, Brooke, and Jill Newton. "Teaching University Mathematics: One Mathematician's Contribution." *NAMS* 62.9(2015): 1062–64.

Mazliak, Laurent. "The Ghosts of the École Normale." *SS* 30.3(2015): 391–412.

McAllister, Alex M., and Diana White. "History of Mathematics: Seeking Truth and Inspiring Students." *NAMS* 62.2(2015): 172–74.

Mcmullen, Tyler. "It Probably Works." *CACM* 58.11(2015): 50–54.

Merrotsy, Peter. "The Tower of Hanoi and Inductive Logic." *Australian Senior Mathematics Journal* 29.1(2015): 16–24.

Merow, Katharine. "A Toast! To Type 15." *MH* 23.2(2015): 10–11.

Merttens, Ruth. "Textbooks from Shanghai and Singapore: A National Debate." *The Mathematical Gazette* 99(2015): 391–401.

Misfeldt, Morten, and Mikkel Willum Johansen. "Research Mathematicians' Practices in Selecting Mathematical Problems." *ESM* 89(2015): 357–73.

Moeller, Korbinian, Samuel Shaki, Silke M. Göbel, and Hans-Christoph Nuerk. "Language Influences Number Processing: A Quadrilingual Study." *Cognition* 136(2015): 150–55.

Morar, Florin-Stefan. "Reinventing Machines: The Transmission History of the Leibniz Calculator." *British Journal for the History of Science* 48.1(2015): 123–46.

Moretti, Alessio. "Was Lewis Carroll an Amazing Oppositional Geometer?" *HPL* 35.4 (2015): 383–409.

Morris, Bradley J., and Amy M. Masnick. "Comparing Data Sets: Implicit Summaries of the Statistical Properties of Number Sets." *Cognitive Science* 39(2015): 156–70.

Naccarato, Emilie, and Gulden Karakok. "Expectations and Implementations of the Flipped Classroom Model in Undergraduate Mathematics Courses." *IJMEST* 46.7 (2015): 968–78.

Naresh, Nirmala. "A Stone or a Sculpture? It Is All in Your Perception." *IJSME* 13(2015): 1567–88.

Nescolarde-Selva, Josué Antonio, José Luis Usó-Domenech, and Miguel Lloret-Climent. "Ideological Complex Systems: Mathematical Theory." *Complexity* 21.2(2015): 47–65.

Nirode, Wayne. "Exploring New Geometric Worlds." *MT* 109.2(2015): 119.

Nolan, Caroline, and Sandra Herbert. "Introducing Linear Functions: An Alternative Statistical Approach." *MERJ* 27(2015): 401–21.

Norton, Anderson. "The Wonderful Gift of Mathematics." *Mathematics Educator* 24.1 (2015): 3–20.

Nosrati, Mona. "Temporal Freedom in Mathematical Thought: A Philosophical-Empirical Enquiry." *JMB* 37(2015): 18–35.

Nyman, Rimma. "Indicators of Student Engagement: What Teachers Notice During Introductory Algebra Lessons." *IJMTL* (2015): 1–17.

O'Daly, Emer. "The Art of Roughness." *Benoit Mandelbrot: A Life in Many Dimensions*, edited by Michael Frame and Nathan Cohen. Singapore: World Scientific, 2015, 381–97.

O'Halloran, Kay L. "The Language of Learning Mathematics." *JMB* 40A(2015): 63–74.

Ochkov, Valery F., and Andreas Look. "A System of Equations: Mathematics Lessons in Classic Literature." *JHM* 5.2(2015).

Odic, Darko, Matthew Le Core, and Justin Halberda. "Children's Mappings between Number Words and the Approximate Number System." *Cognition* 138(2015): 102–21.

Olsen, James R. "Five Keys for Teaching Mental Math." *MT* 108.7(2015): 543–47.

Ornes, Stephen. "The Whole Universe Catalog." *Scientific American* 313.1(2015): 68–75.

Oxley, James. "Writing a Teaching Statement." *NAMS* 62.1(2015): 59–61.

Pais, Alexandre. "Symbolising the Real of Mathematics Education." *ESM* 89(2015): 375–91.

Palha, Sonia, Rijkje Dekker, and Koeno Gravemeijer. "The Effect of Shift-Problem Lessons in the Mathematics Classroom." *IJSME* 13(2015): 1583–1623.

Partanen, Anna-Maija, and Raimo Kaasila. "Sociomathematical Norms Negotiated in the Discussions of Two Small Groups Investigating Calculus." *IJSME* 13(2015): 927–46.

Passeau, Alexander. "Knowledge of Mathematics without Proof." *British Journal for the Philosophy of Science* 66(2015): 775–99.

Peng, Roger. "The Reproducibility Crisis in Science." *Significance* 12(2015): 30–32.

Periton, Cheryl. "The Medieval Counting Table Revisited." *BSHMB* 30.1(2015): 35–49.

Petrocchi, Alessandra. "A New Theoretical Approach to Sample Problems and Deductive Reasoning in Sanskrit Mathematical Texts." *BSHMB* 30.2(2015): 2–19.

Pierce, Pamela, and Robert Wooster. "Conquer the World with Markov Chains." *MH* 22.4(2015): 18–21.

Polotskaia, Elena, Annie Savar, and Viktor Freiman. "Duality of Mathematical Thinking When Making Sense of Word Problems." *EJMSTE* 11.2(2015): 251–61.

Prodromou, Theodosia. "Teaching Statistics with Technology." *Australian Mathematics Teacher* 71.3(2015): 32–40.

Proulx, Jérôme. "Mental Mathematics with Mathematical Objects Other Than Numbers: The Case of Operations on Functions." *JMB* 39(2015): 156–76.

Pudwell, Lara, and Eric Rowland. "What's in YOUR Wallet?" *MI* 37.4(2015): 54–60.

Reed, Daniel A., and Jack Dongarra. "Exascale Computing and Big Data." *CACM* 58.7 (2015): 56–68.

Reed, Michael C. "Mathematical Biology Is Good for Mathematics." *NAMS* 62.10(2015): 1172–76.

Reeder, Patrick. "Zeno's Arrow and the Infinitesimal Calculus." *Synthese* 192(2015): 1315–35.

Reinhardt, Lloyd. "Good and Bad Arithmetical Manners." *Analysis* 75.1(2015): 26–28.

Roberts, Siobhan. "Cogito, Ergo Summer." *The New Yorker* August 27, 2015.

Rodríguez, Laura. "Riesz and the Emergence of General Topology: The Roots of 'Topological Space' in Geometry." *AHES* 69(2015): 55–102.

Roero, Clara Silvia. "M. G. Agnesi, R. Rampinelli and the Riccati Family: A Cultural Fellowship Formed for an Important Scientific Purpose, the *Instituzioni analitiche*." *HM* 42(2015): 296–314.

Roth, Wolff-Michael. "Excess of Graphical Thinking: Movement, Mathematics, and Flow." *FLM* 35.1(2015): 2–7.

Rowlett, Peter. "'The Unplanned Impact of Mathematics' and Its Implications for Research Funding." *BSHMB* 30.1(2015): 67–74.

Sahin, Zulal, Arzu Aydogan Yenme, and Ayhan Kursat Erbas. "Relational Understanding of the Derivative Concept through Mathematical Modeling." *EJMSTE* 11.1(2015): 177–88.

Samasa, Gregory P. "Has It Really Been Demonstrated That Most Genomic Research Findings Are False?" *ASt* 69.1(2015): 1–4.

Sanchez, Wendy B., Alyson E. Lischka, Kelly W. Edenfield, and Rebecca Gammill. "An Emergent Framework: Views of Mathematical Processes." *School Science and Mathematics* 115.2(2015): 88–99.

Savic, Milos. "The Incubation Effect: How Mathematicians Recover from Proving Impasses." *JMB* 39(2015): 67–78.

Sengupta-Irving, Tesha, and Noel Enyedy. "Why Engaging in Mathematical Practices May Explain Stronger Outcomes in Affect and Engagement." *Journal of the Learning Sciences* 24(2015): 550–92.

Seo, Byung-In. "Mathematical Writing." *JHM* 5.2(2015): 133–45.

Shang, Jian, et al. "Assembling Molecular Sierpiński Triangle Fractals." *Nature Chemistry* 7(2015): 389–93.

Shell-Gellasch, Amy. "The Shilling Kinematic Models at the Smithsonian." *JHM* 5.1(2015): 167–79.

Shell-Gellasch, Amy, and J. B. Thoo. "Power of Two . . . in Poetry." *MH* 23.2(2015): 28–29.

Shorser, Lindsey. "Manifestations of Mathematical Meaning." *Semiotic and Cognitive Science Essays on the Nature of Mathematics,* edited by Mariana Bockarova, Marcel Danesi, and Rafael Núñez. Munich, Germany: Lincom Europa, 2012, 268–78.

Sidney, Pooja G., Shanta Hattikudur, and Martha W. Alibali. "How Do Contrasting Cases and Self-Explanation Promote Learning? Evidence from Fraction Division." *Learning and Instruction* 40(2015): 29–38.

Simoson, Andrew J. "Life Lessons from Leibniz." *MH* 22.4(2015): 5–7, 29.

Siu, Man Keung. "Mathematics: What Has It to Do with Me?" *JHM* 5.2(2015).

Skow, Bradford. "Are There Genuine Physical Explanations of Mathematical Phenomena?" *British Journal for the Philosophy of Science* 66(2015): 69–93.

Smeding, Annique, Céline Darnon, and Nico W. Van Yperen. "Why Do High Working Memory Individuals Choke? An Examination of Choking under Pressure Effects in Math from a Self-Improvement Perspective." *Learning and Individual Differences* 37(2015): 176–82.

Smith, Sheldon R. "Incomplete Understanding of Concepts: The Case of the Derivative." *Mind* 124.496(2015).

Steel, Daniel. "Acceptance, Values, and Probability." *SHPS* 53(2015): 81–88.

Stephan, Michelle M., et al. "Grand Challenges and Opportunities in Mathematics Education Research." *JRME* 46.2(2015): 134–46.

Stöltzner, Michael. "Hilbert's Axiomatic Method and Carnap's General Axiomatics." *SHPS* 53.1(2015): 12–22.

Suri, Manil. "Don't Expect Math to Make Sense." *The New York Times* March 13, 2015.

Talbert, Robert. "Flipped Calculus: A Gateway to Lifelong Learning in Mathematics." *Best Practices for Flipping the College Classroom*, edited by Julee B. Waldrop and Melody A. Bowdon, New York: Routledge, 2015, 29–43.

Tanswell, Fenner. "A Problem with the Dependence of Informal Proofs on Formal Proofs." *Philosophia Mathematica* 23.3(2015): 295–310.

Tao, Terrence. "On Addition" (https://plus.google.com/u/0/+TerenceTao27/posts/6diqmz1JQrB).

Tarran, Brian. "How Machines Learned to Think Statistically." *Significance* 12.1(2015): 8–15.

Thomas, Amanda, and Alden J. Edson. "How Common Is the Common Core?" *MT* 108.5(2015): 382–86.

Thomas, Rachel. "Kissing the Curve: Manifolds in Many Dimensions." *Plus Magazine* June 23, 2015.

Tjoe, Hartono. "Giftedness and Aesthetics: Perspectives of Expert Mathematicians and Mathematically Gifted Students." *Gifted Child Quarterly* 59.3(2015): 165–76.

Toeplitz, Otto. "The Problem of University Courses on Infinitesimal Calculus and Their Demarcation from Infinitesimal Calculus in High Schools." *Science in Context* 28.2(2015): 297–310.

Tolle, Penelope P. "Changing Classroom Instruction: One Teacher's Perspective." *MT* 108.8(2015): 616–21.

Torrence, Eve. "Start with Art." *MH* 23.2(2015): 34.

Trefethen, Lloyd N. "Computing Numerically with Functions instead of Numbers." *CACM* 58.10(2015): 91–97.

Tsang, Jessica M., Kristen P. Blair, Laura Bofferding, and Daniel L. Schwartz. "Learning to 'See' Less than Nothing: Putting Perceptual Skills to Work for Learning Numerical Structure." *Cognition and Instruction* 33.2(2015): 154–97.

Usiskin, Zalman. "The Relationship between Statistics and Other Subjects in the K-12 Curriculum." *Chance* 28.3(2015): 4–18.

Ulivi, Elisabetta. "Masters, Questions and Challenges in the Abacus Schools." *AHES* 69(2015): 651–70.

Utts, Jessica. "The Many Facets of Statistics Education: 175 Years of Common Themes." *ASt* 169.2(2015): 107.

Vallée-Tourangeau, Gaëlle, and Marlène Abadie. "Interactivity Fosters Bayesian Reasoning without Instruction." *Journal of Experimental Psychology: General* 144.3(2015): 581–603.

Van Benthem, Johan. "Fanning the Flames of Reason: A Personal View of Logic." *Nieuw Archief voor Wiskunde* 5/16.1(2015): 52–61.

Veličković, Vladica. "What Everybody Should Know about Statistical Correlation." *ASc* 103.1(2015): 26–29.

Verburgt, Lukas M. "Remarks on the Idealist and Empiricist Interpretation of Frequentism: Robert Leslie Ellis versus John Venn." *BSHMB* 29.3(2015): 184–95.

Verschaffel, Lieven, Fien Depaepe, and Erik De Corte. "Mathematics Education." *International Encyclopedia of the Social and Behavioral Sciences*, edited by James D. Wright, vol. 14, Amsterdam: Elsevier, 2015, 816–21.

Wagner, Katie, Katherine Kimura, Pierina Cheung, and David Barner. "Why Is Number Word Learning Hard? Evidence from Bilingual Learners." *Cognitive Psychology* 83(2015): 1–21.

Wagner, Roy. "Citrabhānu's Twenty-One Algebraic Problems in Malayalam and Sanskrit." *HM* 42(2015): 263–79.

Wainer, Howard, and Donald B. Rubin. "Visual Revelations: Causal Inference and Death." *Chance* 28.2(2015): 58–64.

Walkington, Candace, and Matthew Bernacki. "Students Authoring Personalized 'Algebra Stories': Problem-Posing in the Context of Out-of-School Interests." *JMB* 40(2015): 171–91.

Wallace, Dorothy. "Moderating Competition in the World of Ideas." *Numeracy* 8.1(2015).

Wallace, Dorothy. "Quantitative Literacy on a Desert Island." *Numeracy* 8.2(2015).

Walton, Douglas. "The Basic Slippery Slope Argument." *Informal Logic* 35.3(2015): 273–311.

Wardhaugh, Benjamin. "A 'Lost' Chapter in the Calculation of π: Baron Zach and MS Bodleian 949." *HM* 42(2015): 343–51.

Wasserman, Nicholas H. "A Random Walk: Stumbling across Connections." *MT* 108.9(2015): 686–95.

Watson, Jane M., and Lyn D. English. "Introducing the Practice of Statistics: Are We Environmentally Friendly?" *MERJ* 27(2015): 585–613.

Weber, Keith, and Juan Pablo Mejía-Ramos. "On Relative and Absolute Convictions in Mathematics." *FLM* 35.2(2015): 15–21.

Weinberg, Aaron, Tim Fukawa-Connelly, and Emily Wiesner. "Characterizing Instructor Gestures in a Lecture in a Proof-Based Mathematics Class." *ESM* 90(2015): 233–58.

Wertheimer, Michael. "The Mathematics Community and the NSA: Encryption and the NSA Role in International Standards." *NAMS* 62.2(2015): 165–67.

Wickramasinghe, Indika, and James Valles. "Can We Use Polya's Method to Improve Students' Performance in Statistics Classes?" *Numeracy* 8.1(2015).

Wilczek, Frank. "A Puzzling Solution for Math Education." *The Wall Street Journal* January 8, 2016.

Wilder, Sandra, Annette Lang, and Max Monegan. "The Man Who Counted: A Collection of Integrated Adventures." *TMIA* 34(2015): 102–14.

Wildstrom, Jacob D. "Design and Serial Construction of Digraph Braids." *Journal of Mathematics and the Arts* 9.1–2(2015): 17–26.

Wilkerson-Jerde, Michelle Hoda, and Uri J. Wilensky. "Patterns, Probabilities, and People: Making Sense of Qualitative Change in Complex Systems." *Journal of the Learning Sciences* 24.2(2015): 204–51.

Wilkinson, Alec. "The Pursuit of Beauty." *The New Yorker* February 2, 2015.

Williams, Wendy M., and Stephen J. Ceci. "National Hiring Experiments Reveal 2:1 Faculty Hiring Preference for Women on STEM Tenure Track." *Proceedings of the National Academy of Sciences* 112.17(2015): 5360–65.

Witten, Thomas. "Of Bagels and Burgers." *Nature Physics* 11.2(2015): 95–96.

Wuorinen, Charles. "Music and Fractals." *Benoit Mandelbrot: A Life in Many Dimensions*, edited by Michael Frame and Nathan Cohen. Singapore: World Scientific, 2015, 501–6.

Yin, Shen, and Okyay Kaynak. "Big Data for Modern Industry: Challenges and Trends." *Proceedings of the IEEE* 103.2(2015): 143–46.

Yopp, David A., Rob Ely, and Jennifer Johnson-Leung. "Generic Example Proving Criteria for All." *FLM* 35.3(2015): 8–13.

Zalamea, Federico. "The Mathematical Description of a Generic Physical System." *Topoi* 34(2015): 339–48.

Zazkis, Dov, Keith Weber, and Juan Pablo Mejía-Ramos. "Two Proving Strategies of Highly Successful Mathematics Majors." *JMB* 39(2015): 11–27.

Zazkis, Rina, and Boris Koichu. "A Fictional Dialogue on Infinitude of Primes: Introducing Virtual Duoethnography." *ESM* 88(2015): 163–81.

Zeljić, Marijana. "Modelling the Relationships between Quantities: Meaning in Literal Expressions." *EJMSTE* 11.2(2015): 431–42.

Zhang, Xiaofen, M. A. (Ken) Clements, and Nerida F. Ellerton. "Conceptual Mis(understandings) of Fractions: From Area Models to Multiple Embodiments." *MERJ* 27(2015): 233–61.

Zhang, Xiaofen, M. A. (Ken) Clements, and Nerida F. Ellerton. "Enriching Student Concept Images: Teaching and Learning Fractions through a Multiple-Embodiment Approach." *MERJ* 27(2015): 201–31.

Notable Book Reviews and Review Essays

Below are references to substantive book reviews, some reaching beyond the subject of the book they discuss. To save printed space, in some cases I omit the full titles—and in all cases I omit the publishers of the books. Readers interested in reviews of mathematics books or on books on mathematics can consult the reviews website of the Mathematical Association of America (http://www.maa.org/press/maa-reviews) and Robin Whitty's web page (http://www.theoremoftheday.org/Resources/Bibliography.htm). The list below is selective (just as all others presented here).

Arianrhod, Robyn, reviews *Enlightening Symbols* by Joseph Mazur. *NAMS* 62.2(2015): 148–51.

Artigue, Michéle, reviews *The Mathematics Teacher in the Digital Era* edited by Alison Clark-Wilson, Ornella Robutti, and Nathalie Sinclair. *ESM* 90(2015): 357–62.

Carroll, Tom, reviews *My Life and Functions* by Walter K. Hayman. *LMSN* 449(2015): 36–38.

Catalano, Michael T., reviews *Naked Statistics* by Charles Wheelan. *Numeracy* 8.1(2015).

Chen, Jiang-Ping Jeff, reviews *Taming the Unknown: A History of Algebra from Antiquity to the Early Twentieth Century* by Victor J. Katz and Karen Hunger Parshall. *CMJ* 46.2(2015): 149–52.

Cohen, Marion, reviews *Mathematicians on Creativity* edited by Peter Borwein, Peter Liljedahl, and Helen Zhai. *AMM* 122.6(2015): 613–16.

Drasin, David, reviews *My Life and Functions* by Walter K. Hayman. *NAMS* 62.5(2015): 517–20.

Ferrara, Francesca, reviews *The Mathematics Teacher in the Digital Era* edited by Alison Clark-Wilson, Ornella Robutti, and Nathalie Sinclair. *ESM* 90(2015): 215–20.

Folland, Gerald B., reviews *Hidden Harmony—Geometric Fantasies: The Rise of Complex Function Theory* by Umberto Bottazzini and Jeremy Gray. *AMM* 122.2(2015): 183–88.

Folland, Gerald B., reviews *A History in Sum: 150 Years of Mathematics at Harvard (1825–1975)* by Steve Nadis and Shing-Tung Yau. *AMM* 122.5(2015): 508–10.

Franklin, James, reviews *How Not to Be Wrong* by Jordan Ellenberg. *The New Criterion* January 2015.

Garrett, Kristina C., reviews *Catalan Numbers* by Richard P. Stanley. *CMJ* 46.3(2015): 228–32.

Goff, Christopher, reviews *Animating Popular Mathematics* by Simon Singh. *NAMS* 62.1(2015): 40–44.

Grabiner, Judith, reviews *Enlightening Symbols* by Joseph Mazur. *JHM* 5.2(2015).

Greenwald, Sarah J., and Jill E. Thomley review *The Mαth βook* by Clifford A. Pickover. *NAMS* 62.4(2015): 384–85.

Grosholz, Emily R., reviews *Love and Math* by Edward Frenkel. *JHM* 5.1(2015).

Harper, Frances K., et al., review *Mathematics and the Body* by Elizabeth de Freitas and Nathalie Sinclair. *ESM* 90(2015): 221–30.

Harris, Michael, reviews *Birth of a Theorem* by Cédric Villani. *AMM* 122.10(2015): 1018–22.

Hartshorne, Robin, reviews *Alexandre Grothendieck* edited by Leila Schneps. *AMM* 122.9 (2015): 907–11.

Henle, Michael, reviews *Manifold Mirrors: The Crossing Paths of the Arts and Mathematics* by Felipe Cucker. *AMM* 122.9(2015): 912–15.

Hersh, Reuben, reviews *How Humans Learn to Think Mathematically* by David Tall. *AMM* 122.3(2015): 292–96.

Hersh, Reuben, reviews *Why Greatness Cannot Be Planned* by Kenneth O. Stanley and Joel Lehman. *JHM* 5.2(2015).

Holt, Jim, reviews *Mathematics without Apologies* by Michael Harris. *New York Review of Books* December 5, 2015.

Kapraff, Jay, reviews *Mathematics in 20th Century Literature and Art* by Robert Tubbs. *LMSN* 443(2015): 49–51.

Klyve, Dominic, reviews *Enlightening Symbols: A Short History of Mathematical Notation and Its Hidden Powers* by Joseph Mazur. *CMJ* 46.1(2015): 67–72.

Malkevitch, Joseph, reviews *The Mathematical Coloring Book* by Alexander Soifer. *AMM* 122.7(2015): 708–12.

Manaster, Alfred, reviews *What's Math Got to Do with It?* by Jo Boaler. *NAMS* 62.9(2015): 1077–79.

McArthur, Jane E., reviews *Argumentation and Health* edited by Sara Rubinelli and Francisca A. Snoeck Henkemans. *Informal Logic* 35.3(2015): 446–49.

O'Farrell, Anthony G., reviews *Mathematics and the Real World* by Zvi Artstein. *LMSN* 446(2015): 39–40.

O'Rourke, Joseph, reviews *Genius at Play: The Curious Mind of John Horton Conway* by Siobhan Roberts. *CMJ* 46.4(2015).

Pfaff, Thomas J., reviews *Street-Fighting Mathematics* by Sanjoy Mahajan. *Numeracy* 8.2(2015).

Ricchezza, Victor J., and H. L. Vacher review *Developing Quantitative Literacy Skills in History and the Social Sciences* by Kathleen W. Craver. *Numeracy* 8.2(2015).

Schaefer, Edward F., reviews *The Mathematics of Encryption* by Margaret Cozzens and Steven J. Miller. *AMM* 122.1(2015): 83–88.

Siegmund-Schultze, Reinhard, reviews *The Scholar and the State: In Search of Van der Waerden* by Alexander Soifer. *NAMS* 62.8(2015): 924–29.

Traves, Will, reviews *Perspectives in Projective Geometry* by Jürgen Richter-Gebert. *AMM* 122.4(2015): 398–402.

Tunno, Ferebee, reviews *Probability Theory in Finance* by Seán Dineen. *AMM* 122.8(2015): 809–12.

Tunstall, Samuel L., reviews *Case Studies for Quantitative Reasoning* by Bernard L. Madison et al. *Numeracy* 8.2(2015).

Walker, James S., reviews *Creating Symmetry* by Frank Farris. *NAMS* 62.11(2015): 1350–53.

Whitty, Robin, reviews *Birth of a Theorem* by Cédric Villani. *LMSN* 448(2015): 37–39.

Notable Interviews

In this subsection I reference recent printed interviews with mathematical people. I omit the original titles (if other than "Interview"). Readers interested in other sources of recent interviews can consult Shecky Riemann's *MathTango* blog (list of interviews hosted at http://math-frolic.blogspot.com/p/blog-page.html). Other recent interviews are featured on the Mathematical Association of America website: http://www.maa.org/maa.org/centennial-interviews.

Atkinson, Anthony C., and Barbara Bogacka. Interview with Tadeusz Caliński. *SS* 30.3(2015): 423–42.

Carlin, Bradley P., and Amy H. Herring. Interview with Alan Gelfand. *SS* 30.3(2015): 413–22.

Cox, Sonja. Interview with Martin Hairer. *Nieuw Archief voor Wiskunde* 16.2(2015): 133–34.

Daróczy, Zoltan. Interview with János Aczél. *Aequationes Mathematicae* 89.1(2015): 1–16.

Fisher, N. I. Interview with Jerry Friedman. *SS* 30.2(2015): 268–95.

Genest, Christian. Interview with Herbert Tate. *ME* 12.1–3(2015): 404–16.

Honner, Patrick. Interview with Steven Strogatz. *MH* 21.3(2014): 8–11.

Jackson, Allyn. Interview with Robert L. Bryant. *NAMS* 62.3(2015): 260–62.

Klarreich, Erica. Interview with Neil Sloane. *Quanta Magazine Online* August 6, 2015.

Lin, Thomas. Interview with Yitang Zhang. *Quanta Magazine Online* April 2, 2015.

Ooguri, Hirosi. Interview with Edward Witten. *NAMS* 62.5(2015): 491–506.

Raussen, Artin, and Christian Skau. Interview with Yakov Sinai. *NAMS* 62.2(2015): 152–60.

Rice, John A. Interview with Richard A. Olshen. *SS* 30.1(2015): 118–32.

Richeson, David. Interview with Timothy Gowers. *MH* 23.1(2015): 10–14.

Rosenberg, William F. Interview with Nancy Fluornoy. *SS* 30.1(2015): 133–46.

Slovic, Scott. Interview with Sandra Steingraber. In *Numbers and Nerves*, edited by Scott and Paul Slovic (Oregon University Press, 2015), 192–99.

Törner, Günter. Interview with Alan Schoenfeld. *ME* 11.3(2014): 745–52.

Zheng, Gang, Zhaohai Li, and Nancy L. Geller. Interview with Robert C. Elston. *SS* 30.2(2015): 258–67.

Notable Journal Issues

The following selective list is ordered alphabetically by the unabbreviated title of the publication.

"American Statistical Association's 175th Anniversary." *ASt* 69.2(2015).

"Statistics and the Undergraduate Curriculum." *ASt* 69.4(2015).

"Toward Computational Social Science: Big Data in Digital Governments." *Annals of the American Academy of Political and Social Science* 659 (2015).

"Works by John Milnor." *BAMS* 52.4(2015).

"Nurturing Statistical Thinking before College." *Chance* 28.3&4(2015).

"Learning Analytics, Educational Data Mining and Data-Driven Educational Decision Making." *Computers in Human Behavior* 47(2015).

"Statistical Reasoning: Learning to Reason from Samples." *ESM* 88.3(2015).
"Logical Issues in the History and Philosophy of Computer Science." *HPL* 36.5(2015).
"The Languages of Scientists." *History of Science* 53.4(2015).
"Computer Security." *IEEE Annals of the History of Computing* 47.2(2015).
"History of Computing in Latin America." *IEEE Annals of the History of Computing* 37.4 (2015).
"Technology and Its Integration into Mathematics Education." *International Journal for Technology in Mathematics Education* 22.2–3(2015).
"Video-Based Research on Teacher Expertise." *IJSME* 13.2(2015).
"Governing by Numbers" *Journal of Education Policy* 20.3(2015).
"The Language of Learning Mathematics" *JMB* 40A(2015).
"Mind the Gap! Studies on the Development of the Rational Number Concept." *Learning and Instruction* 37(2015).
"Turmoil and Transition: Tracing Émigré Mathematicians in the Twentieth Century." *MI* 37.1(2015).
"Identity in Mathematics Education." *MERJ* 27.1(2015).
"The Economics of Risk." *ME* 12.1–3(2015).
"Assessment." *Numeracy* 8.1(2015).
"Mathematical Depth." *Philosophia Mathematica* 23.2(2015).
"Flipped Classroom." *PRIMUS* 25.8–10(2015).
"Using Inquiry-Based Learning in Mathematics for Liberal Arts Courses." *PRIMUS* 25.3(2015).
"Einstein." *Scientific American* 313.3(2015).
"Integrated History and Philosophy of Science in Practice." *SHPS* 50(2015).
"History and Philosophy of Infinity." *Synthese* 192.8(2015).
"Logic and Relativity Theory." *Synthese* 192.7(2015).
"Computer Aided Assessment of Mathematics" *TMIA* 34.3(2015).
"Evidence-Based Continuous Professional Development." *ZDM* 47.1(2015).
"Enactivist Methodology in Mathematics Education Research." *ZDM* 47.2(2015).
"Geometry in the Primary School." *ZDM* 47.3(2015).
"Numeracy." *ZDM* 47.4(2015).
"Inhibitory Control in Mathematical Thinking, Problem Solving and Learning." *ZDM* 47.5(2015).
"Design Research with a Focus on Learning Processes." *ZDM* 47.6(2015).
"Scaffolding and Dialogic Teaching in Mathematics Education." *ZDM* 47.6(2015).

Acknowledgments

Thanks to the authors of the pieces included in this book, for writing the articles and for cooperating in the reprinting in this form; similar thanks to the original publishers and to the copyright holders of the illustrations.

Vickie Kearn at Princeton offered me indispensable support and advice at key decision points during the process of selecting the final content. Three anonymous readers/rapporteurs graded and opined on an extended collection of pieces I advanced at an early stage. Nathan Carr oversaw the production process, and Paula Bérard copyedited the manuscript. Other people were involved in the making of this book, some of whom I do not know. Thank you to all. I take upon myself the responsibility for the shortcomings you might find in the final result.

This series is the most scholarly of the *Best* . . . series on the market. I am comfortable with that; mathematics is special; we should not dilute the content in order to give the readers a false sense of facility. For the broad span we give to this series, access to an excellent academic library is needed. I commend once again the services I enjoyed at the Cornell University Library—although I lost important privileges lately, after I finished my doctorate.

Many thanks to my wife Fangfang and son Leo, who have put up with my daily eloping into the libraries and have kept us full of hope despite living hundreds of miles apart. Special thanks to my daughter Ioana, for overcoming heartbreaking vicissitudes for yet one more year.

Credits

"Mathematics and Teaching" by Hyman Bass. Originally published in *I, Mathematician,* edited by Peter Casazza, Steven G. Krantz, and Randi D. Ruden. Washington, DC: Mathematical Association of America, 2015, pp. 129–139. Reprinted with permission. Copyright © 2015 Mathematical Association of America. All Rights Reserved.

"In Defense of Pure Mathematics" by Daniel S. Silver. Originally published in *American Scientist* 103.6(2015): 418–421. Reprinted by permission of Sigma Xi, The Scientific Research Society.

"G. H. Hardy: Mathematical Biologist" by Hannah Elizabeth Christenson and Stephan Ramon Garcia. Originally published in *Journal of Humanistic Mathematics* 5.2(2015). Reprinted by permission of the authors.

"The Reasonable Ineffectiveness of Mathematics" by Derek Abbott. © 2013 IEEE. Reprinted with permission from, *Proceedings of the IEEE* 101.10(2013): 2147–2153.

"Stacking Wine Bottles Revisited" by Burkard Polster. Reprinted with kind permission from Springer Science+Business Media: *Mathematical Intelligencer* 37.2(2015): 43–51. Copyright © 2015 Springer Science+Business Media New York. DOI 10.1007/s00283-014-9509-6

"The Way the Billiard Ball Bounces" by Joshua Bowman. Originally published in *Math Horizons* 22.3(2015): 18–22. Reprinted with permission. Copyright © 2015 Mathematical Association of America. All Rights Reserved.

"The Intersection Game" by Burkard Polster. Originally published in *Math Horizons* 22.4(2005): 8–11. Copyright © 2015 Mathematical Association of America. All Rights Reserved.

"Tonight! Epic Math Battles: Counting vs. Matching" by Jennifer J. Quinn. Originally published in *Math Horizons* 22.3(2005): 5–9. Reprinted with permission. Copyright © 2015 Mathematical Association of America. All Rights Reserved.

"Mathematicians Chase Moonshine's Shadow" by Erica Klarreich. This article was originally published on March 12, 2015 in *Quanta Magazine*, an editorially independent publication of the Simons Foundation.

"The Impenetrable Proof" by Davide Castelvecchi. Reprinted by permission from Macmillan Publishers Ltd: *Nature* 526(Oct. 8, 2015): 179–181. Copyright 2015.

"A Proof That Some Spaces Can't Be Cut" by Kevin Hartnett. This article was originally published on January 13, 2015 in *Quanta Magazine*, an editorially independent publication of the Simons Foundation.

"Einstein's First Proof" by Steven Strogatz. Originally published in *New Yorker Online* November 19, 2015, http://www.newyorker.com/tech/elements/einsteins-first-proof-pythagorean-theorem. Copyright © 2015 Condé Nast. All Rights Reserved. Reprinted by permission.

"Why String Theory Still Offers Hope We Can Unify Physics" by Brian Greene. Originally published in *Smithsonian Magazine* (Jan 2015): http://www.smithsonianmag.com/science-nature/string-theory-about-unravel-180953637/?all. Reprinted by permission of the author.

"The Pioneering Role of the Sierpinsky Gasket" by Tanya Khovanova, Eric Nie, and Alok Puranik. Originally published in *Math Horizons* 23.1(2015): 5–9. Copyright ©2015 Mathematical Association of America. All Rights Reserved.

"Fractals as Photographs" by Marc Frantz. Originally published in *Math Horizons* 23.1(2015): 18–21. Copyright © 2015 Mathematical Association of America. All Rights Reserved.

"Math at the Met" by Joseph Dauben and Marjorie Senechal. Reprinted with kind permission from Springer Science+Business Media: *The Mathematical Intelligencer* 37.3(2015): 41–54. Copyright © 2015 Springer Science+Business Media New York. DOI 10.1007/s00283-015-9571-8

"Common Sense about the Common Core" by Alan H. Schoenfeld. Originally published in *The Mathematics Enthusiast*, vol. 11, no. 3, pp. 737–744, http://www.math.umt.edu/TMME/vol11no3/Schoenfeld_13.pdf. Reprinted by permission of the author.

"Explaining Your Math: Unnecessary at Best, Encumbering at Worst" by Katharine Beals and Barry Garelick. Originally published in *The Atlantic Online* November 11, 2015, http://www.theatlantic.com/education/archive/2015/11/math-showing-work/414924/. Reprinted by permission of the authors.

"Teaching Applied Mathematics" by David Acheson, Peter R. Turner, Gilbert Strang, and Rachel Levy. Orginally published in *The Princeton Companion to Applied Mathematics*, edited by Nicholas J. Higham. Princeton, NJ: Princeton University Press, 2015, pp. 933–943.

"Circular Reasoning: Who First Proved that C Divided by d Is a Constant?" by David Richeson. Originally published in *College Mathematics Journal* 46.3(2015): 162–171. Copyright © 2015 Mathematical Association of America. All Rights Reserved.

"A Medieval Mystery: Nicole Oresme's Concept of *Curvitas*" by Isabel M. Serrano and Bogdan D. Suceavă. Originally published in *Notices Amer. Math. Soc.* 62, no.9 (October 2015): 1030–1034. © 2015 American Mathematical Society. Reprinted by permission of the American Mathematical Society.

"The Myth of Leibniz's Proof of the Fundamental Theorem of Calculus" by Viktor Blåsjö. Originally published in *Nieuw Archief voor Wiskunde* 16.1(2015): 6–50. Reprinted by permission of the author.

"The Spirograph and Mathematical Models from Nineteenth Century Germany" by Amy Shell-Gellasch. Originally published in *Math Horizons* 22.4(2015): 22–25. Copyright © 2015 Mathematical Association of America. All Rights Reserved.

"What Does 'Depth' Mean in Mathematics?" by John Stillwell. Originally published in *Philosophia Mathematica* 23.2(2015): 215–232. Reprinted by permission of the author.

"Finding Errors in Big Data" by Marco Puts, Piet Daas, and Ton de Waal. Originally published in *Significance* 45.3(2015): 26–29. © 2015 The Royal Statistical Society. Reprinted by permission of John Wiley & Sons and the authors.

"Programs and Probability" by Brian Hayes. Originally published in *American Scientist* 103.5(2015): 326–329. Reprinted by permission of the author.

"Lottery Perception" by Jorge Almeida. Reprinted by permission of the Applied Probability Trust. First published in *Mathematical Spectrum* 47.3(2014/2015): 105–107. Copyright © Applied Probability Trust 2015.

"Why Acknowledging Uncertainty Can Make You a Better Scientist" by Andrew Gelman. Originally published as "Working through Some Issues" in *Significance* 12.3(2015): 33–35. © 2015 The Royal Statistical Society. Reprinted by permission of John Wiley & Sons and the author.

"For Want of a Nail: Why Unnecessarily Long Tests May Be Impeding the Progress of Western Civilization" by Howard Wainer and Richard Feinberg. Originally published in *Significance* 45.2(2015): 17–21. © 2015 The Royal Statistical Society. Reprinted by permission of John Wiley & Sons and the authors.

"How to Write a General Interest Mathematics Book" by Ian Stewart. Originally published in *The Princeton Companion to Applied Mathematics*, edited by Nicholas J. Higham. Princeton, NJ: Princeton University Press, 2015, pp. 906–912.